Introduction to Behavioral Genetics

A series of books in psychology

Editors: *Richard C. Atkinson*
               Jonathan Freedman
               Richard F. Thompson

G. E. McClearn   J. C. DeFries

University of Colorado

Introduction to Behavioral Genetics

W.H. Freeman and Company

San Francisco

591.15
M 126

Copyright © 1973 by W. H. Freeman and Company

No part of this book may be reproduced by any mechanical, photographic, or electronic process, or in the form of a phonographic recording, nor may it be stored in a retrieval system, transmitted, or otherwise copied for public or private use without written permission of the publisher.

Printed in the United States of America

Library of Congress Cataloging in Publication Data

McClearn, G. E. 1927-
Introduction to behavioral genetics.

  Bibliography: p.
  1. Behavior genetics.   I. DeFries, J. C., 1934-   joint author.
  II Title.   [DNLM: 1. Genetics, Behavioral.   BF701 M126i
1973]
QH457.M3   591.1'5   73-8862
ISBN 0-7167-0835-3

10   9   8   7   6   5   4   3   2   1

Preface

Although several excellent monographs and collections of articles pertinent to behavioral genetics are currently available, our experience with teaching courses in the subject has revealed a genuine need for an introductory textbook. Like the authors of this book, students come to behavioral genetics with diverse backgrounds. Most psychology majors have studied behavior and statistics, but have received no formal training in genetics. Most biology undergraduates, on the other hand, have had instruction in basic biology and genetics, but little or no exposure to psychology and statistics. We have endeavored to resolve this problem by writing a text that covers the entire range, realizing that some material will be repetitious to almost everyone.

In Chapter 1, an historical survey, we introduce a few of the basic principles of genetics. In Chapter 2, these principles are considered in more detail, but not belabored. Throughout the rest of the book, additional details and concepts of genetics are introduced. The more elementary concepts of probability and statistics are presented in Chapter 3. Students who have had formal course work in statistics may find this chapter a useful review (as students who have studied genetics may find the first two chapters). For those with no previous training in statistics, mastery of the information in Chapter 3 will be essential to an understanding of the later material. The remainder of the text presents a sampling of the literature of behavioral genetics. We have attempted to be representative in our coverage, but by no means exhaustive or encyclopedic. We have frequently cited our own work, not because we regard it as being of special significance, but simply because we know it best.

Not all aspects of genetics are covered in this volume. We have included only those topics that we regard as being most pertinent to the current status of behavioral genetics. Thus, although some concepts are treated in much greater detail (those of quantitative genetics, for example) than they are in any text on general genetics, this book should not be considered to be a substitute for an introductory genetics text. Any revision of this book may require a change in emphasis. Such a change will serve as an index of the growth and maturation of the field, one which is still in its infancy.

We have relied heavily upon the work of others in preparing this text. Special mention should be made of the extensive use of publications by Falconer (1960), Hsia (1968), Lerner (1968), and Stern (1973). Portions of a chapter by McClearn (1963) that are of a historical nature have been reproduced here with the generous permission of Alfred A. Knopf, Inc.

I. I. Gottesman and I. M. Lerner critically read an earlier version of the manuscript, and we thank them for their many useful suggestions. We wish to acknowledge the highly competent services of Agnes Conley, who not only typed the entire manuscript but also managed many of the editorial problems. Thanks also go to R. Gerry Miles who labored long and hard on the bibliography and to Patti McNeely for secretarial help. Finally, we wish to acknowledge mentors and colleagues, J. F. Crow, E. R. Dempster, D. S. Falconer, and R. W. Touchberry, whose continuing influences are, we sincerely hope, reflected in the following pages.

January 1973

G. E. McClearn
J.C. DeFries

Contents

Introduction to Behavioral Genetics

Chapter 1

Historical perspective

To illustrate a point concerning the inheritance of gestures, Darwin quoted an interesting case that had been brought to his attention by Galton.

> A gentleman of considerable position was found by his wife to have the curious trick, when he lay fast asleep on his back in bed, of raising his right arm slowly in front of his face, up to his forehead, and then dropping it with a jerk so that the wrist fell heavily on the bridge of his nose. The trick did not occur every night, but occasionally.

Nevertheless, the gentleman's nose suffered considerable damage, and it was necessary to remove the buttons from his nightgown cuff in order to minimize the hazard.

> Many years after his death, his son married a lady who had never heard of the family incident. She, however, observed precisely the same peculiarity in her husband; but his nose, from not being particularly prominent, has never as yet suffered from the blows. . . . One of his children, a girl, has inherited the same trick. (1872, pp. 33–34)

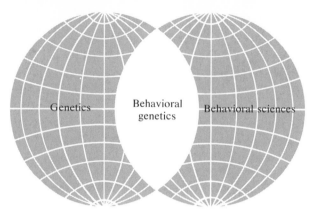

Figure 1.1
*Behavioral genetics as the intersection between genetics
and the behavioral sciences.*

Probably everyone could cite some examples, perhaps less quaint than Galton's, in which some peculiarity of gait, quality of temper, degree of talent, or other trait is characteristic of a family, and such phrases as "a chip off the old block," "like father, like son," and "it runs in the family" give ample evidence of the general acceptance of the notion that behavioral traits may be inherited, as are physiological ones. The central purpose of this chapter is to consider the history of scientific inquiry into these matters, with major emphasis on developments during the latter half of the nineteenth and the early twentieth century.

Origin of an Interdiscipline

Research that may now be categorized as being within the realm of behavioral genetics has been conducted for many years. However, behavioral genetics has only recently achieved the status of a distinct interdiscipline. A monograph, *Behavior Genetics*, by John L. Fuller and W. Robert Thompson, published in 1960, seems to deserve the credit for defining the field, whose growth since that date has been exponential.

As indicated in Figure 1.1, behavioral genetics is simply the intersection between genetics and the behavioral sciences. Behavioral geneticists are currently applying the various techniques of genetic analysis (classical, physiological, molecular, cytological, developmental, population, quantitative, evolutionary) to various behavioral characters in order to learn more about them. The characters under study are truly diverse, varying with the species studied, the behavioral orientation of the investigator (e.g., psychology, ethology, sociology, anthropology, political science) and ranging from the simple to the complex. Behavioral genetics, however, is not a one-way street. Some investigators are also concerned with the impact of behavior upon the genetic composition of populations—that is, the role of behavior as a force in evolution.

Although the fertile hybrid of behavioral genetics has only recently emerged, its roots are ancient.

Ancient Concepts

Pinpointing the earliest expression of a view concerning any subject matter is usually extremely difficult, if not impossible. The present topic is no exception, but as a matter of general interest, we may note that its origins must be very remote indeed. The concept that "like begets like" has had great practical importance in the development of domesticated animals, which have almost certainly been bred for behavioral as well as for morphological characteristics. By extrapolation, it might be suggested that a glimpse of the notion of inheritance, including inheritance of behavioral traits, may have appeared in human thought as early as 8000 B.C., to which date the domestication of the dog has been traced.

The workings of inheritance have been of great interest to men throughout recorded history, and many interesting conjectures were made (Zirkle, 1951). One of the most familiar of the early statements is that of Theognis, who, in the sixth century B.C., commented on contemporary mores:

> We seek well-bred rams and sheep and horses and one wishes to breed from these. Yet a good man is willing to marry an evil wife, if she bring him wealth: nor does a woman refuse to marry an evil husband who is rich. For men reverence money, and the good marry the evil, and the evil the good. Wealth has confounded the race. (Roper, 1913, p. 32)

By implication at least, Theognis believed that such marriages with "evil" spouses would not generate "well-bred" offspring.

The Spartans, as is well known, took direct and positive action to eliminate those who were not "well-bred," by the practice of infanticide, which was designed to eliminate those of unsound soul as well as those of defective body, for as Roper points out, "To the Greeks, believing only in the beauty of the spirit when reflected in the beauty of the flesh, the good body was the necessary correlation of the good soul" (1913, p. 19).

In *The Republic* Plato suggested a course of action whereby the principles of inheritance of behavior could be used to develop an ideal society:

It necessarily follows . . . from what has been acknowledged, that the best men should as often as possible form alliances with the best women, and the most depraved men, on the contrary, with the most depraved women; and the offspring of the former is to be educated, but not of the latter, if the flock is to be of the most perfect kind. . . . As for those youths, who distinguish themselves, either in war or other pursuits, they ought to have rewards and prizes given them, and the most ample liberty of lying with women, that so, under this pretext, the greatest number of children may spring from such parentage. . . . As respects, then, the children of worthy persons, I think, they should carry them to some retirement, to certain nurses dwelling apart in a certain quarter of the city; but as for the children of the more depraved, and such of the rest as may be maimed or lame, they will hide them, as is right, in some secret and obscure place. (Davis, 1849, p. 144)

The age of parents was also seen as an important factor, and Plato suggested that men should procreate when thirty to fifty-five years of age, and women should bear children when between the ages of twenty and forty. If, by chance, children should be conceived past the prime periods, they should be left exposed to the elements at birth.

Aristotle offered less counsel on these matters than did Plato, but he had some definite ideas concerning the proper age of parents and the optimal season of the year for procreation. Eighteen and thirty-seven were the recommended ages, for women and men respectively, to begin reproduction.

It is extremely bad for the children when the father is too young; for in all animals whatsoever the parts of the young are imperfect, and are more likely to be productive of females than males, and diminutive also in size; the same thing of course necessarily holds true in men; as a proof of this you may see in those cities where the men and women usually marry very young, the people in general are very small and ill framed. . . . And thus much for the time which is proper for marriage; but moreover a proper season of the year should be observed, as many persons do now, and appropriate the winter for this business. (Ellis, 1912, pp. 233–234)

Thus we see that both Plato and Aristotle, who contributed so much to subsequent philosophical thought, attached great importance to the circumstances surrounding mating, including the nature of the parents themselves. As we shall see later, ancient concepts such as theirs were often ill founded.

Biological thought during the ensuing centuries was dominated by Aristotle's pronouncements on natural history, and by the teachings of Galen, a Roman, concerning anatomy. Progress in understanding biological phenomena was virtually halted during the general stagnation of secular pursuits that typified the Middle Ages. Then came the Renaissance.

The Renaissance

Leonardo da Vinci's study of anatomy in connection with art, well enough known to require no elaboration here, serves to characterize the broad-ranging inquisitiveness of the Renaissance scholars. Less well known is a family incident that reveals a deep conviction about the workings of heredity. Leonardo was an illegitimate child resulting from a liaison between Piero, a notary from the village of Vinci, and a peasant girl named Caterina. As a modern biographer of Leonardo puts it:

> There was an interesting and deliberate attempt to repeat the experiment. Leonardo had a step-brother, Bartolommeo, by his father's third wife. The step-brother was forty-five years younger than Leonardo, who was already a legend when the boy was growing up and was dead when the following experiment took place. Bartolommeo examined every detail of his father's association with Caterina and he, a notary in the family tradition, went back to Vinci. He sought out another peasant wench who corresponded to what he knew of Caterina and, in this case, married her. She bore him a son but so great was his veneration for his brother that he regarded it as profanity to use his name. He called the child Piero. Bartolommeo had scarcely known his brother whose spiritual heir he had wanted thus to produce and, by all accounts, he almost did. The boy looked like Leonardo, and was brought up with all the encouragement to follow his footsteps. Pierino da Vinci, this experiment in heredity, became an artist and, especially, a sculptor of some talent. He died young. (Ritchie-Calder, 1970, pp. 39–40)*

Another example of thought on the subject of heredity can be cited from the works of the renowned humanist, Montaigne. Distressed by an ailment that had also afflicted his father, Montaigne expressed amazement that the disease could be transmitted in a "drop of seed."

> 'Tis to be believed that I derive this infirmity from my father, for he died wonderfully tormented with a great stone in his bladder; he was never sensible of his disease till the sixty-seventh year of his age; and before that had never felt any menace or symptoms of it, either in his reins, sides, or any other part, and had lived, till then, in a happy, vigorous state of health, little subject to infirmities, and he continued seven years after, in this disease, dragging on a very painful end of life. I was born above five and twenty years before his disease seized him, and in the time of his most flourishing and healthful state of body, his third child in order of birth: where could his propension to this

---

*Reproduced courtesy of Simon and Schuster, Inc., and William Heinemann, Ltd., from *Leonardo and the Age of the Eye*, Copyright 1970.

malady lie lurking all that while? And he being then so far from the infirmity, how could that small part of his substance wherewith he made me, carry away so great an impression for its share? and how so concealed, that till five and forty years after, I did not begin to be sensible of it? being the only one to this hour, amongst so many brothers and sisters, and all by one mother, that was ever troubled with it. He that can satisfy me in this point, I will believe him in as many other miracles as he pleases. (1588)

Neither the wonderment of Montaigne nor the applied husbandry of the da Vinci family can be said to have had a pivotal effect on the subsequent development of biological thought. However, there were other concurrent developments that did. Vesalius' exhaustive work on human anatomy, published in 1543, was based on detailed and painstaking dissection of human bodies, and demonstrated that the teachings of Galen contained serious errors. Harvey's momentous discovery of the circulation of blood followed, after a considerable interval, in 1628. This finding was of far-reaching importance, for it opened the way to a mechanistic as opposed to a vitalistic viewpoint, and thus to empirical research on the phenomena of life.

The Era of Darwin and Galton

The pace of biological research quickened, and many fundamental developments in technique and theory ensued in the following century. One of the cornerstones of biology was laid by Karl von Linné (better known as Linnaeus), who, in 1735, published *Systema Naturae* in which he established a system of taxonomic classification of all known living things. In so doing, Linnaeus emphasized the separateness and distinctness of species, and the view that species were fixed and unchanging became the prevailing one. This notion, of course, fit the Biblical account of creation. Not everyone was persuaded, however, that species were unchangeable. For example, Erasmus Darwin, in the latter part of the eighteenth century, suggested that plant and animal species appear to be capable of improvement or degeneration. A more influential view on this subject was promoted by Jean Baptiste Lamarck, who argued that the deliberate efforts of an animal could result in modifications of the body parts involved, and that the modification so acquired could be transmitted to the animal's offspring. For example:

We perceive that the shore bird, which does not care to swim, but which, however, is obliged . . . to approach the water to obtain its prey, will be continually in danger of sinking in the mud, but wishing to act so that its body shall not fall into the liquid, it will contract the habit of extending and lengthening its feet. Hence it will result in the generations of these birds which continue to live in this manner, that the individuals will find themselves raised as if on stilts, on long naked feet: namely, denuded of feathers up to and often above the thighs. (Packard, 1901, p. 234)

Changes of this sort presumably could accumulate and eventually the characteristics of the species could change. Lamarck was not the first to think that changes acquired in this manner could be transmitted to the next generation, but he crystallized and popularized the notion, and it has come to be called Lamarckism.

The strict and literal interpretation of the account in Genesis of the creation of the earth and its inhabitants was being challenged most seriously on the basis of geological evidence. The discovery of fossilized bones of animals deep in strata beneath the earth's surface proved difficult to accommodate to Bishop Ussher's calculations that the earth had been created in 4004 B.C. A theory of "catastrophism" had been put forward to account for these fossils. The Deity was regarded to have created and extinguished life on many successive occasions with catastrophes like floods and violent upheavals. Hence, the burial of the bones. Many geologists questioned whether catastrophic events needed to be invoked to account for the geological record, however. A school of "uniformitarians" believed that the processes at work in the past were the same as those at work in the present, and thus the accumulation of strata upon strata of rock required millions of years rather than the six thousand odd available from Bishop Ussher's postulated date of creation. A leader of this uniformitarian school of thought, and one of the dominant intellects of the time, was Charles Lyell (see Eiseley, 1959). Lyell published the first volume of his *Principles of Geology* in 1830, and one of the early copies found its way into the baggage of a young man about to embark upon what was probably the most important voyage in the history of scientific thought.

Erasmus Darwin's grandson Charles (Figure 1.2) had been a student of medicine at Edinburgh but was so unnerved by the sight of blood during surgery that he gave up further medical study and went to Cambridge, where, although a student of mediocre accomplishment, he received a degree in 1831. He appeared to be destined for a career as a clergyman when suddenly, and unexpectedly, through the recommendation of one of his old professors, he was nominated for the unpaid post of naturalist aboard H.M.S. *Beagle*, a survey ship of the Royal Navy about to embark on a long voyage. It was not uncommon for a naturalist to be taken on trips of this kind, and the young and devout captain of the *Beagle*, Captain Fitz Roy, was pleased at the prospect, for he expected a naturalist to be able to produce yet more data in support of "natural theology." A central theme of natural theology was the so-called "argument from design" in which the adaptation of animals and plants to the circumstances of their lives was taken as evidence of the Creator's wisdom. Such exquisite design, so the argument went, implied a designer. As exploration opened up hitherto unexplored parts of the world, new evidence of the Designer's works was uncovered, and it was with this end in mind that Captain Fitz Roy welcomed the young Darwin. During the next five years, in which Darwin experienced chronic seasickness, tropical

Figure 1.2
Charles Darwin as a young man.
(Courtesy of Trustees of the British
Museum [Natural History].)

fever, volcanic eruptions, earthquakes, and tidal waves, and the high adventure of encounters with rebels and life with Argentine gauchos, he filled many notebooks full of observations on fossils, primitive men, and various species of animals and their remarkable and detailed adaptation to their environments. A particularly compelling observation was the presence of thirteen species of finch in a very small area on the Galápagos Islands. The principal differences among these finches were in their beaks, and these differences were exactly appropriate for the different eating habits of the species. Somehow, thought Darwin, these birds had all been derived from a common ancestral group. "Seeing this gradation and diversity of structure in one small, intimately related group of birds, one might really fancy that from an original paucity of birds in this archipelago, one species had been taken and modified for different ends" (Darwin, 1896, p. 380). However, influenced by Lyell's book, and his own observations on geology as well as biology, Darwin was not inclined to make the argument in favor of design. Upon his return to England, Darwin began work on several reports summarizing his observations on coral reefs, on barnacles, and on other matters, while he gradually and systematically marshalled the evidence that species had evolved one from another, and pondered the possible mechanisms through which this evolution could occur. He shared his developing theory with a few friends, including Lyell himself, an eminent botanist named Hooker, and the young T. H. Huxley, and gradually convinced some but not all of them of the merit of his theory.

Realizing the kind of opposition that a theory that contradicted the Biblical account of creation would encounter, Darwin hesitated; he planned a monumental work in which he would present an overwhelming mass of evidence. His friends warned him that he should publish a brief version immediately, lest someone anticipate him, but he continued to work slowly and carefully on amassing the evidence and anticipating the objections. He did, however, take time to write out short sketches in correspondence with his friends and confidants. Finally, in 1858 the blow fell. A young man named Alfred Wallace had sent Darwin a manuscript for his comments. In it, with much less evidence than Darwin's in hand, he had arrived at essentially the same theory that Darwin had been developing for more than two decades. Darwin was greatly concerned over the course of action he should take. As he said in a letter to Lyell:

> I should be extremely glad now to publish a sketch of my general views in about a dozen pages or so; but I cannot persuade myself that I can do so honourably. Wallace says nothing about publication, and I enclose his letter. But as I had not intended to publish any sketch, can I do so honourably, because Wallace has sent me an outline of his doctrine? I would far rather burn my whole book, than that he or any other man should think that I had behaved in a paltry spirit. (Darwin, 1858)

Lyell and Hooker took initiative and resolved the issue by arranging for the simultaneous presentation of a sketch Darwin had prepared in 1844 and Wallace's current paper at a meeting of the Linnean Society in 1858.

With the theory now out into the open, Darwin began work on what he called an abstract. This "abstract," published in 1859 under the title *On The Origin of Species by Means of Natural Selection, or the Preservation of Favoured Races in the Struggle for Life,* proved to be one of the most influential books ever written.

The elements of Darwin's theory can be stated as follows. Within any species, many more individuals are born each generation than survive to maturity. Great variation exists among the individuals of a population. At least in part, these individual differences are due to heredity. If the likelihood of·surviving to maturity and reproducing is influenced, even to a slight extent, by a particular trait, offspring of the survivors and reproducers should manifest slightly more of the trait than did the generation from which their parents came. Thus, bit by bit the characteristics of a population can change, and over a sufficiently long period the cumulative changes would be so great that in retrospect we would call the later and the earlier populations different species. The catch phrase that came to characterize this theory was "survival of the fittest." It would more appropriately be called "differential reproduction of the fittest" since mere survival is necessary but not sufficient. As Darwin himself put it:

> Owing to this struggle for life, any variation, however slight and from whatever cause proceeding, if it be in any degree profitable to an individual of any species, in its infinitely complex relations to other organic beings and to external nature, will tend to the preservation of that individual, and will generally be inherited by its offspring. The offspring, also, will thus have a better chance of surviving, for, of the many individuals of any species which are periodically born, but a small number can survive. (1869, p. 61)

This principle was called natural selection, and it is clear that Darwin considered behavioral characteristics to be just as subject to natural selection as are physical traits. In the *Origin of Species* an entire chapter is given to the discussion of instinctive behavior patterns, and in a later book, *The Descent of Man and Selection in Relation to Sex*, detailed consideration is given to comparisons of mental powers and moral senses of animals and man, and to the development of intellectual and moral faculties in man. In these discussions Darwin was satisfied that he had demonstrated that the difference between the mind of man and the mind of animals "is certainly one of degree and not of kind" (1873, p. 101) — an essential point, since one of the strongest objections to the theory of evolution was the qualitative gulf that was supposed to exist between the mental capacities of man and of lower animals. All the behavioral traits cited in support of this idea must be, by implication, inherited, since, for Darwin, it is only the heritable traits that have long-range evolutionary significance.

In an explicit summary statement, based largely on observations of "family resemblance," Darwin said:

> So in regard to mental qualities, their transmission is manifest in our dogs, horses, and other domestic animals. Besides special tastes and habits, general intelligence, courage, bad and good temper, etc., are certainly transmitted. With man we see similar facts in almost every family; and we now know through the admirable labors of Mr. Galton that genius, which implies a wonderfully complex combination of high faculties, tends to be inherited; and, on the other hand, it is too certain that insanity and deteriorated mental powers likewise run in the same families. (1873, Vol. I, pp. 106–107)

It was most crucial for the evolutionary theory that heritable variation be present in each generation, or evolution could not continue. But, by the then commonly accepted principle that characteristics merged or blended in offspring, it was apparent that variability of a trait would be roughly halved in each generation, and would rapidly diminish to a trivial level, were it not replenished in some manner. Darwin (1868) devoted much attention to the causes of variability and concluded that changes in the conditions of life in some way altered the reproductive systems of animals in such a manner that their offspring were more variable than they would have been under stable conditions. Ordinarily, this increased variability would be random — natural selection would then preserve those deviants that by chance happened to be the better adapted as a consequence of their deviation. Sometimes, however, particularly if continued for a number of generations, an environmental condition might induce *systematic* change — the environment directly inducing changes making organisms more adapted to it.

Another source of variability was presumed to be the effects of use and of disuse:

> Increased use adds to the size of a muscle, together with the bloodvessels, nerves, ligaments, the crests of bone to which these are attached, the whole bone and other connected bones. So it is with various glands. Increased functional activity strengthens the sense-organs. Increased and intermittent pressure thickens the epidermis; and a change in the nature of the food sometimes modifies the coats of the stomach, and increases or decreases the length of the intestines. Continued disuse, on the other hand, weakens and diminishes all parts of the organization. Animals which during many generations have taken but little exercise, have their lungs reduced in size, and as a consequence the bony fabric of the chest, and the whole form of the body, become modified. (1868, Vol. II, p. 423)

Likewise, with respect to behavioral characteristics, "some intelligent actions . . . after being performed during many generations, become converted into instincts, and are inherited" (1873, Vol. I, p. 36) and "It is not improbable that virtuous tendencies may through long practice be inherited" (1873, Vol. II, p. 377).

It should be noted that Darwin was not completely satisfied with the doctrine that characters acquired by use, disuse, or environmental modification could be transmitted to subsequent generations (see Fisher, 1958). Yet such a mechanism seemed to be necessary to explain some of the facts. As we shall see, a vigorous controversy has persisted over this theory, which is generally described as Lamarckian.

The *Origin of Species* caused a violent reaction. Fierce denunciation came from those whose sensibilities were shocked by this contradiction of the Biblical account of creation. There was opposition, too, from other scientists, whose favorite theories were challenged by the new conceptions.

In the United States, for example, the eminent Louis Agassiz of Harvard University was a confirmed anti-evolutionist; he was opposed by Asa Gray, also of Harvard University (and a recipient of one of Darwin's prepublication "sketches" of his theory), and a lively controversy followed at Harvard and elsewhere.

Darwin's health had failed, probably because of long-term consequences of a tropical infection he contracted while on the *Beagle*, and this, combined with his great timidity with respect to public encounters, left the actual debating of the issue in Britain largely up to his friends. This distinguished group, which included Wallace, Lyell, Hooker, and Huxley, was certainly equal to the task. In what must be the most famous confrontation in the history of biology, Huxley found himself opposed to Bishop Wilberforce at a meeting of the British Association for the Advancement of Science in 1860. Wilberforce, who was an extremely effective public speaker, had been carefully coached by an anti-Darwinian named Richard Owen. After some stirring oratory on the matter, Wilberforce turned to Huxley and inquired whether it was through his grandfather or grandmother that he claimed to be descended from the apes. A letter of Huxley's describes his response:

> When I got up I spoke pretty much to the effect—that I had listened with great attention to the Lord Bishop's speech but had been unable to discover either a new fact or a new argument in it—except indeed the question raised as to my personal predilections in the matter of ancestry—That it would not have occurred to me to bring forward such a topic as that for discussion myself, but that I was quite ready to meet the Right Rev. prelate even on that ground. If then, said I, the question is put to me would I rather have a miserable ape for a grandfather or a man highly endowed by nature and possessing great means and influence and yet who employs those faculties and that influence for the mere purpose of introducing ridicule into a grave scientific discussion— I unhesitatingly affirm my preference for the ape. (Montagu, 1959, pp. 2–3)

Huxley's rejoinder caused quite an uproar. In the hubbub one particularly plaintive voice was heard. Robert Fitz Roy, now an admiral and the contributor of a paper on meteorology to an earlier session of the Association meetings, waved a Bible, and denounced the views of his old friend and shipmate. After order was restored, Hooker was called upon to speak, and he too spoke persuasively in favor of the Darwinist view. In many ways this meeting represented a turning point, for after two of the most eminent scientists in Britain publicly defended the evolutionary theory, thinking men everywhere had to take it seriously.

Galton's Contributions

Among the supporters and admirers of Darwin at this time was another one of Erasmus Darwin's grandsons, Francis Galton (Figure 1.3). Already Galton had established something of a reputation as a geographer, explorer,

Figure 1.3
Francis Galton, in 1840, from
a portrait by O. Oakley. (Courtesy
of the Galton Laboratory.)

and inventor. He had, for example, invented a printing electric telegraph, a type of periscope, and a nautical signalling device by the time that *Origin of Species* was published. The effect that this book had upon him is revealed in a letter that he later wrote to Darwin:

> I always think of you in the same way as converts from barbarism think of the teacher who first relieved them from the intolerable burden of their superstition. I used to be wretched under the weight of the old-fashioned arguments from design; of which I felt though I was unable to prove to myself, the worthlessness. Consequently the appearance of your *Origin of Species* formed a real crisis in my life; your book drove away the constraint of my old superstition as if it had been a nightmare and was the first to give me freedom of thought. (Pearson, 1924, Vol. I, Plate II)

The *Origin of Species* directed Galton's immense curiosity and talents to biological phenomena and he soon developed what was to be a central and abiding interest for the rest of his life: the inheritance of mental characteristics.

Hereditary Genius. In 1865 two articles by Galton, jointly entitled "Hereditary Talent and Character," were published in *Macmillan's Magazine*. Four years later a greatly expanded discussion was published under the title, *Hereditary Genius: An Inquiry into its Laws and Consequences*.

The general argument presented in this work is that among the relatives of persons endowed with high mental ability is to be found a greater number of other extremely able individuals than would be expected by chance;

furthermore, the closer the family relationship, the higher the incidence of such superior individuals.

Applying Quetelet's "law of deviation from an average," at the time a fairly recent development, but later to become familiar as the normal curve, Galton distinguished fourteen levels of human ability, ranging from idiocy through mediocrity to genius.

No satisfactory way of quantifying natural ability was available, so Galton had to rely upon reputation as an index. By "reputation" he did not mean notoriety from a single act, nor mere social or official position, but "the reputation of a leader of opinion, of an originator, of a man to whom the world deliberately acknowledges itself largely indebted" (1869, p. 37). The designation "eminent" was applied to those individuals who constituted the upper 250-millionths of the population (i.e., one in 4,000 persons would attain such a rank), and it was with such men that the discussion was concerned. Indeed, the majority of individuals presented in evidence were, in Galton's estimation, the cream of this elite group, and were termed "illustrious." These were men whose talents ranked them one in a million.

On the basis of biographies, published accounts, and direct inquiry, Galton evaluated the accomplishments of eminent judges, statesmen, peers, military commanders, literary men, scientists, poets, musicians, painters, Protestant religious leaders, and Cambridge scholars, and their relatives. (Oarsmen and wrestlers of note were also examined to extend the range of inquiry from brain to brawn.) In all, nearly 1,000 eminent men were identified in the 300 families examined. With the over-all incidence of eminence only 1 in 4,000, this result clearly illustrated the tendency for eminence to be a familial trait.

Taking the most eminent man in each family as a reference point, the other individuals who attained eminence (in the same or in some other field of endeavor) were tabulated with respect to closeness of family relationship. Briefly stated, the results showed that eminent status was more likely to appear in close relatives, with the likelihood of eminence decreasing as the degree of relationship becomes more remote.

Galton recognized the possible objection that relatives of eminent men would share social, educational, and financial advantages, and that the results of his investigation might be interpreted as showing the effectiveness of such environmental factors. To demonstrate that reputation is an indication of *natural* ability, and not the product of environmental advantages, three arguments were presented. First, Galton stressed the fact that many men had risen to high rank from humble family backgrounds. Second, it was noted that the proportion of eminent writers, philosophers, and artists in England was not less than that in the United States, where education of the middle and lower socioeconomic classes was more advanced. The educational advantages in America had spread culture more widely, but had not produced more persons of eminence. Finally, a comparison was made between the success of adopted kinsmen of Roman Catholic Popes, who were given great social advantages, and the sons of eminent men, and the latter were judged to be the more distinguished group.

In Galton's view, men of mediocre talents might be suppressed by environmental obstacles, but inherited genius will out, regardless of adversity, and no amount of social or educational advantage can serve to raise a man to eminence unless he possesses inherited natural ability.

Pioneering Research in Psychology and Statistics.   In order to further his researches, it was necessary for Galton to have means for assessing mental characteristics. In a prodigious program of research, he developed apparatus and procedures for measuring auditory thresholds, visual acuity, color vision, touch, smell, judgment of the vertical, judgment of length, weight discrimination, reaction time, and memory span. In addition, a questionnaire technique was employed to investigate mental imagery, and association of ideas was studied by introspection. One particularly intriguing, although not especially successful, investigation involved the use of composite portraiture, whereby the photographs of a number of individuals could be superimposed to yield their common features. These composite photographs were then used in an effort to determine what relationship, if any, existed between the facial characteristics of certain groups and various attributes of their intelligence, personality, morality, and health.

The problems of properly expressing and evaluating the data obtained from such researches were formidable, and Galton also turned his remarkable energies to statistics, pioneering in the development of the concepts of the median, percentiles, and correlation.

It was, of course, desirable to have data from large numbers of individuals, and various stratagems were employed to this end. For example, Galton arranged for an "Anthropometric Laboratory for the measurement in various ways of Human Form and Faculty" to be located at an International Health Exhibition. Some 9,337 people paid four-pence each for the privilege of being measured for various bodily and sensory characteristics! On another occasion a contest was sponsored in which awards of £7 were given to those submitting the most careful and complete "Extracts from their own Family Records." Thus did Galton obtain a large number of pedigrees that he could examine for evidence of human inheritance.

Twins and the Nature–Nurture Problem.   Of special relevance to the present topic is Galton's introduction (1883) of the twin-study method to assess the effectiveness of *nature* (inheritance) and *nurture* (environment). The essential question in this examination of twins was whether twins who were alike at birth became more dissimilar as a consequence of any dissimilarities in their nurture, and conversely, whether twins unlike at birth became more similar as a consequence of similar nurture. Galton acknowledged two types of twins: those arising from separate eggs, and those arising from separate germinal spots on the same egg, yet he did not distinguish between the two types in his discussion, except as they fell into his "alike at birth" or "unlike at birth" categories. Gathering his evidence from answers to questionnaires and biographical and autobiographical material, Galton observed that, among 35 pairs of twins who had been very much alike at birth, and who had been reared under highly similar conditions, the similarities within the twinships persisted after the members had grown to adulthood and gone more-or-less separate ways.

From 20 pairs of originally dissimilar twins, there was no compelling evidence that any had become more alike through being exposed to similar environments.

> There is no escape from the conclusion that nature prevails enormously over nurture when the differences of nurture do not exceed what is commonly to be found among persons of the same rank of society and in the same country. My fear is, that my evidence may seem to prove too much, and be discredited on that account, as it appears contrary to all experience that nurture should go for so little. (1883, p. 241)

Galton's Work in Perspective.   The ten years between *Origin of Species* and *Hereditary Genius* had not been sufficient for the idea of man as an animal to be completely accepted. For many of those who accepted Darwin, of course, Galton was a natural and logical extension: man differs from animals most strikingly in mental powers; man has evolved as have other animals; evolution works by inheritance; mental traits are heritable. For those whose faith in the special creation of man remained firm, Galton was unacceptable, atheistic, and reprehensible.

Even among those not arguing primarily on theological grounds, there were wide differences of opinion about the proper frame of reference for the study of man. In psychiatric theorizing, for example, some views were based upon the concept that human behavior is determined by biological processes, and that no adequate theory of mental functioning or malfunctioning could disregard man's fundamentally animal nature. On the other hand, there were those who chose to regard the "psyche" as capable of investigation in and of itself, with organic processes ignored as irrelevant (see White, 1948).

There were also scholars whose inquiries stemmed, not from interest in psychiatric problems, but from a general desire to understand "mind." This philosophical approach was dominated by the British philosophers, whose emphasis was clearly on experience and thus on "nurture," having been inspired by Locke's seventeenth-century *tabula rasa* dictum that ideas are not inborn, but come from experience. The role of experience was also emphasized by experimental psychology, which is usually dated from Wilhelm Wundt's establishment in 1879 of the Psychologisches Institut at Leipzig. In spite of the fact that Wundt had come to psychology from physiology, his approach was not biological in the same sense as Galton's, and the goal at Wundt's institute was the identification, through introspection, of components of consciousness. Individual differences, which formed the very heart of Galton's investigations, were nuisances in this search for principles having general application to all. One notable exception to this general trend was provided by an American named J. McK. Cattell, who, as a student of Wundt, insisted on studying individual differences. After Cattell left Leipzig, he worked for a while with Galton, and had his belief in the importance of individual differences strengthened and confirmed. Cattell had an important influence on the development of American psychology, and inspired some of the earliest experimental work in behavioral genetics.

From the foregoing it may be seen that Galton's work was neither completely in step nor completely out of step with the times. As it happened, Galton lived during the period of greatest intellectual turmoil in biology. His work was both a product and a causal factor of the advances made. Galton was not the first to insist upon the importance of heredity in traits of behavior. We have seen explicit statements on this matter by the ancient Greeks. Nor was Galton the first to place his conclusions in an evolutionary context. Spencer had introduced an "evolutionary associationism" in 1855 (Boring, 1950, p. 240). But it was Galton who championed the idea of inheritance of behavior, who vigorously consolidated and extended it. In effect, we may regard Galton's inspired efforts as the founding of behavioral genetics.

Theories of Inheritance

For Darwin and Galton the idea that characteristics were transmitted from generation to generation was an essential concept. There was substantial evidence of the importance of heredity, but its laws had proved extremely resistant to analysis. In particular, a vast amount of data had been accumulated from plant and animal breeding. Many offspring bore a closer resemblance to one parent than to the other; also common were situations in which the appearance of offspring was intermediate between the two parents. But two offspring from the very same parents could be quite unlike. As Lush described the situation considerably later, the first rule of breeding was that "like produces like," while the second rule was that "like does not always produce like" (Lush, 1951, p. 496).

Pangenesis. The theory of heredity that seemed to explain most adequately the confusion of facts at the time was the "provisional hypothesis of pangenesis" as described by Darwin. On this view, the cells of the body, besides having the power, as is generally admitted, of growing by self-division, throw off free and minute atoms of their contents, that is gemmules. These multiply and aggregate themselves into buds and the sexual elements. (1868, p. 481)

Gemmules were presumably thrown off by each cell throughout its course of development. In embryogenesis and later development, gemmules from the parents, originally thrown off during various developmental periods, would come into play at the proper times, and thus direct the development of a new organ like that from which they had arisen.

If a body part were modified by use or disuse, the gemmules then cast off by the cells of that part would also be modified, and thus acquired characteristics could be transmitted to the offspring. Of specific interest to our present topic, we may note Darwin's statement:

With respect to mental habits or instincts, we are so profoundly ignorant on the relation between the brain and the power of thought that we do not know whether an inveterate habit or trick induces any change in the nervous system; but when any habit or other mental attribute, or insanity, is inherited, we must believe that some actual modification is transmitted; and this implies, according to our hypothesis, that gemmules derived from modified nerve-cells are transmitted to the offspring. (1868, p. 472–473)

Galton took issue with some of the features of pangenesis, and performed a study that was a direct attempt to determine if gemmules from one breed of rabbit would affect the progeny of another breed after blood transfusions between breeds. This experiment, which, incidentally, was performed in collaboration with Darwin, had a quite negative outcome. Galton doubted the veracity of the concept of the inheritance of acquired characteristics. A substantial, but on the whole friendly, disagreement grew up between Darwin and Galton on these issues, with the latter publishing extensively on a revised theory. Galton's revision foresaw many of the later developments, but to Gregor Mendel must go the credit for providing the basic answer to the riddle of inheritance.

Mendel's Experiment and Theory.   Mendel (Figure 1.4) was an Augustinian monk who conducted his critical researches on pea plants in the garden of a monastery at Brunn, Moravia.

Much of the information concerning heredity available at the time had been derived from "plant-hybridization" experiments in which plants of different species were crossed. Among the difficulties of this approach, two of the most important were that the offspring of many such crosses were sterile or of low fertility, with the result that succeeding generations were

Figure 1.4
Gregor Johann Mendel, 1822–1884.
A picture taken at the time of his research.
(Courtesy of V. Orel, Mendel Museum,
Brno, Czechoslovakia.)

impossible or difficult to obtain, and that the features which had been investigated were generally too complex for clear analysis. Mendel's success can be attributed in large part to his circumventing these problems. By crossing different varieties within the same species, Mendel obtained viable and fertile offspring, and thus was able to proceed to hybrids of the second generation. By concentrating his attention on simple qualitative characters, he was able to make a thorough analysis, uncluttered by problems of measurement or distinction of categories. Curiously, Mendel's greatest innovation seems to have been his insistence on counting all the progeny rather than being content with a verbal summary of the typical result. This was, of course, made convenient by the choice of characters that could easily be assigned to one of two alternative classes.

In all, some seven morphological characters were investigated, and uniform results were obtained with respect to all. In the first-generation hybrid offspring (later named the first filial generation, abbreviated as F_1) of crosses between plants differing in any one of the characters, all plants were uniform, and like one of the parents. That parental character which appeared in the F_1 was called *dominant*; the parental character which was not expressed was called *recessive*. From crosses between two F_1 parents, plants showing the dominant trait and plants showing the recessive trait were found among the offspring (the second filial generation, or F_2) in a definite 3:1 ratio, but no plants were found that were intermediate. Furthermore, it was found that the recessive plants "bred true" — when self-pollinated, all of their offspring showed the recessive character. One-third of the dominant plants also bred true, but two-thirds yielded both types of progeny.

To account for these results, Mendel postulated that each parent possessed two elements which determined the particular trait. Each parent transmitted one of its elements to an offspring. If the parents differed in a characteristic, an element contributed by the one parent might be dominant over that contributed by the other parent, and then the offspring would resemble the former. Nonetheless, the recessive element would not be contaminated in any way by its association with the dominant element. When the individual offspring in turn had offspring, it would pass on the element that it had received from each parent to one-half of its progeny — and the recessive element transmitted would not differ in any way from the recessive element received from the original parent. This was Mendel's first law, the law of *segregation*. Thus, the *gametes* (male and female germ cells, or sex cells) were regarded as pure and essentially inviolable. Now, when such a hybrid offspring (F_1) is self-pollinated (or when two such hybrids are cross-pollinated), the male and female gametes, which each contain only one of the elements, unite at random. Thus if A_1 represents the dominant element and A_2 the recessive, each hybrid is A_1A_2, but each gamete produced by the hybrid will be either A_1 or A_2. When two hybrids are crossed, yielding an F_2 generation, the following combinations can occur: A_1A_1, A_1A_2 and A_2A_2, and these will occur in a $1:2:1$ ratio. Because of dominance, the A_1A_1 will not be distinguishable from the A_1A_2, except by examination of their offspring, so that the observable character will be displayed in a $3:1$ ratio.

The second major law was the law of *independent assortment*. This principle was discovered when parents differing in two or more characteristics were crossed. For example, a plant having yellow, round peas was crossed with one having green, wrinkled peas. The first-generation hybrid plants uniformly had yellow, round peas, for these elements are dominant. In the generation resulting from the self-pollination of the F_1 plants, the characteristics were combined at random. The elements for yellow and round were not bound together simply because they had been associated in that combination in the "grandparents." The elements, indeed, were sorted out at random, hence the name "independent assortment."

Mendel's results and theory were read to the Brunn Society of Natural Science in 1865, and were later published in the proceedings of the society. The crucial experiments had, therefore, been done and reported prior to Darwin's most complete statement of pangenesis, and before *Hereditary Genius*. But Darwin and Galton were not alone in overlooking Mendel's ideas. For thirty-four years, the "Versuche über Pflanzen-Hybriden" (Mendel, 1866) remained almost completely unheeded.

In 1900, three investigators — Correns, de Vries, and von Tschermak — almost simultaneously "rediscovered" Mendel's work, and a period of intensive research was inaugurated in which the Mendelian results were confirmed and extended. Some modifications ensued. Not all elements displayed dominance; there were cases in which the hybrid offspring were intermediate to the parents. Nonetheless, the elements emerged unchanged in later generations. The "purity of gametes" held in spite of the initial appearance of blending. Furthermore, it was found that the law of independent assortment did not hold absolutely. Sometimes assortment was not at random, but elements determining characters tended to stick together in the gametes produced by an individual in the same relationship as in the gametes that produced the individual.

The vigorously developing area of research was given the name *"genetics"* by Bateson in 1905, and the name *"gene"* was proposed for the Mendelian elements by Johannsen in 1909. At the same time, Johannsen made a fundamental distinction between *genotype*, which is the genetic composition of the individual, and *phenotype*, the apparent, visible, measurable characteristic. The importance of this distinction is that it makes clear that the observable trait is not a perfect index of the individual's genetic properties. Given a number of individuals of the same genotype, we might nonetheless expect differences among them — differences caused by environmental agents. Thus, two beans might be from the same "pure line," and have identical genotypes for size, yet one might be larger than the other because of such differences in "nurture," as soil conditions. Nevertheless, their genotypes would remain unaffected, and the beans of the plants grown from these two beans would be of the same average size. The inheritance of "acquired" characters obviously has no place in this scheme.

Modifiability of the Genes.   Mendel's conclusion concerning the "purity of the gametes" and Johannsen's demonstration that environmental modification of a phenotype does not alter the genotype present a view of the genes as being highly stable and well insulated from the effects of environment. There were, however, many observations which showed that the stability of the genes is a relative matter. On occasion, a given gene might undergo a more-or-less permanent change, called a *mutation*. The reasons for such alteration in genes are still incompletely understood, but significant advances have been made since the discovery in 1927 by Muller that irradiation of *Drosophila* increases the rate of gene mutation. Since this discovery of the *mutagenic* effect of X-rays, other agents for experimentally inducing mutations have been discovered, including certain chemical compounds and extreme temperatures, and the mutability of the hereditary material of other species has been demonstrated. Thus, certain environments can bring about changes in genotype, but the phenomenon differs greatly from the old notion of inheritance of acquired characters. Under that scheme, either the environment was thought capable of bringing about systematic changes or else the organism, by use or disuse of body parts, was thought to cause a change, making itself better adapted to the environment, with the adaptation being transmissible to subsequent generations. Mutations, however, induced by X-ray and other mutagenic agents as well as those occurring "spontane-ously," are apparently random — the mutation might affect eye color or wing shape or any of a large number of such characteristics, but the organisms are not necessarily better adapted as a result. In fact, the mutations that occur seem much more likely to be deleterious than advantageous to the organism.

The capability of experimentally inducing mutations has proved to be of marked value in genetic research, and has contributed greatly to the eluci-dation of the molecular structure of genes and of the biochemistry of gene action.

Progress in the understanding of mutations has also been of importance to evolutionary theory. It may be recalled that Darwin took great pains in considering the possible sources of heritable variation, and somewhat reluc-tantly concluded that Lamarckian mechanisms are among the important factors. Contemporary evolutionary theory views mutation as the ultimate source of the genetic variability upon which natural selection operates.

The Extension of Mendelian Theory to Quantitative Characteristics. Throughout the early period of enthusiastic research following the redis-covery of Mendel's laws, Galton's biometrical approach to problems of the inheritance of continuously varying characteristics (i.e., those not readily assigned to a few distinct classes or categories) had been pursued vigorously, notably by Pearson. Rather than finding mutual support in each other's work, the Mendelians and the biometricians came into acute conflict. It was difficult for the Mendelians to reconcile continuous variation with the type of qualitative, discrete difference, mediated by particulate genes, with which they had worked. The biometricians, on the other hand, supported the blending hypothesis, and were inclined to regard the Mendelian type of inheritance as an unimportant exception to the general rule. With justifica-tion, they pointed to the obvious importance of the smoothly continuous, quantitative characteristics, such as height, weight, intelligence, and so on. It was apparent, to the biometricians, at least, that the type of thing investi-gated by Mendelians — causing qualitative differences, and usually abnor-malities — could not possibly account for such continuous distributions.

The groundwork for the resolution of this conflict had been provided, in fact, by Mendel himself, when he suggested that a certain characteristic might be due to two or three elements. General acceptance of this idea, however, was not forthcoming until the work with plants of Nilsson-Ehle (1908) and of East and collaborators (East and Hayes, 1911; Emerson and East, 1913). These researchers showed that if it was assumed that a *number* of gene pairs, rather than just one pair, each exerted a small and cumulative effect upon the same character, and if the effects of environment were taken into consideration, the final outcome would be an apparently continuous distribution of the characteristic instead of dichotomous categories such as had been featured in the typical Mendelian researches. This was quite different from the blending hypothesis for, in this *multiple-factor* hypothesis, the hereditary determiners were not presumed to vary continuously in nature from individual to individual, thus determining a gradation of the characteristic in the population. Rather, the genes were acknowledged to occur in discretely alternative states (typically two, sometimes more), but when a number of such discrete units bear upon the same character, the final outcome approximates a continuous distribution, just as the simultaneous tossing of a number of coins that can each express only one of two "states"— heads or tails—can have a large number of possible outcomes. Elaborate statistical development of this notion was provided by Fisher (1918) and by Wright (1921), and this work presented convincing demonstrations that the biometrical results in fact follow logically from this multiple-factor extension of Mendel's theory. The blending hypothesis was gradually discarded, and as early as 1914 Bateson could remark, "The question is often asked whether there are not also in operation systems of descent quite other than those contemplated by the Mendelian rules . . . none have been demonstrated" (Mather, 1951, p. 111).

Studies on Eminent and "Degenerate" Families

Although most of the major advances in genetics resulted from research on plants and lower animals, many studies were made of the inheritance of a wide variety of characteristics in humans. Many of the characteristics studied were abnormalities of one kind or another, and in time, hereditary defects were identified in almost all organ systems (see Gates, 1946). Almost any organ or tissue, and particularly components of receptor, effector, and associative systems, may play a role in behavior, so that in a very real sense the discovery of the genetic basis for colorblindness, deaf-mutism, and certain forms of ataxia, for example, have been contributions to behavioral genetics. At present, however, we shall emphasize the early developments in respect to those traits that fall within the customary definitions of intelligence, aptitude, mental deficiency, psychosis, neurosis, and personality.

Eminent and Royal Families.   Several extensive surveys subsequently used Galton's procedure of investigating the accomplishments of relatives of notable people. Royal families provided particularly convenient source material, owing to the easy availability of their genealogical records (Woods, 1906; Gun, 1930a, 1930b). One disadvantage of this line of inquiry, however, was the sometimes dubious correspondence of legal and biological parentage. For example, Gun, in discussing King James I, remarks, "his characteristics have but little resemblance to those of any of his ancestors. This

fact was so obvious that from an early period doubts arose as to his parentage, some considering that he was the son of Mary by David Rizzio, while
others contended that he was a changeling" (1930b, p. 195). Nevertheless,
Gun was convinced that the family histories clearly showed the inheritance
of certain traits. Thus the Stewarts were said to be characterized by tact-
less obstinancy, which "ran like a thread down the direct male line" (1930b,
p. 201). The Tudors, on the other hand, were thought to be hereditarily
endowed with love of learning.

The "Jukes."   Dugdale, on the other hand, had concerned himself with
the other end of the scale of social merit. As a member of the executive committee of the Prison Association of New York, Dugdale was named a committee of one to inspect county jails. In one county he was impressed by
finding six of the prisoners to be related. Undertaking an intensive survey
of this family, he was able to trace the lineage back to six sisters, to whom
he gave the pseudonymous label "Jukes." One of the six had left the country
and was not traceable. The remaining five had provided a most striking
posterity, characterized by criminality, immorality, pauperism, and feeble-
mindedness. Dugdale was primarily a social reformer, and was rather cau-
tious in assigning the causal role in this pedigree of sordidity to nature or to
nurture. That there was a social problem was clear enough. Dugdale (1877)
estimated the cost to the state, in welfare relief, institutional care, and so on
to exceed one million dollars over a seventy-five-year period. In 1911 some
of Dugdale's original manuscripts were found, which gave the real name of
the Juke family. Estabrook (1916), acting upon this information, was able to
trace the family history over the forty years that had ensued since the first
study. Estabrook summarized his study as follows:

> For the past 130 years they have increased from 5 sisters to a family which
> numbers 2,094 people, of whom 1,258 were living in 1915. One half of the
> Jukes were and are feeble-minded, mentally incapable of responding normally to
> the expectations of society, brought up under faulty environmental conditions
> which they consider normal, satisfied with the fulfillment of natural passions
> and desires, and with no ambition or ideals in life. The other half, perhaps nor-
> mal mentally and emotionally, has become socially adequate or inadequate,
> depending on the chance of the individual reaching or failing to reach an environ-
> ment which would mold and stimulate his inherited social traits. . . . Heredity,
> whether good or bad, has its complemental factor in environment. The two
> determine the behavior of the individual. (1916, p. 85)*

This conclusion was reasonably modest, assigning importance to both
heredity and environment, but the findings of the study were enthusiasti-

*Reproduced courtesy of Carnegie Institution.

cally endorsed by the more ardent hereditarians, and came to be regarded as proof of "morbid inheritance."

Various criticisms have been leveled at the Juke studies, and at a number of similar studies that followed. Perhaps the most cogent objection raised was that members of the families shared similar environments as well as a common lineage. Thus, although the more-or-less anecdotal evidence could be accepted as presenting a dismaying picture of human degradation, there was no means of determining the relative contributions of environment and heredity.

The "Kallikaks." In 1912 Goddard published an account of a family that, in his view, provided a clear-cut resolution of the problem of disentangling nature and nurture. This family consisted of two branches, each of which could be traced back to the same man. According to the report, "Martin Kallikak" (again, a pseudonym), while a soldier in the Revolutionary War, had an affair with a feeble-minded girl whom he met in a tavern. When the girl gave birth to a son, she named him "Martin Kallikak, Jr." After the war, Martin, Sr. returned home, married a girl of good family, and began the other branch of the family. Among 480 descendants of the illicit affair, a very "Juke-like" picture was presented. Among the descendants of the marriage, almost all were normal, good members of society.

These results were taken to demonstrate that feeble-mindedness, which was regarded as the root of all the family difficulties, was inherited. A discussion of Mendelian principles was provided in the report, but judgment was reserved on whether feeble-mindedness is a unit character, caused by a single gene.

The investigation of the Kallikaks was carried out largely by a field worker interviewing members of the family and people who knew members of the family. In discussing the general methodology, Goddard stated that, although the evidence was occasionally ambiguous, and judgment had to be withheld, the field worker could usually decide easily the mentality of the persons interviewed. He also defended the assessment of the intellect of deceased individuals by interview of acquaintances, which was part of the procedures used in the study.

Criticism of Family Studies. As with the Jukes, the Kallikak findings were widely hailed in some circles, and vigorously criticized in others. In 1942 Goddard wrote a defense of the study, replying to some of the principal critics. To the criticism that assessment by a field worker was unreliable, Goddard replied that the field worker was well trained, and from familiarity with patients in mental institutions, could adequately judge mental level. Furthermore, if there were doubt regarding a person's mental level, it was recorded as "undetermined." To the objection that the evidence that Martin, Sr. was the father of Martin, Jr. was scant, and would not be acceptable in a court case, Goddard simply replied, "A strange statement.

Courts have always accepted such evidence and still do. In this case there was not even a doubt" (1942, p. 575).

These answers were not very satisfying, and one of the strongest critics, Scheinfeld (1944), retorted in detail. Particularly, he remained unconvinced that the evidence for the younger Martin's paternity—"a single short sentence, *unaccompanied by any documentation or supporting evidence*" (p. 262)—could serve the purposes of a scientific investigation. If this particular point is not adequately demonstrated, of course, the whole study becomes meaningless. Scheinfeld also remained unimpressed by the unsupported claim of accuracy in diagnosing the mental attributes of the living, not to mention the dead, Kallikaks.

But even if the above could all be allowed, there remained another fundamental, and indeed a vitiating, problem. This concerns Goddard's failure to consider seriously the possibility that differences in environment might have been strong factors in creating at least some of the disparity between the two Kallikak branches. "This possibility he dismissed lightly by saying that the bad Kallikaks . . . are not open to this argument," and "that we are dealing with a problem of true heredity no one can doubt" (Scheinfeld, 1944, p. 262). Such a major issue cannot be so easily disposed of, and, in fact, the impossibility of separating genetic and environmental effects renders the whole study pointless.

The objections raised to the studies of eminent and degraded families are telling, and by current standards we must judge that, whatever their worth as sociological documents, these studies merely serve to confuse the problem of determining the relative influence of nature and nurture.

Pedigree Studies on Mental Defects

Aside from the large-scale efforts just described, there were numerous smaller pedigree studies of many families with relatively fewer individuals studied per family. In a review of the literature to 1912, Davenport (1912) was in fact able to present data on musical ability, artistic composition, literary composition, mechanical skill, calculating ability, memory, temperament, handwriting, pauperism, narcotism, criminality, and feeble-mindedness.

Most of these studies are susceptible to the same type of criticism as that applied to the Kallikak study, but in the subsequent research on one of the topics, feeble-mindedness, the pedigree approach achieved its most substantial success as applied to the problems of behavioral genetics.

One of the most influential of the early publications was that of Tredgold (1908). In examining the family histories of some 200 cases of "every grade and variety of amentia," he concluded that there were two basic causal factors: intrinsic (hereditary) and extrinsic (environmental), and he regarded the former to be of "immense importance," accounting for some 80 percent of the cases. The roles of parental age and intoxication at the time of conception were specifically examined and judged to be of trivial importance. Other factors, however, were thought to be very effective in bringing about deterioration in the germ cells. After discussing the Mendelian hypothesis that gametes are unaffected by environment, Tredgold rejected it as being inconsistent with the experiences of physicians.

> With regard to the causation of amentia, I believe that there are certain diseases which bring about a deterioration of the germ plasm. The chief of these are alcoholism and consumption [tuberculosis]. . . . In consequence, there results a pathological change in that part of the offspring which is at once the most elaborate, the most vulnerable, and of most recent development — namely, the cerebral cortex. This change consists in a diminished control of the higher, and increased excitability of the lower, centres, and is manifested clinically as neurasthenia, hysteria, migraine, and the milder forms of epilepsy. We may say that a neuropath has been created. Should the adverse environment continue, or should such a person marry one similarly tainted, then the nervous instability becomes accentuated in the following generation, and insanity, the graver forms of epilepsy, and early dementia, make their appearance. (1908, pp. 36–37)

Thus the various traits mentioned, ranging from neuroses through insanity to profound mental deficiency, were regarded to be all the outcome of successive stages in a hereditary deterioration set in action by some environmental factor.

Two years after his study of the Kallikaks was published, Goddard (1914) presented an extensive collection of pedigrees of mentally defective patients at the Vineland Training School. The Binet-Simon Measuring Scale of Intelligence was administered to a number of the inmates, but the remainder of the pedigrees were primarily studied by fieldworker interviews. After studying the pedigrees, Goddard concluded that, of 327 families investigated, the mental defectiveness was certainly inherited in 164, and probably inherited in another 34 cases. The remaining cases were described as due to accident (57), having no determined cause (8), unclassified (27), and neuropathic (37). The latter group was composed of patients whose families had little or no history of feeble-mindedness per se (apart from the institutionalized patient), but in which many other conditions, such as alcoholism, paralysis, suicidal tendency, nervousness, and so on, were prevalent. For Tredgold this was the typical picture in inherited mental deficiency. Goddard thought that the feeble-mindedness in these families was probably *not* transmissible, and suggested that some might be due to adverse influences on the mother's "power of nutrition."

In what Goddard called the hereditary cases, he concluded:

Since our figures agree so closely with Mendelian expectation and since there are few if any cases where the Mendelian formula does not fit the facts, the hypothesis seems to stand: viz. normal-mindedness is, or at least behaves like, a unit character; is dominant and is transmitted in accordance with the Mendelian law of inheritance.

The writer confesses to being one of those psychologists who find it hard to accept the idea that the intelligence even *acts like a unit character.* But there seems to be no way to escape the conclusion from these figures. (1914, p. 556)

In the ensuing years, a number of further pedigree studies were published. Gates reviewed the evidence to 1933, and concluded:

It may be stated that feeble-mindedness is generally of the inherited, not the induced, type; and that the inheritance is generally recessive. Most often a single recessive gene appears to be involved; but, as with other abnormalities, occasionally the inheritance is of a different type (1933, p. 265).

By this time the Mendelian principles dominated the conceptual approach to the problem, but were still not universally accepted. Tredgold (1937), for example, in a revision of his earlier text, considered some of the evidence that feeble-mindedness was a recessive condition, and acknowledged that this might be the case in certain special types of defect, but maintained that for mental defect in general, it had not been demonstrated.

We also find evidence of the still-lingering Mendelian-biometrician dispute in a 1930 lecture by Pearson (published in 1931):

Attempts have been made on very inadequate data, most inadequately handled, to fit insanity and feeble-mindedness into the Mendelian theory. Of these attempts I shall hardly find time to say anything in this lecture; in my opinion they fail hopelessly, for they overlook essentially the fact that insanity and feeble-mindedness are far from being simple unit characters. The boundary between sanity and insanity is a perfectly indefinite one. . . . There is no mental test which will separate the normal from the feeble-minded child, the measurements of intelligence show no breaks from one end of the scale to the other. (1931, p. 366)

To support the last point, Pearson presented his analysis of Jaederholm's data, which showed that when intelligence test scores of normal children and of children classed as mentally deficient were superimposed, the result was a smooth continuous distribution. There was no gap, no separation into two discrete groups. Pearson saw the problem as even more complicated, for not only was it impossible to separate clearly the feeble-minded from the normally intelligent, but feeble-mindedness was also confounded with other defects. In a conclusion reminiscent of Tredgold, Pearson stated: "In feeble-minded stocks mental defect is interchangeable with imbecility, insanity, alcoholism, and a whole series of mental (and often physical) anomalies" (1931, p. 379).

Here, indeed, is a serious problem. If the different phenotypes cannot be adequately distinguished, how can a pedigree study possibly yield any valuable result?

At about the same time, Crew (1932) reviewed the status of work on the genetics of mental defect, and he, as did Pearson, called attention to the continuous distribution of intelligence test scores.

Pearson had concluded that the Mendelian approach was doomed to failure because of the absence of a clear dichotomy. Crew did not question the applicability of Mendelian theory, but emphasized that there were probably many different genetic types of mental defect, and that genetic analysis would need to consider the various types separately. Furthermore, he stressed that the various types need not be subject to the same type of genetic action — some might be dominant, some recessive, some due to multiple factors.

As a matter of fact, there had been an increasing attention to this possibility with a growing tendency to investigate distinct syndromes, and, especially in those conditions in which there was gross nervous system damage, there were encouragingly good "Mendelian" results (see Gates, 1946; Böök, 1953).

Behavioral Genetics and Psychology

To this point, little has been said of the relationship that early studies in behavioral genetics had to psychology in general. To a considerable extent, of course, developments in behavioral genetics were directed by contemporary trends in psychology. The reciprocal influence, that of behavioral genetics upon developments within psychology as a whole, was limited by the predominantly environmentalistic orientation that characterized psychological theory.

There have been vigorous opponents to any suggestion that a man's genotype could have any determining effect upon his intelligence, personality, emotional stability, or any other "mental or moral" characteristic. Around the turn of the century an intense debate began, which has come to be known as the nature-nurture controversy.

In all controversies of this type, apparently, the motivations of the opposing teams are diverse and various, and this is clearly true of the nature-nurture debate. For some, religious convictions may have played a predominant role in shaping opinions. Political attitudes were also undoubtedly involved. Are not all men created equal? This was a self-evident truth to the signers of the Declaration of Independence. Arguments that some men are inherently wiser than others have appeared to some to be inimical to the democratic ideal, and to imply the rightness of a rule by the elite.

But the most important factor was no doubt the development of the "behavioristic" point of view, which assumed a dominating role in the developing discipline of psychology, particularly in America. With J. B. Watson as the prime mover, behaviorism developed as a protest against all forms of introspective psychology. Mental states, consciousness, mind, will, imagery—all became taboo. Stimulus and response were the only acceptable explanatory terms.

The instinct doctrine, which had been brought to its culmination by McDougall (1908), was attacked by behaviorists as being redundant and circular. Instincts had been thought of as inherited patterns of behavior in contrast to learned behavior, and with the rejection of instincts, the whole notion of heredity influencing behavior was cast into discard. The burden of explaining individual differences fell completely to environmental factors.

> So let us hasten to admit—yes, there are heritable differences in form, in structure. . . . These differences are in the germ plasm and are handed down from parent to child. . . . But do not let these undoubted facts of inheritance lead us astray as they have some of the biologists. The mere presence of these structures tells us not one thing about function. . . . Our hereditary structure lies ready to be shaped in a thousand different ways—the same structure— depending on the way in which the child is brought up. . . .
>
> Objectors will probably say that the behaviorist is flying in the face of the known facts of eugenics and experimental evolution—that the geneticists have proven that many of the behavior characteristics of the parents are handed down to the offspring. . . . Our reply is that the geneticists are working under the banner of the old "faculty" psychology. One need not give very much weight to any of their present conclusions. We no longer believe in faculties nor in any stereotyped patterns of behavior which go under the names of "talent" and inherited capacities. . . .
>
> Our conclusion, then, is that we have no real evidence of the inheritance of traits. I would feel perfectly confident in the ultimately favorable outcome of careful upbringing of a *healthy, well-formed* baby born of a long line of crooks, murderers and thieves, and prostitutes. Who has any evidence to the contrary? (Watson, 1930, pp. 97–103)

Then came the familiar and frequently quoted challenge:

> I should like to go one step further now and say, "Give me a dozen healthy infants, well-formed, and my own specified world to bring them up in and I'll guarantee to take any one at random and train him to become any type of

specialist I might select—doctor, lawyer, artist, merchant-chief and, yes, even beggar-man and thief, regardless of his talents, penchants, tendencies, abilities, vocations, and race of his ancestors." I am going beyond my facts and I admit it, but so have the advocates of the contrary and they have been doing it for many thousands of years. (p. 104)*

Woodworth has pointed out that this extreme environmentalism was not a necessary consequence of the behavioristic philosophical position, and suggests that Watson's stand was taken, in part at least, "to shake people out of their complacent acceptance of traditional views" (1948, p. 92). For whatever reason Watson sought to exorcise genetics from psychology, he succeeded to a remarkable degree, and the position taken in his book entitled *Behaviorism* soon became the "traditional view" that was "complacently accepted" by the majority of psychologists.

This majority view was not without opposition. In fact, since Watson's pronouncement, no single year has passed without publication of some evidence showing it to be wrong. Collectively, these researches have demonstrated the important role of the genotype in many kinds of organism and in many varieties of behavior pattern.

Before considering this growing body of literature, it is necessary to examine the basic principles of transmission genetics in greater detail.

Chapter 2

Brief overview of transmission genetics

Although genetics teachers can dream up very complicated problems, the basic principles of Mendelian genetics are elegantly simple. This is not to say, of course, that the field of genetics is without its complexities. The search for detailed understanding of the transmission of hereditary factors and their mode of action has led investigators far into the realms of cytology, embryology, physiology, biochemistry, biophysics, and mathematics. However, the fundamentals of Mendelism were established without knowledge of the physical or chemical nature of the hereditary material, and it is still convenient to introduce the principles of genetics by treating the hereditary determinants as hypothetical factors. In this chapter, a brief overview of the principles of Mendelian genetics will be followed by a consideration of the physical basis of heredity and the Mendelian basis for individual differences.

Mendelian Rules

As described in Chapter 1, Mendel deduced that each individual possesses two hereditary factors, now called genes, which determine a particular character. One gene was derived from the individual's father, the other from its mother. Each gene may exist in two or more alternative forms (*alleles*). For simplicity, only two alleles of the "*A* gene" will be considered here, A_1 and A_2. Later in this chapter, multiple alleles will be discussed.

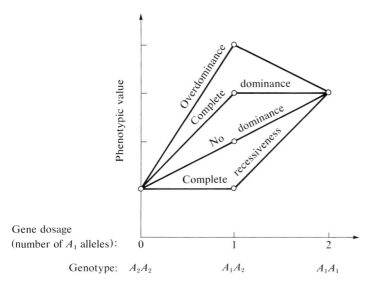

Figure 2.1
Graphical representation of four different types of gene expression.

When two alleles are considered, three kinds of individuals may occur: A_1A_1, A_1A_2, and A_2A_2. Individuals possessing two identical alleles (A_1A_1 or A_2A_2) are known as *homozygotes*; individuals with unlike alleles (A_1A_2) are *heterozygotes*.

The distinction between hereditary constitution, or genotype, and the observed character, or phenotype, is of considerable importance because the relationship between genotype and phenotype is not invariant. Four different types of gene expression are represented in Figure 2.1. The height of the vertical axis represents the phenotypic effect of the various genotypes; the horizontal axis indicates the number of A_1 alleles in the genotype. Arbitrarily, A_1 is used to symbolize the "increasing" allele, i.e., genotypes with one or two A_1 alleles have higher phenotypic values than A_2A_2 homozygotes. If A_1 is completely dominant to A_2, the phenotypic values of A_1A_1 and A_1A_2 are equal. If, however, A_1 is completely recessive to A_2, the phenotypic values of A_1A_2 and A_2A_2 are equal. If there is overdominance, the phenotypic value of A_1A_2 exceeds that of both A_1A_1 and A_2A_2; however, if there is no dominance, the phenotypic value of A_1A_2 is exactly intermediate to that of A_1A_1 and A_2A_2. Most of the characters Mendel and other early geneticists studied exhibited complete dominance; however, dominance is not always complete. Thus, an infinite variety of gene expression is possible.

In addition, a gene may influence many different characters, this manifold effect of the gene being referred to as *pleiotropism*. When different characters influenced by the same gene are examined, the types of gene expression

Father   Mother

Parental genotypes:

Gametes:

Offspring genotypes:

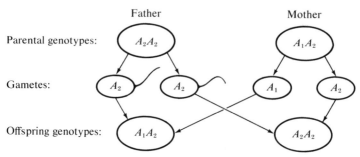

Figure 2.2
Offspring of the mating of a heterozygote and a homozygote.

may differ. Thus, a gene that controls the activity of an enzyme may exhibit no dominance for this character, yet may be completely dominant for some other character (e.g., production of a pigment in the skin).

In the process of reproduction, parents transmit only a sample of half of their genes to their offspring. This allocation is such that the sex cells, or gametes, contain one gene from each gene pair carried by the parent.

Parents who are homozygous, say A_2A_2, can only transmit A_2 to their offspring. Parents who are heterozygous (A_1A_2), however, produce two different kinds of gametes with equal frequency, those bearing A_1 and those bearing A_2. Thus, as indicated in Figure 2.2, two kinds of offspring (A_1A_2 and A_2A_2) will occur with equal frequency from the mating of an A_2A_2 homozygote and an A_1A_2 heterozygote.

When two parents are heterozygous for the same gene pair, both *mono-hybrid* individuals will produce two kinds of gametes in equal frequency. If sperm and eggs produced by these parents unite at random, i.e., if A_1-bearing sperm, for example, have no greater affinity for A_1-bearing eggs than for A_2-bearing eggs, which is almost always found to be the case, the kinds and frequencies of offspring produced are easily determined from a table such as that presented in Table 2.1. From this table it may be seen that three different genotypes (A_1A_1, A_1A_2, and A_2A_2) will occur among offspring of such matings, and that their relative frequencies will be $\frac{1}{4}$, $\frac{1}{2}$, and $\frac{1}{4}$, respectively. Put another way, the expected frequency is $1A_1A_1:2A_1A_2:1A_2A_2$. If there is complete dominance for the character, the observed ratio among a number of such progeny will be $3A_1-:1A_2A_2$, where A_1- indicates individuals possessing at least one dominant A_1 allele, i.e., A_1A_1 or A_1A_2. It may be seen from these calculations that it is immaterial whether a heterozygote received its A_1 allele from its mother or father; all A_1A_2 individuals represented in Table 2.1 would be genotypically indistinguishable.

A basic Mendelian principle is that during gamete formation members of a gene pair separate "cleanly" with no residual effect upon each other. Thus, an A_1 allele derived from an A_1A_1 parent is no different from one

Table 2.1

Offspring from a monohybrid cross

		Sperm	
		$\frac{1}{2}A_1$	$\frac{1}{2}A_2$
Eggs	$\frac{1}{2}A_1$	$\frac{1}{4}A_1A_1$	$\frac{1}{4}A_1A_2$
	$\frac{1}{2}A_2$	$\frac{1}{4}A_1A_2$	$\frac{1}{4}A_2A_2$

Table 2.2

Offspring from a dihybrid cross

		Sperm			
		$\frac{1}{4}A_1B_1$	$\frac{1}{4}A_1B_2$	$\frac{1}{4}A_2B_1$	$\frac{1}{4}A_2B_2$
	$\frac{1}{4}A_1B_1$	$\frac{1}{16}A_1A_1B_1B_1$	$\frac{1}{16}A_1A_1B_1B_2$	$\frac{1}{16}A_1A_2B_1B_1$	$\frac{1}{16}A_1A_2B_1B_2$
	$\frac{1}{4}A_1B_2$	$\frac{1}{16}A_1A_1B_1B_2$	$\frac{1}{16}A_1A_1B_2B_2$	$\frac{1}{16}A_1A_2B_1B_2$	$\frac{1}{16}A_1A_2B_2B_2$
Eggs	$\frac{1}{4}A_2B_1$	$\frac{1}{16}A_1A_2B_1B_1$	$\frac{1}{16}A_1A_2B_1B_2$	$\frac{1}{16}A_2A_2B_1B_1$	$\frac{1}{16}A_2A_2B_1B_2$
	$\frac{1}{4}A_2B_2$	$\frac{1}{16}A_1A_2B_1B_2$	$\frac{1}{16}A_1A_2B_2B_2$	$\frac{1}{16}A_2A_2B_1B_2$	$\frac{1}{16}A_2A_2B_2B_2$

transmitted from an A_1A_2 parent. As indicated in Chapter 1, this was Mendel's first law, the law of segregation. Members of different gene pairs, however, assort independently. Consider an individual heterozygous for two gene pairs ($A_1A_2B_1B_2$), where the A and B genes influence different characters. Such *dihybrid* individuals produce four different kinds of gametes in equal frequency: $\frac{1}{4}A_1B_1:\frac{1}{4}A_1B_2:\frac{1}{4}A_2B_1:\frac{1}{4}A_2B_2$. This outcome shows that segregation of these two gene pairs is independent, i.e., the probability that a gamete contains A_1B_1, for example, is equal to the product of their separate probabilities ($\frac{1}{2} \times \frac{1}{2} = \frac{1}{4}$). As indicated in Chapter 1, this is Mendel's second law, the law of independent assortment.

When two dihybrid individuals mate, each produces four kinds of gametes with equal frequency. Thus, we may determine the kinds and expected frequency of resulting offspring from a 4×4 table such as that presented in Table 2.2. This, however, results in a table with 16 cells, with some of the same genotypes appearing in more than one cell. For example, there are two cells that each contain $\frac{1}{16}A_1A_1B_1B_2$. A more expedient way to obtain the desired results is to utilize the law of independent assortment. If members of different gene pairs assort independently, the probability that an individual is A_1A_1 should be $\frac{1}{4}$, regardless of whether it is B_1B_1, B_1B_2, or B_2B_2. Thus, we should be able to multiply the genotypic frequencies, each

Table 2.3

Summary of offspring genotypes and their frequencies resulting from a dihybrid cross, utilizing law of independent assortment

		A genotypes		
		$\frac{1}{4}A_1A_1$	$\frac{1}{2}A_1A_2$	$\frac{1}{4}A_2A_2$
	$\frac{1}{4}B_1B_1$	$\frac{1}{16}A_1A_1B_1B_1$	$\frac{1}{8}A_1A_2B_1B_1$	$\frac{1}{16}A_2A_2B_1B_1$
B genotypes	$\frac{1}{2}B_1B_2$	$\frac{1}{8}A_1A_1B_1B_2$	$\frac{1}{4}A_1A_2B_1B_2$	$\frac{1}{8}A_2A_2B_1B_2$
	$\frac{1}{4}B_2B_2$	$\frac{1}{16}A_1A_1B_2B_2$	$\frac{1}{8}A_1A_2B_2B_2$	$\frac{1}{16}A_2A_2B_2B_2$

gene pair considered separately, to obtain their joint frequencies. Such a multiplication is summarized in Table 2.3. Nine cells are contained in such a table, one representing each of the nine possible genotypes and its corresponding frequency.

As indicated in Chapter 1, Mendel crossed strains of peas that differed in two characters to produce a dihybrid F_1 generation and an F_2 generation was subsequently obtained. The expected results of such a breeding system are indicated in Figure 2.3. From this figure it may be seen when two gene pairs affect different characters, and when one allele is completely dominant over the other in each pair, a classic 9:3:3:1 Mendelian ratio of offspring genotypes results.

Many other possibilities exist, however. Consider, for example, two different gene pairs affecting the same character. The effects of "*A* genes" may simply add to those of "*B* genes" or, on the other hand, nonadditive interactions between different gene pairs may occur. Such interactions are known as *epistasis*. In Figure 2.4, hypothetical examples of no epistasis (left) and epistasis (right) are graphed. In Figure 2.4, left, the difference between the phenotypic values of individuals of B_2B_2 and B_1- genotypes is the same, regardless of whether the other gene pair is A_2A_2, A_1A_2, or A_1A_1. In Figure 2.4, right, however, this is not the case. Many examples of such dependency of the expression of one gene pair on the constitution of another gene pair have been described.

Physical Basis of Heredity

Mendel was convinced that his "elements" were material units located in the gametes, but with the state of knowledge of cytology at the time, it was not possible for him to specify their physical nature in any greater detail. It was fortunate that, for the purposes of establishing the basic Mendelian laws, the "elements" or genes could be treated as hypothetical constructs,

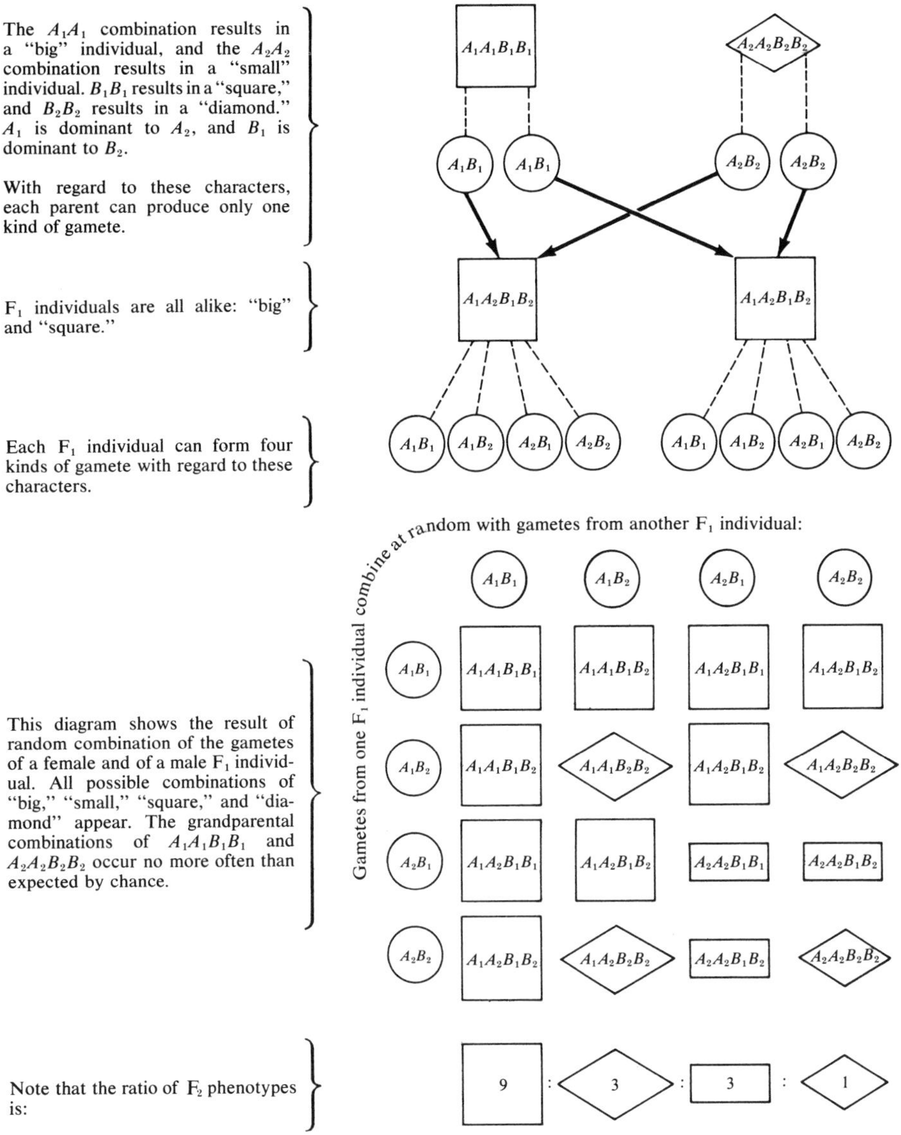
The A_1A_1 combination results in a "big" individual, and the A_2A_2 combination results in a "small" individual. B_1B_1 results in a "square," and B_2B_2 results in a "diamond." A_1 is dominant to A_2, and B_1 is dominant to B_2.
With regard to these characters, each parent can produce only one kind of gamete.
$A_1A_1B_1B_1$
$A_2A_2B_2B_2$
A_1B_1
A_2B_2
F_1 individuals are all alike: "big" and "square."
$A_1A_2B_1B_2$
Each F_1 individual can form four kinds of gamete with regard to these characters.
A_1B_1 A_1B_2 A_2B_1 A_2B_2
Gametes from one F_1 individual combine at random with gametes from another F_1 individual:
This diagram shows the result of random combination of the gametes of a female and of a male F_1 individual. All possible combinations of "big," "small," "square," and "diamond" appear. The grandparental combinations of $A_1A_1B_1B_1$ and $A_2A_2B_2B_2$ occur no more often than expected by chance.
$A_1A_1B_1B_1$ $A_1A_1B_1B_2$ $A_1A_2B_1B_1$ $A_1A_2B_1B_2$
$A_1A_1B_1B_2$ $A_1A_1B_2B_2$ $A_1A_2B_1B_2$ $A_1A_2B_2B_2$
$A_1A_2B_1B_1$ $A_1A_2B_1B_2$ $A_2A_2B_1B_1$ $A_2A_2B_1B_2$
$A_1A_2B_1B_2$ $A_1A_2B_2B_2$ $A_2A_2B_1B_2$ $A_2A_2B_2B_2$
Note that the ratio of F_2 phenotypes is:
9 : 3 : 3 : 1

Figure 2.3
A schematic illustration of the law of independent assortment. Two hypothetical characters are considered, each of which is expressed in a dominant fashion. (After McClearn, 1963, p. 159.)

and no precise knowledge of their location or structure was necessary. There was, naturally, considerable speculation, but a real breakthrough in understanding the physical nature of the determiners of heredity awaited critical developments in the field of cytology.

(a)   (b)

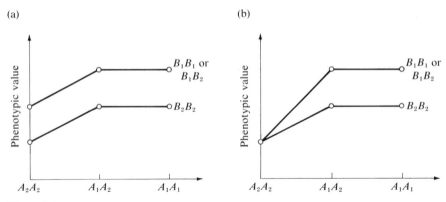

Figure 2.4
Illustration of no epistasis (a) *and epistasis* (b).

Cytological Discoveries. The study of the cell and its contents had progressed rapidly since the general acceptance, in the mid-nineteenth century, of the doctrine that cells are the structural and functional units of living organisms. Aided in no small degree by advances in the chemistry of dyes, cytologists were able to develop means of staining the contents of cells to render them more visible for study. It was soon found that a portion of the cell, the *nucleus,* contains a number of small rod-shaped bodies that are called *chromosomes* (colored bodies) because of their capacity to be stained by particular dyes. The number of chromosomes, with some exceptions that need not concern us now, is the same in all somatic cells of an organism, and all individuals of a species have the same number. The number of chromosomes, however, varies greatly from species to species. It was known that in the process of growth the cells divide into two "daughter cells," each of which then later divides into two more, and so forth. Study of the chromosomes revealed that a remarkable series of changes takes place during cell division. The major features of this process of *mitosis* are illustrated in Figure 2.5. Prior to the splitting of the cell, the chromosomal material doubles and spindle fibers become attached to the *kinetochore* or *centromere* of each chromosome. During the cell division, half of the material goes into one daughter cell, half into the other. The spindle fibers appear to be associated with this allotment. As different chromosomes are somewhat distinctive in shape and size, it was possible to determine that each daughter cell receives an equivalent chromosomal complement. This distinctiveness of chromosomes also permitted the observation that chromosomes are present in pairs. In Figure 2.5, the dark chromosomes may be taken to indicate paternal origin and the light ones, maternal origin. Thus, two pairs of *homologous* chromosomes are shown.

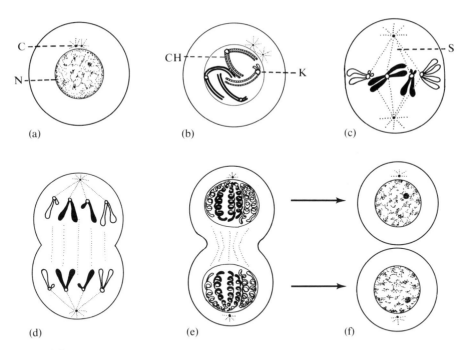

Figure 2.5
Cell division and mitosis. C = *centriole;* N = *nucleus;* CH = *chromosome;* K = *kinetochore;*
S = *spindle. (From Stern, Principles of Human Genetics, 3rd Ed. W. H. Freeman and
Company. Copyright © 1973, p. 19.)*

Quite independently of knowledge of the Mendelian laws, evidence was
obtained that the chromosomes are, in some way, concerned with heredity,
and it was concluded that one *set* of chromosomes is contributed by each
parent. The process by which this is accomplished (*meiosis,* Figure 2.6)
consists essentially of the splitting of a cell into two without the prior doubling of chromosomal number that is found in mitosis. One member of each
pair of homologous chromosomes is drawn into each daughter cell before
the division is complete. The set included in any one gamete, however, is
not necessarily the same set which the individual had received from its
mother or from its father. A reshuffling takes place, so that an individual
transmits to its offspring some of the chromosomes it received from its
mother along with some received from its father.

Chromosomes and Genes. This interesting behavior of the chromosomes was seen to parallel the behavior of Mendel's "elements": two elements, paired chromosomes; one element in each gamete, one of each pair

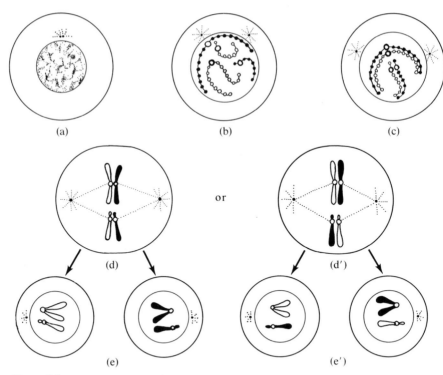

Figure 2.6

Meiosis, simplified: two pairs of chromosomes. Dark: Paternal chromosomes. Light:
Maternal chromosomes. (a) Nucleus in a premeiotic germ cell. (b-c) Pairing of
homologous chromosomes. (d-d') The two alternative arrangements of the chromosome
pairs on a meiotic spindle. (e-e') The four different types of reduced chromosome
constitutions of the gametes. (From Stern, Principles of Human Genetics, 3rd Ed.
W. H. Freeman and Company. Copyright © 1973, p. 87.)

of chromosomes in each gamete. On this, and other evidence, it was suggested that the genes are in fact particulate physical bodies residing at
specific *loci* on the chromosomes.

The advances in understanding of the chromosomal basis of heredity
also allowed explanation of exceptions to the law of independent assortment, which had been noted. It was evident that there are more genes than
there are chromosomes, and that therefore each chromosome must contain
a number of genes. If two characteristics under study are determined by
genes on the same chromosome, it is clear that these genes may not assort
independently. Such *linkage* was experimentally demonstrated, but it was
also discovered that linkage is not permanent, or unbreakable. During one
stage of gamete formation, homologous chromosomes line up pair by pair.
Each member of each pair separates into two, except at the centromere.
The adjacent members of this *tetrad* frequently come into contact and ex-

change parts. This mutual exchange is usually done with such precision that equivalent sections are traded — each of the members participating in the exchange receiving the same loci that it gives.

Figure 2.7 is a diagrammatic illustration of this process for one pair of chromosomes only. It should be remembered that the same events may be occurring at the same time for all other chromosome pairs. In Figure 2.7(a) are shown the two members of the chromosome pair. The maternal chromosome, carrying the genes *A*, *b*, and *C*, is shown as white, and the paternal chromosome with the genes *a*, *B*, and *c* is shown as gray. At one stage in meiosis each of the chromosomes can be seen to be duplicate, as shown in Figure 2.7(b). In Figure 2.7(c) the adjacent members are shown as crossed over one another. During this stage the chromosomes may break and rejoin, yielding the configuration of Figure 2.7(d). Following two meiotic divisions, each one of these four members will be transmitted to one gamete. Consider only the *A-a* and *B-b* loci for the moment. As shown in Figure 2.7(e), one gamete will carry the genes *A* and *b* as in the grandmother, one will carry *a* and *B* as in the grandfather, and the other two will carry *A* with *B* and *a* with *b*. For these last, recombination has taken place. Crossing over of this kind does not always occur at the same place, and the probability that recombination will occur is a function of the distance between the particular genes. In Figure 2.7, for example, the crossing over has not affected the relationship between the *A-a* and the *C-c* loci. All gametes are either *AC* or *ac*, as in the grandparental combinations, since the crossover did not occur between these loci. Crossing over could occur between the *A* and *C* loci, but would be less frequent than between *A* and *B*. Because of this, the crossover gametes frequently occur less often than the noncrossover, and this forms an exception to the law of independent assortment. Genes located on different chromosomes do, of course, assort at random.

Autosomes and Sex Chromosomes. Detailed examination of the chromosomes revealed that one pair was exceptional, in that the members of the pair were of obviously different size and shape. Eventually, it was possible to relate this atypical pair of chromosomes to sex determination. Whereas the situation differs in various species, the mechanism in mammals, including man, is briefly as follows: Females have two similar-sized chromosomes that are called X. Males have one X and a smaller chromosome called Y. Females obviously can form only X-bearing eggs, but males form both X- and Y-bearing sperm. If an egg is fertilized by a Y-bearing sperm, the zygote will be male; if fertilization is by an X-bearing sperm, the zygote will be female.

Genes located on the *sex chromosomes* give phenotypic results that differ from the usual Mendelian results of genes carried on *autosomes* (chromosomes other than sex chromosomes), primarily because the Y chromosome appears to be relatively barren. In humans, for example, colorblindness is due to a gene carried on the X chromosome. As it acts

42

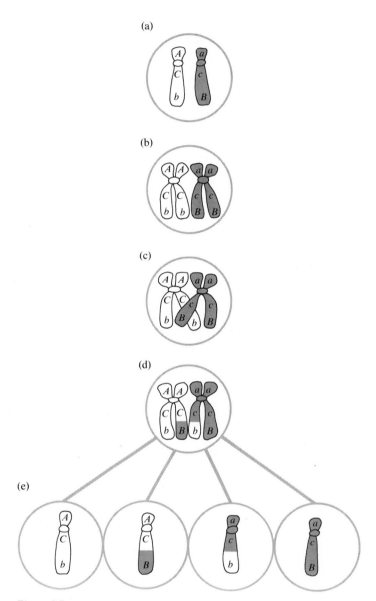

Figure 2.7

Diagrammatic illustration of crossing over — the mutual exchange of material by homologous chromosomes. (After McClearn, 1963, p. 163.)

as a recessive in females, a woman will be colorblind only if homozygous. Frequently, the recessive gene will be paired with a dominant one, and such a heterozygous female will have normal color vision. In males, however, there is no corresponding locus on the Y chromosome, so that a single recessive gene, present on the X, will be expressed. Thus, colorblindness and other X-linked conditions are much more frequent in men than in women.

Recent developments in the study of human chromosomes and the relationship between chromosomal anomalies and behavior will be considered in Chapter 7. From the foregoing brief account of chromosomal transmission, however, it is apparent that the processes of segregation, independent assortment, and crossing over facilitate a considerable amount of reshuffling of genes. Let us consider this potential for generating genetic variation in more detail.

Mendelian Basis of Individual Differences

Individuals may be grouped into distinct classes when genes segregating at only a few loci are observed. When many loci are considered, it becomes obvious that each individual is truly unique. As indicated earlier in this chapter, when only two alleles at one autosomal locus are considered, an individual must be one of three possible genotypes: A_1A_1, A_1A_2, or A_2A_2. When N such loci are considered, where N may represent any whole number (N = 1, 2, 3, etc.), the number of possible genotypes becomes 3^N. However, more than two alternative forms of a gene may occur. The example of two alleles per locus is merely the special case of a multiple allelic series in which the number of alleles (m) is two. Let us now examine the more general case. The number of different genotypes that may be expressed by m alleles at one locus is equal to the sum of the number of different homozygotes and the number of different heterozygotes. With m alleles at the locus, m different homozygotes may occur $(A_1A_1; A_2A_2; A_3A_3; \ldots ; A_mA_m)$. The number of different heterozygotes that may occur is equal to the number of combinations that may result when m things are considered two at a time, i.e.,

$$\text{number of heterozygotes} = \frac{m!}{2!(m-2)!} = \frac{m(m-1)}{2},$$

where $m! = m(m-1)(m-2) \ldots$ (1). Thus, the total number of different genotypes that may occur at one such autosomal locus is as follows:

$$\text{number of genotypes} = m + \frac{m(m-1)}{2}$$

$$= \frac{2m + m^2 - m}{2} = \frac{m^2 + m}{2} = \frac{m(m+1)}{2}.$$

With m alleles at each of N such loci, the number of different genotypes that may occur is equal to

$$\left[\frac{m(m+1)}{2}\right]^N.$$

With only four alleles at each of ten loci, the number of different genotypes which may occur is

$$\left[\frac{(4)(5)}{2}\right]^{10} = 10,000,000,000, \text{ i.e., 10 billion.}$$

Thus, when only a relatively few alleles at only ten loci are considered, the number of possible different genotypes is considerably larger than the present human population of the earth. Complex loci have been studied in which the number of alleles is greater than 200. In addition, although the exact number of genes in man is not known, it is conservatively estimated (Stern, 1973) to be between 10,000 and 100,000, and one-third or more of these loci may be segregating for two or more alleles (Hopkinson and Harris, 1971). Thus, segregation and independent assortment provide a mechanism for generating astronomical numbers of different genotypes.

Much variation may even be found among members of the same family. When parents are heterozygous for the same alleles at a given locus, three genotypes are possible among their offspring. If they are heterozygous for different alleles, four genotypes are possible. Of course, if one parent is homozygous and the other heterozygous, only two genotypes may occur. Finally, if both are homozygous (having either the same or different alleles), only one genotype is possible. Consider the following cross:

$$A_1A_2B_1B_2C_1C_1D_1D_2E_1E_2F_1F_2G_1G_1 \times A_1A_2B_3B_4C_2C_2D_1D_2E_2E_2F_1F_3G_1G_2.$$

Among the offspring of such a mating, $(3)(4)(1)(3)(2)(4)(2) = 576$ different genotypes may occur. Most parents, however, probably differ at many more loci than indicated by this mating. Thus, the number of possible genotypes, even among members of the same family, is very large. When we consider environmental effects, plus the likelihood of mutation, even members of identical twin pairs would never be exactly alike, phenotypically; thus, each individual is truly unique. Segregation, independent assortment, crossing over, and mutation not only provide a basis for biological individuality, they also provide the raw material essential for evolution, the subject of Chapters 9 and 10.

In this chapter, a brief overview of the basic principles of transmission genetics and chromosomal mechanics has been presented. In such a short space, of course, nothing approaching completeness can be claimed. Before proceeding to behavioral examples, some basic concepts in elementary probability theory and statistics will be considered.

Chapter 3

A very short course in probability theory and statistics

Genes are transmitted from one generation to the next according to the laws of probability. One of the major factors in Mendel's success was his understanding of basic probability and his ability to relate this theory to experimental results. In this chapter, some elementary concepts of probability theory will be presented. This will be followed by a discussion of statistical methods that are particularly useful in genetic analysis. This coverage is intended to provide the student with only a "cook book" level of comprehension. For those desiring more than a superficial knowledge of the subject, excellent introductory statistics texts are available (see, e.g., Snedecor and Cochran, 1967; Hays, 1963).

Probability

Probability may be defined as a number between zero and one that measures the likelihood of the occurrence of an event, i.e.,

$$0 \leq P(A) \leq 1,$$

where $P(A)$ is the probability that an event A will occur. $P(A)$ may be estimated empirically or it may be predicted from *a priori* information. If we conduct an experiment and observe event A n times during a total of N trials, then n/N is our best estimate of the likelihood that A will be observed on some future trial.

For example, in 100 consecutive live births at a local hospital, we may observe 53 boys and 47 girls. Therefore, we may estimate the probability that an unborn child will be a boy as follows:

$$P(\text{Boy}) = \frac{n}{N} = \frac{53}{100} = 0.53.$$

However, we also know from our consideration of segregation in Chapter 2 that one-half of the sperm should contain an X chromosome and the other half a Y chromosome. When an X-bearing sperm fertilizes an egg (all of which are X bearing), a female zygote will result, and when a Y-bearing sperm fertilizes an egg, a male zygote will result. Thus, from this *a priori* information we may predict that P(Boy) is one-half. Of course, in samples of only 100 live births, some deviation from expected would likely occur, merely due to chance.

Multiplication Law. Let P(AB) symbolize the probability that two events, A *and* B, will *both* occur on some future trial. It may be shown that

$$P(AB) = P(A)P(B/A)$$
$$= P(B)P(A/B),$$

where P(B/A) is the *conditional probability* that event B will be observed, given that A has occurred. If A and B are *independent* events, P(AB) = P(A)P(B), i.e., A and B are independent if P(A/B) = P(A) and P(B/A) = P(B). In other words, two events are independent if the occurrence of one event tells us nothing about the likelihood of the occurrence of the other event. By an algebraic division of the above relations, it may be seen that P(A/B) and P(B/A) may be defined as follows:

$$P(A/B) = \frac{P(AB)}{P(B)}$$

$$P(B/A) = \frac{P(AB)}{P(A)}.$$

Addition Law. Let P(A + B) symbolize the probability that event A *or* B will occur. It may be shown that P(A+B) = P(A)+P(B)−P(AB). A and B are *mutually exclusive* if P(AB) = 0; thus, P(A+B) = P(A)+P(B), only when A and B are mutually exclusive. Examination of Figure 3.1 may clarify this relationship. The total fraction of the area occupied by the circles designated A and B in Figure 3.1(a) is simply the sum of the two areas divided by

(a)

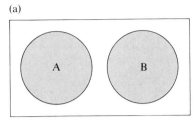

(b)

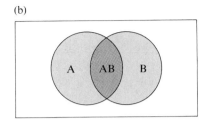

Figure 3.1

Diagrammatic representation of the probability of two events A and B. In (a), events A and B are mutually exclusive. In (b), events A and B are not mutually exclusive.

the total area. If the square were placed on a wall and darts were tossed at it by a blindfolded person, the probability of a dart's landing in circle A or B on any particular toss would be equal to the sum of the separate probabilities, i.e., $P(A + B) = P(A) + P(B)$, where $P(A)$ is the fraction of the area occupied by A, etc. This necessarily follows since there is no overlap of the A and B circles. If we let $P(A)$ symbolize the probability of tossing a dart into the A circle and $P(B)$ symbolize the probability of tossing a dart into the B circle, $P(AB) = 0$, since tossing a dart into both A and B would be impossible in Figure 3.1(a). In Figure 3.1(b), however, the A and B spaces overlap. Thus, if the areas occupied by A and B were merely added in this example, the intersection (AB) would be counted twice. It should be clear that the probability of tossing a dart into either A or B or both should be equal to $P(A) + P(B) - P(AB)$, when the target is arranged as in Figure 3.1(b). Thus, probabilities are strictly additive only when events are mutually exclusive.

In Figure 3.1, it is obvious that all of the area outside the A circle represents the probability of not observing (or not landing in) A. Let us symbolize the probability of not observing A by $P(\overline{A})$. Obviously, $P(A) + P(\overline{A}) = 1$, since A and $\overline{A}$ are mutually exclusive and exhaustive (there are no other possibilities). Thus, $P(\overline{A}) = 1 - P(A)$. In a similar manner, if several trials are observed, event A will either occur at least once or not at all. Thus, P (at least one) $= 1 - P$ (none).

An Example.   In order to illustrate the use of these elementary concepts of probability theory, let us assume that a genetically uniform strain of brown mice, almost 100 percent of whom display convulsive seizures when initially exposed to a loud high frequency sound, is crossed with a genetically uniform black strain that does not have these audiogenic seizures. The gene responsible for black coat color (symbolized here by B_1) is dominant to that for brown (B_2). As discussed in Chapter 5, there is recent evidence which indicates that susceptibility to audiogenic seizures of this type is due to a recessive autosomal allele (symbolized here by S_2). If dihybrid mice $(B_1B_2S_1S_2)$, resulting from the above mating, are *testcrossed* to members of the $B_2B_2S_2S_2$ strain, we might predict the following results on the assumption that segregation at the B locus is independent of segregation at the S locus:

$$0.25 \ B_1B_2S_1S_2 \ (\text{black, seizure resistant}),$$

$$0.25 \ B_1B_2S_2S_2 \ (\text{black, seizure susceptible}),$$

$$0.25 \ B_2B_2S_1S_2 \ (\text{brown, seizure resistant}),$$

$$0.25 \ B_2B_2S_2S_2 \ (\text{brown, seizure susceptible}).$$

One-half of the progeny from such a testcross should be black and one-half should be seizure resistant. Thus, if coat color and seizure resistance are transmitted independently, the frequency of the black, seizure resistant class should be equal to the product of the separate probabilities, i.e.,

$$P(\text{black, seizure resistant}) = P(\text{black})P(\text{seizure resistant})$$

$$= (0.50)(0.50) = 0.25.$$

The actual data from such an experiment, however, might be as follows:

$$0.30 \ B_1B_2S_1S_2 \ (\text{black, seizure resistant}),$$

$$0.20 \ B_1B_2S_2S_2 \ (\text{black, seizure susceptible}),$$

$$0.20 \ B_2B_2S_1S_2 \ (\text{brown, seizure resistant}),$$

$$0.30 \ B_2B_2S_2S_2 \ (\text{brown, seizure susceptible}).$$

These results would indicate that genes at the B and S loci are not assorting independently. This is due to linkage, the closeness of which may be estimated from the sum of the recombination type percentages, i.e., the map distance between the B and S loci is approximately equal to 40 map units. It is clear from the above data that genes at two loci would assort independently, even if they were on the same chromosome, if the map distance between them were 50 map units or greater.

Black and brown coat colors in mice are mutually exclusive. Thus, the probability that a mouse drawn at random from the testcross progeny is either black or brown may be determined as follows:

$$P(\text{black} + \text{brown}) = P(\text{black}) + P(\text{brown}) = 0.50 + 0.50 = 1.$$

Barring mutation, the coat color of such a mouse has to be either black or brown. Coat color and sex, however, are independent. Thus, the probability that this mouse is *either* black or male would be as follows:

$$P(\text{black} + \text{male}) = P(\text{black}) + P(\text{male}) - P(\text{black, male})$$
$$= 0.50 + 0.50 - (0.50)(0.50) = 0.75.$$

The only mice that would not fall into the black or male category are brown females, which would be expected to occur with a frequency of 0.25. Thus, the probability that a mouse drawn at random from such a population is not a brown female is as follows:

$$P(\text{not a brown female}) = 1 - P(\text{brown female})$$
$$= 1 - 0.25 = 0.75,$$

which, of course, is the probability that the mouse is either black or male. Although the probability of observing a brown female on any given "trial" is only 0.25, the probability of observing at least one in a litter of four is as follows:

$$P(\text{at least 1 brown female in a litter of 4}) = 1 - P(\text{none}).$$

$P(\text{none})$ is equivalent to the probability of observing four black *or* male offspring, each with a probability of $\frac{3}{4}$. Since these "events" are independent,

$$P(\text{none}) = \left(\tfrac{3}{4}\right)\left(\tfrac{3}{4}\right)\left(\tfrac{3}{4}\right)\left(\tfrac{3}{4}\right) = \left(\tfrac{3}{4}\right)^4.$$

Thus,

$$P(\text{at least 1 brown female in a litter of 4}) = 1 - \left(\tfrac{3}{4}\right)^4 = 175/256,$$

i.e., the probability is greater than two-thirds.

Composition Law. The probability of an event may also be determined from the sum of the probabilities of its occurrence with other events. From Figure 3.2, it may be seen that the space occupied by A is simply the sum of the AC and A$\overline{C}$ spaces. Thus, $P(A) = P(AC) + P(A\overline{C})$, which, according to the multiplication law, equals $P(C)\,P(A/C) + P(\overline{C})\,P(A/\overline{C})$. For example, colorblindness in man is due to a sex-linked recessive gene and thus is expressed much more frequently in males than in females. Assume that 10

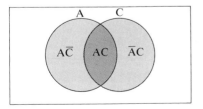

Figure 3.2

The probability of an event A is equal to the probability that it will occur with another event (AC) plus the probability that it will occur with the null set of the other event (A$\overline{C}$).

percent of the males and only 1 percent of the females in a population are colorblind. What is the probability that an individual picked at random from this population, composed of equal numbers of males and females, is colorblind? Let A symbolize the event, being colorblind, and C and $\overline{C}$ symbolize males and females, respectively. As indicated above,

$$P(A) = P(C)P(A/C) + P(\overline{C})P(A/\overline{C}),$$

i.e., the probability that an individual picked at random from the population is colorblind is equal to the probability that the individual is a male, times the probability of colorblindness in males, plus the probability that the individual is a female, times the probability of colorblindness in females. Upon substitution,

$$P(A) = (\tfrac{1}{2})(0.1) + (\tfrac{1}{2})(0.01) = 0.055.$$

This expression may also be used for practical application in genetic counseling. For example, assume a young couple go to a genetic counselor for advice. A brother of the young man is afflicted with phenylketonuria (PKU), a hereditary disease that results in mental retardation as discussed in some detail in Chapters 4 and 6. Individuals with the disease are homozygous for a recessive autosomal allele, symbolized here as P_2; thus, parents of this young man must have both been carriers (P_1P_2) of this recessive allele. The expected genotypic ratio among children of such a marriage is as follows: $\tfrac{1}{4}P_1P_1 : \tfrac{1}{2}P_1P_2 : \tfrac{1}{4}P_2P_2$. Given that the young man is phenotypically normal, the *a priori* probability that he is a carrier is $\tfrac{1}{2}/(\tfrac{1}{4}+\tfrac{1}{2}) = \tfrac{2}{3}$. As the frequency of carriers in the general population is relatively low, the odds are that the woman is P_1P_1. After receiving this information, the young couple decide to have a family of three children. Two children are born to them, both of whom are normal. Then, however, an identical twin sister of the young mother bears a child with PKU, indicating that the mother of the PKU child and her twin sister are both carriers for the disease. In spite of this, the young couple still desire to have a third child. If the husband is a carrier, the probability that the next child will have PKU is one-fourth. However, if he is homozygous P_1P_1, the child would be phenotypically normal.

Let C symbolize the event that the young man is a carrier (P_1P_2) and let *A* symbolize the observation of two normal children already born to the couple. We already know that the *a priori* probability, $P(C)$, is $\tfrac{2}{3}$. What we wish to do, however, is to utilize the additional information that he has already fathered two normal children in assessing the probability that he is a carrier, i.e., $P(C/A)$. This may be determined as follows:

$$P(C/A) = \frac{P(AC)}{P(A)} = \frac{P(C)P(A/C)}{P(AC) + P(A\overline{C})} = \frac{P(C)P(A/C)}{P(C)P(A/C) + P(\overline{C})P(A/\overline{C})}.$$

The wife is known to be a carrier; thus, the probability of her bearing two normal children, assuming that the husband is a carrier $[P(A/C)]$, is equal to $(\frac{3}{4})^2$. However, the probability of her bearing two normal children, assuming that the husband is not a carrier $[P(A/\overline{C})]$, is one. Thus,

$$P(C/A) = \frac{(\frac{2}{3})(\frac{3}{4})^2}{(\frac{2}{3})(\frac{3}{4})^2 + (\frac{1}{3})(1)} = 9/17.$$

In this example, $P(C/A)$ has a very similar value to $P(A/C)$. This, however, is not usually the case.

Given this new reduced probability that the husband is a carrier, what is the probability that their third child would have PKU? Let K symbolize the event that the third child will have PKU.

$$P(K) = P(KC) + P(K\overline{C}) = P(C)P(K/C) + P(\overline{C})P(K/\overline{C})$$
$$= (9/17)(\tfrac{1}{4}) + (8/17)(0) = 0.13.$$

Although this may appear to be a relatively small chance, the couple's risk of having a PKU child is roughly three thousand times higher than that of couples in the population at large.

Binomial Theorem.   When several trials are considered and only one of two alternatives may be expressed in each trial (e.g., head versus tail; boy versus girl; dominant versus recessive), the probability of each possible outcome is described by the following expression:

$$P(\text{observing event A n times during N trials}) = \frac{N!}{n!(N-n)!}a^n b^{N-n},$$

where "a" is the probability of observing event A on any one trial and b is the probability of observing the alternative event, i.e., $b = 1 - a$. For example, in a three-child family, what is the probability of obtaining two girls and one boy?

$$P(\text{2 girls and 1 boy}) = \frac{3!}{2!1!}(\tfrac{1}{2})^2(\tfrac{1}{2})^1 = \frac{(3)(2)(1)}{(2)(1)(1)}(\tfrac{1}{4})(\tfrac{1}{2}) = \tfrac{3}{8},$$

where n is arbitrarily defined as the number of girls.

Multinomial Theorem.   When more than two events may be observed on each of N trials, the probability of observing any particular outcome is described by the multinomial expansion. For simplicity, let us only consider the case in which three different events are possible on any given trial. Generalization to more complicated cases should be obvious. In an F_2

Table 3.1

*Activity scores of two
inbred strains of mice*

A	C57BL
29	155
29	157
44	161
58	199
63	202
83	218

generation resulting from a monohybrid cross, three genotypes are expected to occur in the following ratio: $1A_1A_1 : 2A_1A_2 : 1A_2A_2$. What, however, is the probability of observing exactly such a ratio in a sample of only four F_2 individuals?

$$P(1A_1A_1 : 2A_1A_2 : 1A_2A_2) = \frac{4!}{1!2!1!}(\tfrac{1}{4})^1(\tfrac{1}{2})^2(\tfrac{1}{4})^1 = \tfrac{3}{16}.$$

Thus, the "expected" outcome will occur less than one-fourth of the time in samples of four individuals. This outcome is "expected" in that the probability of its occurrence is higher than that of any other single outcome. Verification of this fact may provide a useful exercise.

Statistical Methods

Statistics may be conveniently partitioned into two areas: (1) parameter estimation, in which true characteristics of populations are estimated from sample values; and (2) hypothesis testing, in which the significance of the departure of observed results from those expected on the basis of some hypothesis is determined. Both areas are of considerable importance to behavioral genetics.

Parameter Estimation. In order to illustrate the estimation of parameters, consider the data presented in Table 3.1. These data are activity scores of mice drawn at random from each of two highly inbred strains, A and C57BL. As will be discussed in Chapter 9, the process of inbreeding results in decreased heterozygosity, and hence, increased genetic uniformity within lines. Differences between lines may be either genetic or environmental in origin. If, however, two highly inbred strains have been reared and tested under similar conditions, an observed difference in some character may be

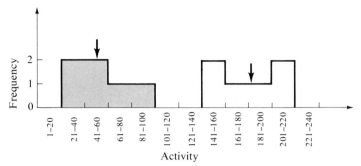

Figure 3.3
Frequency histograms of the activity scores of two inbred strains of mice,
A (shaded) and C57BL. Means are indicated by arrows.

taken as *prima facie* evidence for a heritable difference. Only subsequent breeding tests will reveal the manner in which the character is inherited.

The activity scores in Table 3.1 were obtained by placing the subjects one at a time in a square arena, the floor of which was marked off into a number of square areas, for a period of five minutes. The number of squares entered during this observation period was used as each subject's score. Frequency histograms representing these data are plotted in Figure 3.3. Although there are many ways in which these samples may be characterized, only the two most frequently employed sample values will be calculated here: the sample mean and the sample variance.

The *mean*, or arithmetic average, is a measure of *central tendency*. Let us symbolize the scores listed in Table 3.1 by the letter X. Using a subscript, we may indicate the first score, the second score, etc., for members of the A strain as follows: $X_1 = 29$; $X_2 = 29$; $X_3 = 44$; $X_4 = 58$; $X_5 = 63$; $X_6 = 83$. The mean, of course, is the sum of the scores divided by the sample size (N). Thus, the mean score of subjects in the A sample may be calculated as follows:

$$\overline{X}_A = \frac{X_1 + X_2 + X_3 + X_4 + X_5 + X_6}{6},$$

where $\overline{X}_A$ symbolizes the mean of the A sample. For small samples, this notation presents no particular problem. However, for large samples, such notation would be most cumbersome. Thus, we shall symbolize the sum of N observations by ΣX_i, where

$$\Sigma X_i = X_1 + X_2 + \cdots + X_N.$$

In general, then, the sample mean is calculated as follows:

$$\overline{X} = \frac{\Sigma X_i}{N}.$$

The sum of the scores obtained from the six A subjects is 306. Thus,

$$\overline{X}_A = \frac{306}{6} = 51.$$

The mean score of C57BL subjects in the sample is:

$$\overline{X}_C = \frac{1092}{6} = 182.$$

These means are indicated by arrows in Figure 3.3 and are seen to bisect each of the distributions into two equally balanced parts. Given that no bias was introduced by our sampling procedure, these sample means provide our best estimate of the true means of the populations from which these samples were drawn.

As the name implies, the *variance* is a measure of the variability, or dispersion, in a population. Variance may be defined as the average of the squared deviations from the mean. The population variance (σ^2 or V) is thus calculated as follows:

$$\sigma^2 = \frac{\Sigma(X_i - \mu)^2}{N},$$

where μ is the population mean. For reasons beyond the scope of our present discussion, the sample variance (s^2) should be calculated according to the formula that follows in order to provide an appropriate estimate of the population variance:

$$s^2 = \frac{\Sigma(X_i - \overline{X})^2}{N - 1}.$$

To illustrate the calculation of s^2, the data of Table 3.1 are presented again in Table 3.2, along with corresponding deviations from means and squared deviations. As may be seen, the variance of activity scores in the C57BL sample is somewhat larger than that of the A subjects.

When the sample size is large or when the sample mean is not a whole number, a somewhat different method for calculating the sum of the squared deviations from the mean is more appropriate. It may be shown that

$$\Sigma(X_i - \overline{X})^2 = \Sigma X_i^2 - \frac{(\Sigma X_i)^2}{N},$$

where ΣX_i^2 is the sum of the individually squared observations. For A subjects,

$$\Sigma X_i^2 = (29)^2 + (29)^2 + (44)^2 + (58)^2 + (63)^2 + (83)^2 = 17,840;$$

$$\frac{(\Sigma X_i)^2}{N} = \frac{(306)^2}{6} = 15,606;$$

Table 3.2

Examples of variance estimation from activity scores of two inbred strains of mice

A

X_i	$X_i - \overline{X}$	$(X_i - \overline{X})^2$
29	−22	484
29	−22	484
44	− 7	49
58	+ 7	49
63	+12	144
83	+32	1024
$\Sigma X_i = 306$	$\Sigma(X_i - \overline{X}) = 0$	$\Sigma(X_i - \overline{X})^2 = 2234$
$\overline{X}_A = 51$		$s_A^2 = \dfrac{2234}{5} = 446.8$

C57BL

X_i	$X_i - \overline{X}$	$(X_i - \overline{X})^2$
155	−27	729
157	−25	625
161	−21	441
199	+17	289
202	+20	400
218	+36	1296
$\Sigma X_i = 1092$	$\Sigma(X_i - \overline{X}) = 0$	$\Sigma(X_i - \overline{X})^2 = 3780$
$\overline{X}_C = 182$		$s_C^2 = \dfrac{3780}{5} = 756.0$

thus, $\Sigma(X_i - \overline{X})^2 = 17{,}840 - 15{,}606 = 2{,}234$, in agreement with the result previously obtained in Table 3.2. This "machine method" is particularly expedient when an electronic calculator is available for such computations.

Since variance is the average of the *squared* deviations from the mean, the obtained values are expressed in squared units, rather than in the actual units of measure. In spite of this, as will be seen in Chapter 9, variance has many important applications in genetics. Nevertheless, a measure of variability expressed in actual units, rather than squared units, is useful. Such a measure is provided by the square root of the variance, the so-called *standard deviation*. If our sample has been drawn at random from a population with a *normal distribution* (see Figure 3.4), the sample standard deviation (s) provides a precise measure of dispersion within that population. The population mean in Figure 3.4 is assumed to be zero and the population

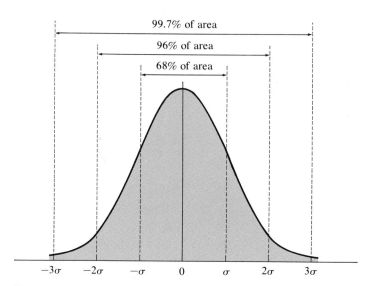

Figure 3.4
The normal distribution curve.

standard deviation, i.e., the population parameter estimated by s, is symbolized by σ. From this illustration it may be seen that approximately $\frac{2}{3}$ of the population would fall within one standard deviation above and below the mean and about 96 percent of the observations would occur within two standard deviations. Thus, we would predict that in a large population of mice of the A strain, approximately $\frac{2}{3}$ of their activity scores would fall within the range of 51 ±21.14, i.e., between 29.86 and 72.14. The precision of such estimation increases as a function of increasing sample size.

When two variables are measured on each subject or when the same variable is measured on pairs of subjects (parents and their offspring, for example), a measure of the covariation of these variables may be obtained. This *covariance* (symbolized by s_{XY} or Cov XY) is perfectly analogous to variance:

$$s_{XY} = \frac{\Sigma[(X_i - \overline{X})(Y_i - \overline{Y})]}{N - 1}.$$

As will be shown in Chapter 9, the concept of covariance is very important in quantitative genetic theory.

A related parameter, also of considerable importance in genetics, is the correlation. The *correlation coefficient* (symbolized by r_{XY}) may be estimated as follows:

$$r_{XY} = \frac{s_{XY}}{(s_X)(s_Y)},$$

Table 3.3

Sample calculation of a correlation coefficient, r_{XY}

	X_i	$X_i - \overline{X}$	$(X_i - \overline{X})^2$	Y_i	$Y_i - \overline{Y}$	$(Y_i - \overline{Y})^2$	$(X_i - \overline{X})(Y_i - \overline{Y})$
	1	-2	4	2	-4	16	$+8$
	2	-1	1	8	$+2$	4	-2
	3	0	0	6	0	0	0
	4	$+1$	1	4	-2	4	-2
	5	$+2$	4	10	$+4$	16	$+8$
Σ:	15	0	10	30	0	40	12

$$s_X^2 = \frac{10}{4} = 2.5 \qquad s_Y^2 = \frac{40}{4} = 10 \qquad s_{XY} = \frac{12}{4} = 3$$

$$r_{XY} = \frac{3}{\sqrt{(2.5)(10)}} = 0.6$$

where s_X and s_Y are the sample standard deviations for the variables X and Y, respectively. Consider the hypothetical data presented in Table 3.3 and plotted in Figure 3.5. It may be noted that Y tends to increase as X increases. The sample correlation, 0.6 in this example, provides a means of assessing the closeness of this association. A correlation of this type may assume any value from -1 to $+1$. A correlation of zero (or near zero) indicates that the two variables are independent; a high correlation (close to $+1$ or -1) indicates a close relationship. The correlation coefficient is "standardized" in the sense that it is not expressed in terms of actual units of measure.

An observed correlation between X and Y, of course, by no means proves the existence of a causal relationship. Such a relationship must be demonstrated by alternative means. In genetics, however, the causal association between genotype and phenotype is clear. When a causal relationship has been established, the correlation coefficient may be used to estimate the variance in one variable (say Y) due to variation in the other, i.e., $r_{XY}^2 s_Y^2$ provides an estimate of the variance in Y that may be attributed to variation in X, where there is some reason to believe that X is at least to some degree a cause of Y. X is sometimes referred to as the *independent variable,* and Y as the *dependent variable.* If it were possible to hold X constant, $1 - r_{XY}^2$ of the variance observed in Y would still remain. For example, when $r_{XY} = 0.6$, 36 percent of the variance in Y may be thought of as being due to variation in X and 64 percent of the variance in Y would remain, even if X were held constant. This may be somewhat clearer after we consider the related concept of regression.

The *regression coefficient* is of particular importance to genetics because it provides a means for making predictions, the requisite of all science.

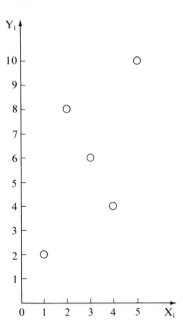

Figure 3.5
Plot of hypothetical data presented in Table 3.3.

Unlike correlation, the regression coefficient is expressed in terms of observed units of measure. In fact, the linear regression of Y on X, symbolized by b_{YX}, measures the number of units on the average that Y changes, corresponding to each unit change in X. The regression of Y on X may be calculated as follows:

$$b_{YX} = \frac{s_{XY}}{s_X^2}.$$

From the data presented in Table 3.3,

$$b_{YX} = \frac{3}{2.5} = 1.2,$$

i.e., for each unit increase in X, Y increased on the average 1.2 units (see Figure 3.6).

We may now utilize the regression coefficient to estimate the value of Y, given information on X. Such prediction may seem unnecessary, given that information has already been obtained on both variables. However, from the sample regression, we may estimate Y for other members of the population, given only information regarding variable X. In addition, this prediction equation may be used to fit a straight line to the observed points, a so-called "least-squares" regression line, i.e., the sum of the squared deviations of

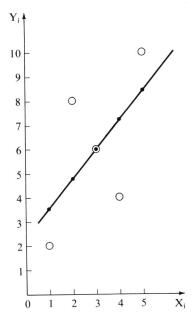

Figure 3.6

Plot of observed values (open circles) and expected values (small black circles) of Y, *corresponding to observed values of* X. *Expected values obtained from the regression equation,* $Y_i = 2.4 + 1.2X_i$.

the observed from the predicted points will be at a minimum. This prediction equation is:

$$\hat{Y}_i = \overline{Y} + b_{YX}(X_i - \overline{X}),$$

where $\hat{Y}_i$ is the expected or predicted value of Y, given information on X. From the data of Table 3.3,

$$\hat{Y}_i = 6 + 1.2(X_i - 3) = 2.4 + 1.2X_i.$$

Using this equation, we may calculate the expected value of Y corresponding to each observed value of X in Table 3.3. These observed and expected values are presented in Table 3.4 and graphed in Figure 3.6.

We may now use these expected values of Y to demonstrate how variance in Y may be partitioned into two parts: (1) that due to variation in X; and (2) that independent of variation in X. Y_i may be partitioned into parts as follows:

$$Y_i = \hat{Y}_i + (Y_i - \hat{Y}_i),$$

i.e., the observed value is equal to the expected value plus the deviation of the observed value from expected. Let us calculate the variance of each part as shown in Table 3.5. From Table 3.5, it may be seen that the variance in Y (s_Y^2) is equal to the sum of the variance of the expected values of

Table 3.4

Observed and expected values of **Y**

X_i	Y_i	$\hat{Y}_i$
1	2	3.6
2	8	4.8
3	6	6.0
4	4	7.2
5	10	8.4

Table 3.5

Calculation of the variance in **Y** *due to regression and due to deviations from regression*

	Y_i	$\hat{Y}_i$	$\hat{Y}_i - \overline{Y}$	$(\hat{Y}_i - \overline{Y})^2$	$Y_i - \hat{Y}_i$	$(Y_i - \hat{Y}_i)^2$
	2	3.6	−2.4	5.76	−1.6	2.56
	8	4.8	−1.2	1.44	+3.2	10.24
	6	6.0	0.0	0.00	0.0	0.00
	4	7.2	+1.2	1.44	−3.2	10.24
	10	8.4	+2.4	5.76	+1.6	2.56
Σ:	30	30.0	0.0	14.40	0.0	25.60

$$s_Y^2 = 10 \qquad s_{\hat{Y}}^2 = \frac{14.4}{4} = 3.6 \qquad s_{Y-\hat{Y}}^2 = \frac{25.6}{4} = 6.4$$

Y $(s_{\hat{Y}}^2)$ and the variance of the deviation of the observed from expected values $(s_{Y-\hat{Y}}^2)$. We may think of $s_{\hat{Y}}^2$ as variance in Y "due to regression of Y on X" and $s_{Y-\hat{Y}}^2$ as variance in Y "due to deviations from regression." This was the meaning of our earlier discussion concerning correlation, where it was indicated that part of the variance in Y may be thought of as being due to variation in X:

$$r_{XY}^2 s_Y^2 = (0.36)(10) = 3.6$$

and

$$(1 - r_{XY}^2) s_Y^2 = (0.64)(10) = 6.4.$$

Thus, rather than calculating each value of $\hat{Y}_i$ and $Y_i - \hat{Y}_i$, the variance in Y due to regression may be simply estimated from the product of r_{XY}^2 and

$s_{\hat{Y}}^2$. This represents the variance of the expected values (points on the regression line) about the population mean (see Figure 3.6). The variance in Y due to deviations from regression (or the variance in Y that would remain if X were held constant), calculated from $(1 - r_{XY}^2)s_Y^2$, is really nothing more than a measure of the variation of the observed value of Y_i about the regression line. This partitioning of variation will be most useful when we begin discussing genetic variance in Chapter 9.

Hypothesis Testing. Tests of statistical significance will be cited frequently throughout the remainder of this text. Many different tests of significance are available, only a few of which will be described here. The choice of which test to use in any particular case depends upon the nature of the data at hand. We shall begin by considering enumeration data.

As will be discussed in Chapter 5, a new neurological mutation in mice was described by van Abeelen and van der Kroon in 1967. Mice from mutant and normal strains were crossed to produce an F_1 generation, all of which were normal. In order to test the hypothesis that the neurological condition was due to a recessive allele at an autosomal locus, F_1 males were mated with F_1 females, yielding the following F_2 ratio: 124 normal : 47 mutant. If the condition were due to homozygosity for an autosomal recessive allele, a ratio of 3 normal to 1 mutant should have been obtained, i.e., of the 171 mice in the F_2, $171/4 = 42.75$ should have been mutant and $(3)(42.75) = 128.25$ should have been normal. The observed results appear to be in relatively good agreement with those expected on the basis of the hypothesis of a recessive allele at a single autosomal locus. How large, however, may the discrepancy between observed and expected be and still be regarded as simply due to chance? Alternatively, at what point do we begin to regard the differences as evidence that the hypothesis should be modified or rejected? The chi-square (χ^2) test may be used to provide answers to these questions. Calculation of chi square, followed by consultation of a chi-square table or chart (see Figure 3.7), yields the approximate probability that, assuming the hypothesis is correct, results would be obtained that deviate from expectation by as much or more than those actually observed. Let us now apply this technique to the data on the mutant strain of mice under consideration.

Chi square is calculated as follows:

$$\chi^2 = \Sigma \left[\frac{(\text{observed} - \text{expected})^2}{\text{expected}} \right].$$

Thus, from the data,

$$\chi^2 = \frac{(124 - 128.25)^2}{128.25} + \frac{(47 - 42.75)^2}{42.75} = 0.56.$$

Before consulting Figure 3.7, one other quantity, the number of *degrees of freedom,* is required. Except for rare exceptions (Crow and Kimura, 1970, p. 496), the number of degrees of freedom for a chi-square test is one less than the number of classes or categories considered. In our example, mice are either normal or mutant. Thus, the number of degrees of freedom is $2 - 1 = 1$. If a mouse is not normal, it must be a mutant in this example. There are no alternatives; thus, there is only one degree of freedom.

To obtain the appropriate probability value from Figure 3.7, find the calculated value of chi square on the horizontal axis and then go up from this point to where the corresponding number of degrees of freedom (N) is reached. The probability value corresponding to the chi-square value and the number of degrees of freedom is read directly from the vertical axis on the left. For a chi-square value of 0.56 with one degree of freedom, the probability value is greater than 0.40. Thus, the probability of observing a deviation as large or larger than that actually observed is greater than 0.40; if the hypothesis is correct, we shall observe deviations at least this large in 40 percent of the experiments we might perform to test it. Because such a deviation would be this frequent, we have no reason to reject the hypothesis of a recessive allele at a single autosomal locus.

Let us consider another mutant neurological character in mice, which was first described by Mary F. Lyon (1958). It is known as *twirler* because affected mice tend to run in tight circles. Preliminary evidence suggested that this character was due to a dominant allele at a single autosomal locus. However, when F_1 mice, symbolized here as $T_1 T_2$, from matings between twirlers and normals were crossed with normal mice $(T_2 T_2)$, the following results were obtained: 466 twirlers: 558 normals. Assuming dominance of the mutant allele, such a testcross should yield a 1:1 ratio, i.e., among the total of 1,024 mice, we would expect to observe 512 twirlers and 512 normals. Could this departure of observed from expected be due to chance alone?

$$\chi^2 = \frac{(466 - 512)^2}{512} + \frac{(558 - 512)^2}{512} = 8.26.$$

From Figure 3.7, again with one degree of freedom, we may see that the probability of obtaining a deviation this large or larger, simply due to chance alone, is between 0.005 and 0.002. Thus, either the hypothesis is wrong or else a very unlikely outcome has been observed. The reason for the shortage of affected mice in the twirler example will be discussed in Chapter 5.

Conventionally, if the probability is less than 0.05, but greater than 0.01, the departure of observed from expected results is regarded as being statistically *significant.* If the probability is less than 0.01, the difference is considered *highly significant.* In such cases, we reject our hypothesis and seek some alternative explanation, realizing, of course, that we will occasionally make the error (1 time out of 20 or less) of rejecting a true hypothesis — a *type-I error.*

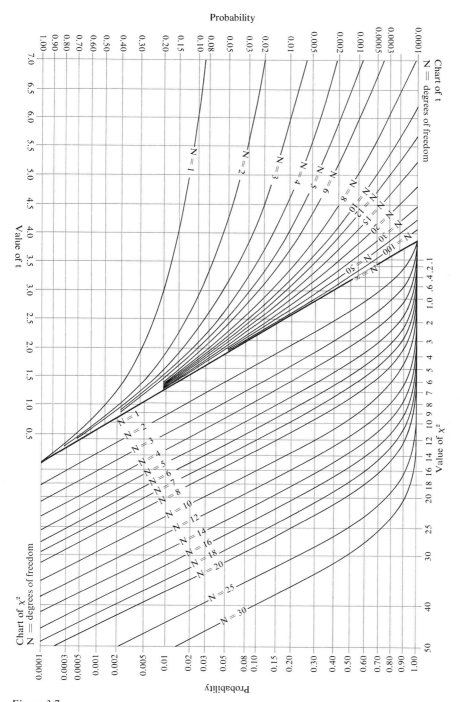

Figure 3.7
Chart of chi-square and t values. (From Crow and Kimura, 1970, p. 516.)

As indicated earlier, the chi-square test is only approximate. The approximation is quite good with large numbers, but may result in considerable error with small numbers. Thus, as a conservative rule, the chi-square test should not be used if the expected number within any class is less than 5. In addition, it is important to remember that only numbers, not proportions or percentages, are to be used in calculating chi-square values. Let us now consider a test of significance appropriate for measurement data.

The t distribution may be used for tests of significance when data are obtained from measurements, rather than counts. If we wish to test the significance of a difference between the means of two groups, t may be calculated as follows:

$$t = \frac{\overline{X}_A - \overline{X}_B}{\sqrt{\frac{(N_A + N_B)[(N_A - 1)s_A^2 + (N_B - 1)s_B^2]}{N_A N_B (N_A + N_B - 2)}}}.$$

As an example, consider again the activity data of A and C57BL mice in Table 3.2. Is the difference between the two strains real, or could it be due to sampling errors? Let us assume a *null hypothesis*: that is, let us hypothesize that there is no difference between the strains, and determine the probability that a difference this large or larger could be obtained by chance alone:

$$t = \frac{182 - 51}{\sqrt{\frac{(6 + 6)[(5)(756.0) + (5)(446.8)]}{(6)(6)(6 + 6 - 2)}}} = 9.23,$$

where the larger mean is arbitrarily symbolized by $\overline{X}_A$ so as to yield a positive t value. The corresponding number of degrees of freedom is $N_A + N_B - 2$, i.e., $6 + 6 - 2 = 10$ in our example. When turned upside down, Figure 3.7 becomes a chart of probability values corresponding to calculated values of t. This t chart is read in exactly the same manner as the chi-square chart. When $t = 9.23$ and the degrees of freedom are 10, we see that the probability value is less than 0.0001. This highly significant result indicates that less than one time in ten thousand would differences as large or larger than observed occur due to chance alone. Thus, we may safely reject our null hypothesis and conclude that the difference is almost certainly real.

The t distribution may also be used to test the significance of correlation and regression coefficients. For example, consider the hypothetical data presented in Table 3.3. Is the obtained value of $r_{XY} = 0.6$ significantly different from zero? We shall assume the null hypothesis, $r_{XY} = 0$, and then determine the probability that a value as large as 0.6 could be obtained by chance alone. For a correlation coefficient,

$$t = r\sqrt{\frac{N - 2}{1 - r^2}},$$

with $N - 2$ degrees of freedom, where N is the number of paired observations. From our example,

$$t = 0.6\sqrt{\frac{5 - 2}{1 - 0.36}} = 1.3.$$

With $t = 1.3$ and only $5 - 2 = 3$ degrees of freedom, the probability of obtaining a value this large or larger due to chance alone lies between 0.20 and 0.30. Using the criterion previously established, we are not justified in concluding from these calculations that the observed correlation is significantly different from zero. This is not surprising in view of the very small sample size in this example. The "power" of tests of significance (the probability of rejecting the null hypothesis) is very low with such small sample sizes. In this case, we would not be justified in accepting the null hypothesis that $r_{XY} = 0$. Instead, additional data should be obtained. With increasing sample size, the chance of accepting a null hypothesis that is actually false—a *type-II error*—decreases.

From the data of Table 3.3, we found that the regression of Y on X, symbolized b_{YX}, was equal to 1.2. Let us assume the null hypothesis that $b_{YX} = 0$ and determine the probability that a regression as large or larger than 1.2 would be obtained by chance alone. For regression coefficients,

$$t = \frac{b_{YX}}{\sqrt{\frac{1}{N-2}\left[\frac{s_Y^2}{s_X^2} - (b_{YX})^2\right]}}.$$

From the data of Table 3.3,

$$t = \frac{1.2}{\sqrt{\frac{1}{3}\left[\frac{10}{2.5} - (1.2)^2\right]}} = 1.3.$$

As with the correlation coefficient estimated from the same data, $t = 1.3$ with 3 degrees of freedom. These results may be succinctly summarized as follows: The sample regression coefficient, $b_{YX} = 1.2$, was found to be not significantly different from zero ($t = 1.3$, $df = 3$, $0.20 \leq p \leq .30$), where df symbolizes degrees of freedom and p is the probability of obtaining a value this large or larger due to chance alone. As with the correlation coefficient, p is sufficiently large that the null hypothesis of $b_{YX} = 0$ would not be rejected. However, with only three degrees of freedom, prudence would dictate reserving judgment until more data were collected.

Many other important tests of statistical significance are currently available. However, the methods briefly discussed in this chapter are sufficient for an understanding of a large part of the behavioral genetics literature. Let us begin sampling this literature by a consideration of single-gene analysis with human subjects.

Chapter 4

Single-gene analysis and human behavior

In many respects, man is an unfavorable organism for genetic analysis: experimental crosses may not be performed; environmental control may not be imposed; the generation interval is relatively long; and the number of offspring per family is relatively small. As Stern (1973) points out, however, these problems are not insurmountable. The human population is very large and thus a rich store of genetic information is potentially available. Although planned Mendelian crosses are not feasible, data may be collected from families in which particular mating types are of interest. Although environmental control may not be imposed, it may be possible to study members of different families which have been reared in more-or-less similar environments. Although man's generation interval is long, data from several generations may nevertheless be available. Finally, although the number of children in human families is relatively small, data from many families may be pooled to provide adequate samples for statistical analysis. That many important advances have occurred in human genetics within the last two decades demonstrates that man, like the mouse, the fruit fly, bread mold, and colon bacteria, is also a favorable organism for genetic analysis.

Single-gene analysis of human behavior is principally concerned with testing the adequacy of single-locus hypotheses. Familial transmission of some character of interest is usually noted and it is then determined whether

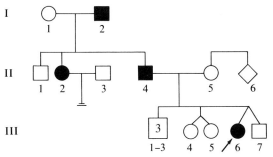

Figure 4.1

A sample pedigree. For explanation, see text. (From Stern, Principles of Human Genetics, 3rd Ed., W. H. Freeman and Company. Copyright © 1973, p. 107.)

the observed pattern of transmission conforms to that expected on the basis of a simple Mendelian model. Both pedigree analysis and gene-frequency analysis may be employed in such research.

Pedigree Analysis

A sample *pedigree* (from *pié de grue,* "crane's foot," a three-line mark denoting succession) is shown in Figure 4.1. *Affected* individuals, i.e., those manifesting the condition under study, are designated by solid symbols; *nonaffected* individuals are indicated by open symbols. Females are represented by circles and males by squares. A diamond is used to represent an individual whose sex is unknown. Parents are joined by a horizontal *marriage line,* offspring being listed below. Members of a *sibship* are connected to a horizontal line that is joined by a perpendicular to their parents' marriage line, with the *sibs* (brothers and sisters) being listed from left to right in order of birth. In this pedigree, each generation is designated by a Roman numeral, and each individual within a generation is denoted by an Arabic numeral. Individuals II-5 and II-6 were sibs, but information concerning their parents was not included in the pedigree. The marriage of individuals II-2 and II-3 resulted in no children. In order to save space, a number enclosed in a large symbol may be used to indicate the number of sibs of like condition. Thus, III-1, III-2, and III-3 were three unaffected males. Twins are indicated by two symbols that are connected either at or just below the sibship line. Individuals III-4 and III-5 were *monozygotic* (identical) twins, indicated by the short vertical line that descends from the sibship line; individuals III-6 and III-7 were *dizygotic* (fraternal) twins. The finding of a family of interest frequently comes only after the discovery of a particular affected individual. This specific individual who first comes to

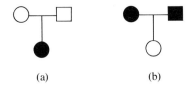

(a)   (b)

Figure 4.2

Hypothetical pedigrees. Condition indicated in (a) *could be due to a recessive autosomal gene, but not a dominant, whereas that in* (b) *could be due to a dominant, but not a recessive.*

the attention of the investigator is referred to as the *index case* (also *propositus* or *proband*). The index case in Figure 4.1, individual III-6, is indicated by an arrow.

Different patterns of transmission are observed for single-gene dominant, single-gene recessive, and sex-linked recessive characters. With autosomal recessive inheritance, individuals that are homozygous recessive, thus expressing the condition, usually have parents who are phenotypically normal. Because inbreeding increases the likelihood of homozygosity, the incidence of genetic diseases due to recessive inheritance is higher among children of *consanguineous* marriages, e.g., cousin marriages. Thus, when a rare character or condition is observed more frequently among offspring of cousin matings than in the population at large, recessive inheritance is suggested. A quite different pattern of transmission is observed for characters caused by genes with completely dominant effects. At least one of the parents of an individual affected with such a character is likewise affected, and about one-half of the children of an affected parent are affected.

For cases of classic dominant or recessive inheritance, pedigree analysis provides a useful means of discriminating between alternative models. Consider the two hypothetical pedigrees presented in Figure 4.2. If the condition indicated in Figure 4.2(a) were due to a dominant gene, the parent who carried the dominant gene would also have been affected. Thus, when both parents of an affected offspring are unaffected, the condition may be due to a recessive gene, but it could not be due to a dominant one. Conversely, the condition indicated in Figure 4.2(b) could be due to a dominant gene, but it could not be due to a recessive one. If it were due to a recessive gene, both parents, who are affected, would have to be homozygous recessive; thus, any children from such a marriage would be homozygous and similarly affected. As discussed later in this chapter, in sex-linked inheritance, the occurrence of the condition in any generation depends upon the sex of the affected parent and that of the offspring.

Genotypes do not always express themselves in exactly the same way in all individuals, probably due to the complexity of the developmental pathways by way of which genes are ultimately expressed in the phenotype (see Chapters 6 and 8). Characters that are not expressed in all individuals who have the appropriate genotype are said to be *incompletely penetrant;* those that are expressed to varying degree (e.g., different degrees of severity) are said to display *variable expressivity.* Although not fully understood, incomplete penetrance and variable expressivity are presumably due to the

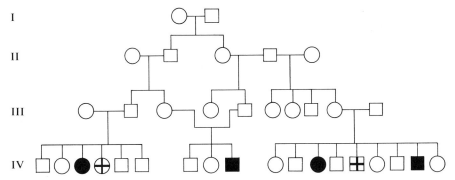

Figure 4.3

Pedigree of phenylketonuria and associated mental deficiency. (From Stern, Principles of Human Genetics, 3rd ed., W. H. Freeman and Company. Copyright © 1973, p. 165. After Følling, Mohr, and Ruud.)

effects of environment or of other genes. In any case, these phenomena may obfuscate the transmission of "simple" dominant or recessive characters.

Let us now consider the pedigree shown in Figure 4.3. Solid symbols denote individuals affected with phenylketonuria (PKU), a condition discussed in greater detail in Chapter 6. Individuals marked with a cross were probably affected, but their condition was not known for certain because they died young. A cousin marriage is indicated in the middle of the pedigree. As these individuals all lived in an isolated group of small islands in Norway, there may actually have been more inbreeding than indicated by this pedigree. Although, as discussed in Chapter 6, PKU may manifest incomplete penetrance, the pattern of transmission indicated in Figure 4.3 nonetheless conforms closely to that expected of an autosomal recessive gene (i.e., affected subjects have normal parents and an increased incidence accompanies inbreeding). Of the 18 children in generation IV, 4 were definitely affected and 2 were probably affected. If both unaffected parents of each sibship were carriers, as must be the case for a recessive condition, then only one-fourth of the children would be expected to be affected. Although this departure from expectation (6 vs. 4.5) is not significant, an excess of affected individuals is frequently observed in pedigree data for reasons described in the following section.

Five pedigrees depicting the transmission of colorblindness are presented in Figure 4.4. Although a number of different forms of colorblindness are known (see Stern, 1973, for a discussion of these various forms), the more common types have a similar genetic basis and will not be differentiated here. The pattern of transmission evident in Figure 4.4 conforms closely to that expected of a sex-linked recessive gene. As discussed in Chapter 2, females have two X chromosomes, whereas males have one X and one Y. The transmission of X chromosomes from one generation to

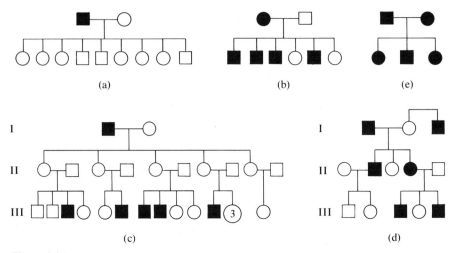

Figure 4.4

*Pedigrees of colorblindness. (a) Part of pedigree no. 406, Nettleship. (b) Pedigree no. 584.
(c) Part of Horner's pedigree. (d) Pedigree, Whisson, 1778. This is the first known pedigree
of colorblindness. (e) Pedigree, Vogt. (From Stern, Principles of Human Genetics, 3rd Ed.
W. H. Freeman and Company. Copyright © 1973, p. 305. a,b, after Bell, Treas. Hum. Inher.,
II, 2, 1926; c, after Gates, Human Genetics, Macmillan; e, after Bauer, Fischer, Lenz.)*

the next is shown in Figure 4.5, from which it may be seen that daughters
inherit their father's X chromosome, but sons do not. Thus, sons cannot
inherit sex-linked conditions from their father. Daughters inherit sex-linked
genes from their father; however, they do not express the condition if it is
recessive unless they receive another such allele from their mother (cf.
Figures 4.4(a) and 4.6). One-half of the sons of these carrier daughters subsequently inherit replicates of the X chromosome bearing the recessive gene
and thus express the same condition as their maternal grandfather (see
Figure 4.4(c)). This is why sex-linked recessive inheritance sometimes
results in the appearance of a "skip-a-generation" phenomenon.

  If a mother is homozygous recessive and a father is normal, all their
daughters will be carriers and all sons will be affected (Figures 4.4(b) and 4.7).
However, if the father is affected and the mother is a carrier, one-half of both
sons and daughters will be affected—compare Figures 4.4(d) and 4.8. Note
again the slight excess of affected sons in generation III of Figure 4.4(c).

  Another example of sex-linked recessive inheritance, Lesch-Nyhan syndrome, will be discussed in Chapter 6.

  The Problem of Ascertainment.   When the criterion for inclusion of
families in a pedigree is that they contain an affected individual, there will be
an excess of affected individuals over the proportion expected simply as a
result of using this system of selection. In order to illustrate this problem
of *ascertainment*, let us consider sex ratio as an example. Among two-child

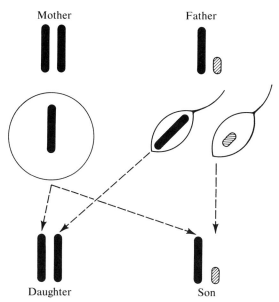

Figure 4.5

Transmission of the X-chromosomes from one generation to the next. (From Stern, Principles of Human Genetics, 3rd Ed. W. H. Freeman and Company. Copyright © 1973, p. 303.)

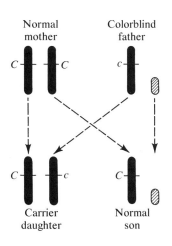

Figure 4.6

Transmission of colorblindness. Normal woman × colorblind man. (From Stern, Principles of Human Genetics, 3rd Ed. W. H. Freeman and Company. Copyright © 1973, p. 304.)

families, we would expect to observe the different possible arrangements in the following proportions:

$$\tfrac{1}{4}BB : \tfrac{1}{2}BG : \tfrac{1}{4}GG,$$

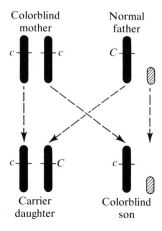

Colorblind
mother

Normal
father

Carrier
daughter

Colorblind
son

Figure 4.7

Transmission of colorblindness. Colorblind woman ×
normal man. (From Stern, Principles of Human
Genetics, 3rd Ed. W. H. Freeman and Company.
Copyright © 1973, p. 304.)

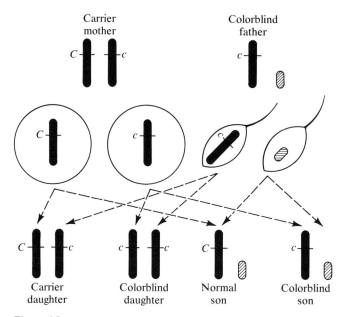

Carrier
mother

Colorblind
father

Carrier
daughter

Colorblind
daughter

Normal
son

Colorblind
son

Figure 4.8

Transmission of colorblindness. Carrier woman × colorblind man.
(From Stern, Principles of Human Genetics, 3rd Ed. W. H.
Freeman and Company. Copyright © 1973, p. 304.)

where B and G symbolize boy and girl and birth order is ignored. Thus, if
all two-child families are randomly sampled, all of the children in BB sib-
ships and one-half of the children in BG sibships $[\frac{1}{4}BB + \frac{1}{2}(\frac{1}{2}BG)]$ or one-
half of the children so ascertained will be boys. However, what will be the

Table 4.1

Probability of various sibship combinations in four-child families when both parents are heterozygous for a recessive allele (D = dominant phenotype and R = recessive phenotype)

Sibship	Probability
4D:0R	$\dfrac{4!}{4!0!}(\tfrac{3}{4})^4(\tfrac{1}{4})^0 = \dfrac{81}{256}$
3D:1R	$\dfrac{4!}{3!1!}(\tfrac{3}{4})^3(\tfrac{1}{4})^1 = \dfrac{108}{256}$
2D:2R	$\dfrac{4!}{2!2!}(\tfrac{3}{4})^2(\tfrac{1}{4})^2 = \dfrac{54}{256}$
1D:3R	$\dfrac{4!}{1!3!}(\tfrac{3}{4})^1(\tfrac{1}{4})^3 = \dfrac{12}{256}$
0D:4R	$\dfrac{4!}{0!4!}(\tfrac{3}{4})^0(\tfrac{1}{4})^4 = \dfrac{1}{256}$

sex ratio if only "affected" families are sampled (let us say families that contain at least one boy)? Such ascertainment will eliminate GG sibships from our sample. Of the 1BB:2BG families sampled, $\tfrac{1}{3}$BB $+ \tfrac{1}{2}(\tfrac{2}{3})$BG $= \tfrac{2}{3}$ of the children so ascertained will be boys. The $\tfrac{1}{2}$B:$\tfrac{1}{2}$G ratio was obtained when ascertainment was complete, whereas a $\tfrac{2}{3}$B:$\tfrac{1}{3}$G ratio resulted when ascertainment was by *truncate selection*, i.e., when an entire group was not included in the sample.

Because genotypes of all parents are not readily available to investigators, most couples who are at risk of producing affected offspring but who have not actually done so are automatically excluded from pedigree data, which brings about inclusion in pedigrees of more affected individuals than would be expected from Mendelian calculations. For example, when both parents are heterozygous for a recessive allele, a ratio of 3D:1R is expected among their progeny, where D and R symbolize dominant and recessive phenotypes. In sibships of four, various combinations may occur: 4D:0R; 3D:1R; 2D:2R; 1D:3R; OD:4R. From the binomial theorem, the probability of each combination may be determined; this has been done and the results are shown in Table 4.1. Thus, when all combinations are included in a computation to determine the proportion of total offspring having the dominant phenotype, we find that $81/256 + (\tfrac{3}{4})(108/256) + (\tfrac{1}{2})(54/256) + (\tfrac{1}{4})(12/256) + (0)(1/256) = 192/256 = \tfrac{3}{4}$ of the progeny will express the dominant phenotype, as we expect from basic Mendelian principles. However, if families are included in a sample only when at least one affected child is present in the sibship, as is always the case when the genotypes of parents are unknown, 4D:0R sibships are excluded from the sample. In this example, 81/256 or

Table 4.2

*Observed and expected proportions from heterozygous
parents of normal to PKU children in sibships with
at least one affected*

Sibs per sibship	No. sibships	Ratio of normal:PKU	
		Observed	Expected
1	6	0:1	0:1
2	7	0.75:1	0.75:1
3	6	0.80:1	1.31:1
4	5	1.50:1	1.73:1
5	7	1.69:1	2.05:1
6	5	1.50:1	2.29:1
7	2	2.50:1	2.46:1
8	3	2.00:1	2.60:1
9	1	3.50:1	2.70:1
10	2	3.00:1	2.77:1
11	1	4.50:1	2.83:1
12	1	3.00:1	2.87:1
13	1	2.25:1	2.90:1

SOURCE: After Stern, *Principles of Human Genetics,* 3rd Ed.
W. H. Freeman and Company. Copyright © 1973, p. 204;
data from Munro, 1947.

almost one-third of the relevant sibships would not be included. As a result, when relevant four-child families are ascertained only by the presence of affected individuals, the expected proportion of individuals with the dominant phenotype will be $[(\frac{3}{4})(108) + (\frac{1}{2})(54) + (\frac{1}{4})(12) + (0)(1)]/175 = 111/175 = 0.634$, instead of 0.75. Thus, instead of observing a 0.75D:0.25R or 3:1 ratio, a ratio of 0.634D:0.366R or 1.73:1 is obtained.

The departure from a 3:1 ratio with truncate selection is a function of the sibship size. With increasing sibship size, the proportion of families not included decreases since the probability that all children will exhibit the dominant phenotype in large sibships becomes small. Thus, the departure from a 3:1 ratio decreases as the size of the sibship increases. In Table 4.2, the expected proportion of normal (dominant) to PKU (recessive) sibs in sibships with at least one affected individual is compared to the proportion actually observed in 47 British families by Munro (1947). It may be seen from this table that the expected ratio approaches 3:1 as the number of children per sibship (n) becomes large. It may also be observed that in spite of considerable fluctuation due to small sample size, the observed ratio also approaches a 3:1 ratio as n increases.

Ascertainment by truncate selection results in an excess of affected individuals because the index case is included in the sample. If the ratio is determined only among the sibs of each index case, however, this systematic bias may be avoided. Application of this correcting method assumes that the sibships are not chosen because of an unusual number of affected members and that complete information is available on each sibship. This *simple sib method* of correcting for ascertainment bias presupposes no specific ratio among the progeny. More elaborate methods are available, however, that assume such ratios. Let us illustrate the use of one of these methods by returning to the PKU data presented in Table 4.2.

We shall test the hypothesis that PKU is transmitted as a simple Mendelian recessive character. Let $P(A)$ symbolize the probability that a child will be affected if both parents are heterozygous, i.e., $P(A) = \frac{1}{4}$. Let $P(\overline{A})$ symbolize the probability that the child is unaffected; thus, $P(\overline{A}) = 1 - P(A) = \frac{3}{4}$. Let $P(A)'$ be the probability that in a sample obtained by truncate selection a child is affected. Among sibships of n members, $[P(\overline{A})]^n$ will have no affected children. In our sample, which includes only families with at least one affected child, $1 - [P(\overline{A})]^n$ of all at-risk families will be included. Thus, in our example, $P(A)'$ may be determined as follows:

$$P(A)' = \frac{P(A)}{1 - [P(\overline{A})]^n} = \frac{\frac{1}{4}}{1 - (\frac{3}{4})^n}.$$

In sibships of 4,

$$P(A)' = \frac{\frac{1}{4}}{1 - (\frac{3}{4})^4} = 0.366,$$

which agrees with our previous calculation of the expected ratio of 0.634D:0.366R in sibships of 4 when both parents are heterozygous for a recessive allele and at least one sib is affected.

Multiplication of $P(A)'$ times the sibship size (n) yields the expected number of affected children per sibship. Multiplication of this product by the number of sibships (x) in our sample results in an expected number of affected individuals per sibship size category. This expected number may be compared to the observed number, thereby permitting a chi-square test of the adequacy of our model. These calculations are presented in Table 4.3. However, because of the small number of individuals expected to be affected in some categories, only data on sibships of 6 and smaller are included in this table. Recall that chi-square tests should not be applied when expected numbers are less than 5. In addition, since sibships of one member necessarily include only 1 affected individual and no normals, this category has also been deleted. The resulting chi-square values, each with 1 degree of freedom, all have accompanying probabilities greater than 0.20. An over-all chi-square test may also be performed by merely summing the individual

Table 4.3

Test of the hypothesis that PKU is caused by a recessive gene at a single autosomal locus (see text for explanation)

Sibship size (n)	Number of sibships (x)	$[P(A)](n)$	Number affected		Number unaffected		χ^2	$p \geq$
			Expected $[P(A)](n)(x)$	Observed	Expected $[1 - P(A)](n)(x)$	Observed		
2	7	1.143	8.001	8	5.999	6	0.00	1.00
3	6	1.297	7.782	10	10.218	8	1.11	0.30
4	5	1.463	7.315	8	12.685	12	0.10	0.70
5	7	1.640	11.480	13	23.520	22	0.30	0.60
6	5	1.825	9.125	12	20.875	18	1.30	0.20

chi-square values. The resulting total chi square of 2.81, with 5 degrees of freedom, has a probability of $0.70 \leq p \leq 0.80$. Thus, these data are highly consistent with the hypothesis that PKU is due to a recessive gene at a single autosomal locus.

Variable Age of Onset. As indicated earlier, genotypes do not always express themselves in exactly the same way in all individuals. In addition, the age at which a given condition is expressed may also differ among individuals. An example of a condition with variable age of onset is *Huntington's chorea*, characterized by loss of motor control, progressive dementia, and if the age of onset is early, mental deficiency. This condition has been traced through many generations of pedigrees and a consistent pattern of transmission is observed: (1) most afflicted patients had a parent who was also afflicted; and (2) approximately one-half of the children of an afflicted parent eventually develop the disease. Since the age of onset is variable, death occasionally precedes the onset of the disease in individuals genetically predisposed to develop it. Nevertheless, this pattern of transmission closely conforms to that expected of a condition caused by a dominant gene. Individuals who develop the disease received the dominant gene from a parent who was also afflicted. Since afflicted individuals are heterozygous, one-half of their children would be expected to receive the dominant gene and thus to develop the disease. The persistence of this insidious dominant lethal gene in the population is apparently due to the fact that the disease is not usually expressed until after the childbearing years. This late age of onset illustrates the principle that hereditary conditions are *not* always manifested at birth.

The distribution of age of onset of Huntington's chorea in 762 patients is plotted in Figure 4.9. Although onset may occasionally occur during the first few years of life, the mean age of onset is between 40 and 45 years. Data of this type are very useful for obtaining *age-corrected* incidence data. As an oversimplified example, assume that only one-half of individuals genetically predisposed to develop a disease actually express the disease by age 40. In such a case, the observed incidence of the disease among individuals at age 40 should be doubled to obtain age-corrected incidence data. Another condition with variable age of onset is *schizophrenia*, discussed in some detail in Chapter 11.

Gene-frequency Analysis

With pedigree data, only affected families are investigated. However, when a character is present in two alternative forms and when both forms are relatively frequent, the adequacy of genetic hypotheses may be tested by subjecting all available family data to genetic analysis. This procedure will be illustrated after some of the basic principles of population genetics are first considered.

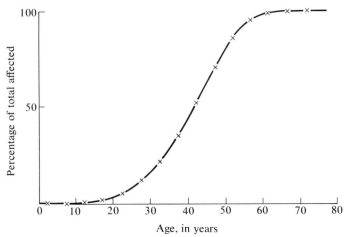

Figure 4.9

Huntington's chorea. Distribution of age of onset in 762 patients.
(From Stern, Principles of Human Genetics, 3rd Ed. W. H. Freeman
and Company. Copyright © 1973, p. 381. After Landzettel,
Unterreiner, and Wendt, Acta Genet. Stat. Med., 9, 1959.)

Gene and Genotype Frequencies. Let us symbolize the frequencies of the A_1A_1, A_1A_2, and A_2A_2 genotypes in a population as **P**, **H**, and **Q**, respectively. These *genotype frequencies* are so defined that $\mathbf{P} + \mathbf{H} + \mathbf{Q} = 1.0$. If the genotypes are distinguishable phenotypically, the observed genotype frequencies permit determination of gene frequencies, which are defined as the fraction or percentage of the alleles of a particular locus that are of a given type in a particular population. Let p symbolize the frequency of the A_1 allele and q the frequency of the A_2 allele. As each individual who is *diploid* (having two sets of chromosomes) must possess two genes at each autosomal locus, $p = (2\mathbf{P} + \mathbf{H})/2 = \mathbf{P} + \frac{1}{2}\mathbf{H}$. That is, all A_1A_1 and $\frac{1}{2}$ of the A_1A_2 individuals' alleles are A_1. As there are only two alleles at the locus under consideration, $q = 1 - p = \mathbf{Q} + \frac{1}{2}\mathbf{H}$.

For example, assume that a population of 100 individuals has been classified according to antigens carried on the surface of their red blood cells. Assume that 50 individuals possess the M antigen, 30 the N antigen and 20 both the M and N antigens. Pedigree studies have revealed that individuals with the M antigen are homozygous $L^M L^M$, N individuals are homozygous $L^N L^N$ and MN individuals are heterozygous $L^M L^N$. Because heterozygotes express both L^M and L^N, the alleles are referred to as being *codominant*. Using p to symbolize the frequency of the L^M allele, the genotype and gene frequencies in this population are obtained as follows:

$$\mathbf{P} = 50/100 = 0.5 \qquad \mathbf{H} = 20/100 = 0.2 \qquad \mathbf{Q} = 30/100 = 0.3$$

$$p = \mathbf{P} + \tfrac{1}{2}\mathbf{H} = 0.5 + \tfrac{1}{2}(0.2) = 0.6 \qquad q = 1 - p = \mathbf{Q} + \tfrac{1}{2}\mathbf{H} = 0.4$$

Table 4.4

Genotype frequencies after
one generation of random mating

		Sperm	
		p A_1	q A_2
Eggs	p A_1	p² A_1A_1	pq A_1A_2
	q A_2	pq A_1A_2	q² A_2A_2

The genotype frequency may be predicted from the gene frequency when the system of mating is known. *Random mating* refers to that situation in which every possible single mating is as likely to occur as any other. If gametes produced by mates unite at random, which is almost always the case, then random mating may be seen to be equivalent to the random union of gametes. The concept of random mating is important in population genetics not because it is assumed to be ubiquitous or even frequent in nature but because it provides us a baseline or sea level from which to operate. Random mating is to population genetics what the concept of standard temperature and pressure is to chemistry and physics. However, it should be noted that there are well-documented instances of random mating in investigations of particular characters.

When gametes unite at random, the genetic composition of the sperm is independent of that of the egg with which it unites. Thus, if the frequency of A_1 in males is p and it is also p in females, the probability that an A_1-bearing sperm will fertilize an A_1-bearing egg is equal to the product of the separate probabilities, i.e., **P** will equal p² after one generation of random mating. The frequencies of the various genotypes, as a function of gene frequency, are summarized in Table 4.4.

From Table 4.4 it may be seen that the genotypic frequencies are described by the square of the *gametic array* after one generation of random mating: $(pA_1 + qA_2)^2 = p^2A_1A_1 + 2pqA_1A_2 + q^2A_2A_2$. This is an extremely important result; it indicates that genotypes will be in these proportions after only one generation of random mating, regardless of the genotypic frequencies in the preceding generation. In addition, since $p_1 = \mathbf{P} + \frac{1}{2}\mathbf{H} = p^2 + pq = p$, where p_1 is the frequency of A_1 after one generation of random mating, it is clear that gene frequencies are not changed by random mating. Thus, with continued generations of random mating, these genotypic proportions will remain stable generation after generation. This *equilibrium* law has been referred to as the Hardy-Weinberg law, since it was independently formulated by Hardy, an English mathematician, and Weinberg, a German physician, in 1908. Recently, however, it has been pointed out that W. E. Castle, an early American geneticist, utilized and even extended this relationship in a paper

Table 4.5

Frequencies of types of matings under random mating

		Males		
		P A_1A_1	**H** A_1A_2	**Q** A_2A_2
	P A_1A_1	P^2 $A_1A_1 \times A_1A_1$	PH $A_1A_1 \times A_1A_2$	PQ $A_1A_1 \times A_2A_2$
Females	**H** A_1A_2	PH $A_1A_2 \times A_1A_1$	H^2 $A_1A_2 \times A_1A_2$	HQ $A_1A_2 \times A_2A_2$
	Q A_2A_2	PQ $A_2A_2 \times A_1A_1$	HQ $A_2A_2 \times A_1A_2$	Q^2 $A_2A_2 \times A_2A_2$

published in 1903. For this reason, Keeler (1968) has suggested that we place in our textbooks a belated recognition of "Castle's law."

Several assumptions are implicit in the above calculations. First, it was assumed that the gene frequencies were equal in males and females in the initial generation. If this assumption were not met, it would require two generations of random mating, rather than one, to achieve an *equilibrium distribution*. It was also assumed that the population was large and that the systematic processes that change the frequency of genes in populations (mutation, selection, and migration) were not operating. We shall consider the consequences of relaxing these assumptions in Chapter 9.

Because of the importance of this equilibrium law, let us derive it by alternative means. When individuals mate at random, the genotype of the male is independent of the female with which he mates. The frequency of genotype A_1A_1, **P**, is equivalent to the probability that the mate of some specified female will be an A_1A_1 male. Similarly, the probability that a randomly chosen female is A_1A_1 is also **P**. Since these "events" are independent with random mating, the probability that a given mating is $A_1A_1 \times A_1A_1$ is equivalent to the product of the separate probabilities, i.e., P^2. The frequencies of the various types of matings are summarized in Table 4.5. Although there are nine cells in Table 4.5, only six different types of matings are indicated. For example, the distribution of genotypes among the offspring of an A_1A_1 male and an A_1A_2 female is exactly the same as that of an A_1A_2 male and an A_1A_1 female. The six types of matings, their frequencies, and frequencies of offspring produced by each are summarized in Table 4.6. From the column totals of Table 4.6, it may be seen that the equilibrium distribution will be achieved after only one generation of random mating, regardless of the genotypic distribution in the parental generation. Recall that nowhere have we assumed that $P = p^2$ in the parental generation. Thus, no matter what the previous mating pattern was, one generation of random mating will generate an equilibrium distribution. The facility of considering random mating as the random union of gametes should now be clear. It was much easier merely to square the gametic array as in Table 4.4 and obtain

Table 4.6

Genotype frequencies after one generation of random mating

Mating	Frequency	Offspring		
		A_1A_1	A_1A_2	A_2A_2
$A_1A_1 \times A_1A_1$	P^2	P^2		
$A_1A_1 \times A_1A_2$	2PH	PH	PH	
$A_1A_1 \times A_2A_2$	2PQ		2PQ	
$A_1A_2 \times A_1A_2$	H^2	$\frac{1}{4}H^2$	$\frac{1}{2}H^2$	$\frac{1}{4}H^2$
$A_1A_2 \times A_2A_2$	2HQ		HQ	HQ
$A_2A_2 \times A_2A_2$	Q^2			Q^2
Total	$(P+H+Q)^2$ $= 1.0$	$(P+\frac{1}{2}H)^2$ $= p^2$	$2(P+\frac{1}{2}H)(Q+\frac{1}{2}H)$ $= 2pq$	$(Q+\frac{1}{2}H)^2$ $= q^2$

the desired results than it was to determine all possible types of matings and then calculate the expected genotypic proportions among the resulting offspring.

When a population is in an equilibrium distribution, it is possible to estimate the frequency of a recessive allele even when there is complete dominance. At equilibrium, $Q = q^2$. Thus, we may estimate q by merely obtaining the square root of the observed frequency of the recessive homozygote. However, if the population is not in or near an equilibrium distribution, such estimation is not valid. For example, from our previous hypothetical data set where $P = 0.5$, $H = 0.2$, and $Q = 0.3$, we found $q = 0.4$. If we had assumed that this population was in an equilibrium distribution, and used the square root of 0.3 as our estimate of q, an erroneous value of 0.55 would have been obtained.

Testing the Adequacy of Genetic Hypotheses.   About 70 percent of the Caucasians in the United States experience a very bitter taste when a solution of phenylthiocarbamide (PTC), also referred to as phenylthiourea, is applied to the tongue, whereas about 30 percent find it virtually tasteless at the same concentration. As will be discussed later, almost everyone can taste PTC when the concentration is very high. We shall assume here that we are considering the ability to taste PTC at a concentration which maximally discriminates between tasters and nontasters. As there are two phenotypes, three mating combinations are possible. Actual data for this character are presented for parents and their children in Table 4.7.

Let us first assume the simplest genetic hypothesis, test it, and then discard it only if a significant departure from expectation is obtained. Thus, we shall retain the most parsimonious model until we are compelled by the

Table 4.7

Data on the inheritance of ability to taste phenylthiocarbamide

Mating	No. of families	Offspring		Fraction of nontasters among offspring
		Tasters	Nontasters	
Taster × Taster	425	929	130	0.123
Taster × Nontaster	289	483	278	0.366
Nontaster × Nontaster	86	5	218	0.978

SOURCE: From Stern, *Principles of Human Genetics*, 3rd Ed. W. H. Freeman and Company. Copyright © 1973, p. 242; after Snyder.

data to consider more complex hypotheses. The simplest genetic model is two alleles at one autosomal locus, with one allele (T) being completely dominant to the other (t). If this is the actual situation for taste sensitivity to PTC, it is clear that nontasters must be tt, since almost all offspring of nontaster × nontaster marriages yield nontaster offspring. The 5 taster children out of 223 could be due to variable gene expression (modifier genes or the environment might occasionally cause individuals of tt genotype to taste PTC), misclassification, or illegitimacy.

If taster individuals are either TT or Tt, the mating combinations listed in the first two rows of Table 4.7 must represent some combination of homozygote and heterozygote matings. If mating is at random with regard to taste sensitivity to PTC, which would seem likely to be true, then the genotypic proportions in the population should be $p^2TT + 2pqTt + q^2tt$. If our model is correct, we should be able to utilize this information to predict the numbers of tasters and nontasters among the children of these mating combinations.

In Table 4.6, the frequencies of the six types of matings that may occur with respect to two alleles at one autosomal locus were listed as a function of **P**, **H**, and **Q**. However, if the population has been mating at random for at least one generation with regard to this character, the following relations should exist: $\mathbf{P} = p^2$; $\mathbf{H} = 2pq$; and $\mathbf{Q} = q^2$. Thus, the frequencies of the six types of matings that may occur with respect to the T locus may be expressed in terms of p and q as shown in Table 4.8. Of the six types of matings listed in Table 4.8, two are matings of tasters with nontasters, corresponding to the middle row of Table 4.7. The expected proportions of taster and nontaster offspring resulting from such matings are shown in Table 4.9.

If tasters and nontasters do not reproduce differentially, then the frequency of matings should be equivalent to the frequency of offspring produced. As shown in Table 4.9, of the total $2p^2q^2 + 4pq^3 = 2pq^2(p + 2q) = 2pq^2(1 + q)$ offspring produced by these types of matings, $2pq^3$ should be

Table 4.8

Frequencies of types of matings when a population is in a Hardy-Weinberg-Castle equilibrium distribution

Mating	Frequency
$TT \times TT$	p^4
$TT \times Tt$	$4p^3q$
$TT \times tt$	$2p^2q^2$
$Tt \times Tt$	$4p^2q^2$
$Tt \times tt$	$4pq^3$
$tt \times tt$	q^4

Table 4.9

Frequencies of PTC taster × nontaster matings and of resulting offspring

Mating		Offspring	
Type	Frequency	Tasters	Nontasters
$TT \times tt$	$2p^2q^2$	$2p^2q^2$	—
$Tt \times tt$	$4pq^3$	$2pq^3$	$2pq^3$

SOURCE: From Stern, *Principles of Human Genetics,* 3rd Ed. W. H. Freeman and Company. Copyright © 1973, p. 243.

nontasters. Thus, the expected fraction of nontasters among offspring of taster × nontaster matings should be:

$$\frac{2pq^3}{2pq^2(1+q)} = \frac{q}{1+q}.$$

Among a sample of 3,643 persons, 70.2 percent were tasters and 29.8 percent were nontasters. Thus, if our simple model is correct, $q^2 = 0.298$ and $q = 0.545$. Upon substitution into the above formula, the expected proportion of nontasters among children of taster × nontaster matings should be:

$$\frac{q}{1+q} = \frac{0.545}{1.545} = 0.353.$$

Table 4.10

Frequencies of PTC taster × taster matings and of resulting offspring

Mating		Offspring	
Type	Frequency	Tasters	Nontasters
$TT \times TT$	p^4	p^4	—
$TT \times Tt$	$4p^3q$	$4p^3q$	—
$Tt \times Tt$	$4p^2q^2$	$3p^2q^2$	p^2q^2

SOURCE: From Stern, *Principles of Human Genetics,* 3rd Ed. W. H. Freeman and Company. Copyright © 1973, p. 243.

Thus, of the 761 children of taster × nontaster matings listed in Table 4.7, $(0.353)(761)=268.6$ would be expected to be nontasters and $(1-0.353)(761) = 492.4$ would be expected to be tasters. As indicated in Table 4.7, the observed numbers are 278 and 483, resulting in a chi-square value of 0.51. With one degree of freedom, the corresponding probability is greater than 0.40; thus, these results are in accordance with those expected on the basis of our model.

Of the six types of matings listed in Table 4.8, three are matings of tasters with tasters. The expected proportion of taster and nontaster offspring from taster × taster matings is shown in Table 4.10. Of the total $p^4 + 4p^3q + 4p^2q^2 = p^2(p^2 + 4pq + 4q^2) = p^2(p + 2q)^2 = p^2(1 + q)^2$ offspring produced by these matings, p^2q^2 should be nontasters. Thus, the expected fraction of nontaster offspring resulting from taster × taster marriages is:

$$\frac{p^2q^2}{p^2(1 + q)^2} = \frac{q^2}{(1 + q)^2}.$$

Comparison of this formula with that obtained above indicates that the frequency of nontaster offspring resulting from taster × taster matings is the square of that expected among offspring of taster × nontaster matings. Thus, the expected proportion of nontasters among children of taster × taster matings is $(0.353)^2 = 0.125$. Of the 1,059 children of taster × taster matings tested in this study, $(0.125)(1,059) = 132.4$ are expected to be nontasters and $(1 - 0.125)(1,059) = 926.6$ are expected to be tasters. From Table 4.7 it may be seen that the corresponding observed values were 130 and 929, yielding a chi-square value of 0.05. With one degree of freedom, a probability value greater than 0.80 is obtained. Thus the data presented in Table 4.7 closely conform to those expected on the basis of an autosomal-locus two-allele model, where T is dominant to t.

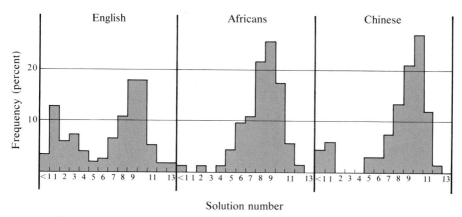

Figure 4.10

Distributions of taste thresholds for phenylthiourea in 155 English males, 74 Africans, and 66 Chinese. The strongest solution (1) had a concentration of 0.13 percent in water, the next (2) half this strength, and so on through 13. (From Barnicot, "Taste deficiency for phenylthiourea in African Negroes and Chinese," Ann. Eugen., 15, 1950. Cambridge University Press.)

Taste sensitivity to PTC and related compounds is now routinely assessed by placing a small drop of a test solution on the tongue of a subject and asking if he can taste it. A variety of concentrations is usually administered so that a taste threshold may be established for each subject. Some people can taste highly dilute solutions of PTC, but others can distinguish only very strong solutions from plain water. Distributions of taste thresholds for 155 English males, 74 Africans, and 66 Chinese are presented in Figure 4.10. Individuals who could taste solution number 13, the most dilute solution, and all stronger solutions were most sensitive and had the lowest taste threshold. Individuals who could not even taste solution number one, the strongest solution, were least sensitive. The resulting threshold distributions tend to be bimodal. Although most individuals may be reliably classified as tasters or nontasters by this procedure, some overlap occurs. Variation within the taster and nontaster categories could be due to segregation at other loci that have relatively small effects on taste sensitivity to PTC, to environmental variation, or both. A higher frequency of nontasters in the English population may also be noted in Figure 4.10.

Scope of the Problem

Only a relatively few of the many known single-gene behavioral effects in man have been discussed in this chapter. As an indication of the number of single-gene conditions that have accompanying behavioral effects, Lindzey,

Loehlin, Manosevitz, and Thiessen (1971) report that among the 1,545 syndromes described in *Mendelian Inheritance in Man* (McKusick, 1968), 112 syndromes due to autosomal recessive genes, 7 due to autosomal dominant genes, and 16 due to sex-linked genes have mental retardation as one of their clinical symptoms. There has even been a recent report (Eldridge, Harlan, Cooper, and Riklan, 1970) which suggests that a recessively inherited condition, torsion dystonia, is associated with superior performance on tests of mental ability. Mental ability, of course, is only one of a very large number of human behavioral characters that may be subjected to single-gene analysis. Thus, although a large number of human behavioral effects due to single genes are known, it is likely that the search for such effects has only just begun.

The mode of action of some single-gene conditions will be discussed in Chapter 6. Before this somewhat different type of genetic analysis is undertaken, however, a discussion of single-gene analysis and behavior in species other than man (with the mouse serving as chief example) will be presented.

Chapter 5

Single-gene analysis and animal behavior

Pedigree analysis and gene-frequency analysis, both of which were applied to human behavioral characters in Chapter 4, may also be used with animal subjects. With animal material, however, additional sorts of analyses are possible. For example, when controlled breeding experiments are feasible, the adequacy of genetic hypotheses may be tested by classical Mendelian crosses. In addition, mutations that have been isolated because of their morphological or physiological effects may be screened for behavioral effects. When employing this latter approach, the analysis of *behavioral pleiotropism*, it is important to measure effects at one locus independently of other gene effects. This may be accomplished by randomizing the genetic background or by employing strains in which only the locus under study is segregating.

Behavioral Pleiotropism

The first study of behavioral pleiotropism was provided by the early geneticist, A. H. Sturtevant, inventor of the chromosome map. As part of a study of mating behavior in *Drosophila*, Sturtevant (1915) put one male fly (either normal or mutant) into a bottle with a normal female and a mutant female

Table 5.1

Matings in male and female "choice" experiments

| Male | Male choice | | Female | Female choice | |
	Female	No. cases		Male	No. cases
Red	Red	54	Red	Red	53
	White	82		White	14
White	Red	40	White	Red	62
	White	93		White	19
Gray	Gray	25	Gray	Gray	60
	Yellow	31		Yellow	12
Yellow	Gray	12	Yellow	Gray	25
	Yellow	30		Yellow	8

SOURCE: After Sturtevant, 1915, pp. 363–364.

and recorded the number of times the male mated with each female; he also recorded observations of the matings of a female (normal or mutant) placed in a bottle with a normal male and a mutant male. A sample of Sturtevant's data on white-eyed mutants with red-eyed (normal) flies and yellow-bodied mutants with gray-bodied (normal) flies is presented in Table 5.1. In general, the data from the experiments in which a male had the choice of two females suggest that both red- and white-eyed males prefer to mate with white-eyed females and that both gray- and yellow-bodied males prefer to mate with yellow females. Sturtevant, however, argued that this higher frequency of mating with mutant females did not really indicate preference by the males; mutant females were observed to be less active and thus were less likely to attempt to escape the courting male.

The results were even more striking in the experiments in which a female had the choice of two males. Although both normal and mutant females appeared to prefer normal males, it was argued that this also did not really indicate preference. Receptive females apparently accepted the first male that courted them; thus, the more sexually active, normal males were more likely to approach and mate receptive females. Mating behavior in *Drosophila* will be discussed in more detail in Chapter 10.

Randomized Genetic Background. A number of investigators have assessed behavioral effects of genes at the albino locus of the house mouse, the most extensive study of which utilized the open-field test. The open-field test was first employed by Hall (1934) to provide an objective index of emotionality in rodents. The test consists of placing a subject in a brightly lit enclosure and observing its behavior. When so placed, the subject may "freeze," defecate and urinate, or it may explore the enclosure. Animals that have relatively low activity and high defecation scores are referred to as "emotional" or "reactive," and those with relatively high activity and low defecation scores are considered "nonemotional" or "nonreactive." The evidence for the validity of this measure has been discussed in some detail (Broadhurst, 1960; Eysenck and Broadhurst, 1964).

It has been known for some time that several albino strains of mice obtain relatively low activity and high defecation scores during open-field tests. However, because inbred strains differ from one another at many loci and because some of these albino strains are closely related, it was generally assumed that the differences observed in open-field tests were due to gene differences other than those determining coat color, i.e., it was assumed that the observed correlation between albinism and open-field behavior was spurious.

An extensive genetic analysis of open-field behavior was initiated by DeFries during the latter part of 1964, one of the objectives of which was to assess the effect of gene substitutions at the c locus. Two highly inbred strains of mice were chosen as the parental strains. One strain (BALB/cJ)* has low open-field activity and high defecation scores and is albino (cc), whereas the other strain (C57BL/6J) has high open-field activity and low defecation scores and is pigmented (CC). These two strains were crossed to produce an F_1 generation and an F_2 generation was subsequently obtained. Members of the F_2 generation were then mated at random to produce an F_3 generation, which was the base population for a selection experiment. In order to assess the effects of albinism on open-field behavior, the means of albino and pigmented subjects may be compared directly in the F_2 and F_3 generations. All genes except those linked to the c locus should be segregating independently of albinism in these generations. In the selected generations, albino and pigmented subjects must be compared within each line. However, means of albino and pigmented mice for each line may be averaged across lines to yield values for each generation. In order to equalize the variances in the pigmented and albino classes (an assumption underlying some tests of statistical significance), square-root transformations (see Table 5.2) were applied to the raw data of this experiment. The resulting mean transformed scores for activity and defecation of albino and pigmented subjects during the F_2 and F_3 generations and eight generations of selection are presented in Table 5.2.

Although there is considerable variation from generation to generation, albinos are consistently less active and have consistently higher defecation scores than pigmented animals within each of the ten generations. This consistency across generations is of considerable significance. If the effects were due to genes linked to the c locus, rather than to genes at the c locus itself, the difference observed in the F_2 and F_3 generations should gradually diminish in subsequent generations as a function of recombination through

---

*Terminal symbols in strain designations refer to particular substrains (see Staats, J., 1964).

Table 5.2

Mean transformed scores for open-field activity and defecation of albino and pigmented mice

Genera-tion	Activity[a]			Defecation[b]			N	
	Albino	Pig-mented	Differ-ence[c]	Albino	Pig-mented	Differ-ence[c]	Albino	Pig-mented
F_2	10.7	12.7	2.0 ± 0.35	1.97	1.76	0.21 ± 0.069	152	485
F_3	10.0	12.3	2.3 ± 0.37	2.24	1.81	0.43 ± 0.062	180	661
S_1	9.9	12.3	2.4 ± 0.50	2.21	1.86	0.35 ± 0.076	113	315
S_2	11.5	13.4	1.9 ± 0.49	1.99	1.71	0.28 ± 0.078	112	268
S_3	12.0	13.5	1.5 ± 0.40	2.03	1.70	0.33 ± 0.070	156	334
S_4	12.6	14.3	1.7 ± 0.40	1.90	1.63	0.27 ± 0.067	172	284
S_5	9.8	12.2	2.4 ± 0.37	2.12	1.80	0.32 ± 0.067	169	256
S_6	12.8	14.9	2.1 ± 0.34	2.37	1.94	0.43 ± 0.075	181	262
S_7	10.3	14.1	3.8 ± 0.34	2.54	1.96	0.58 ± 0.082	186	219
S_8	12.7	15.2	2.5 ± 0.32	2.41	1.97	0.44 ± 0.081	174	284

SOURCE: DeFries, 1969, p. 66.

[a] Mean activity scores were obtained from transformed data, where each subject's score is the square root of the total activity over the two day test.

[b] Mean defecation scores were obtained from transformed data, where $x = (\text{total boluses} + 1/2)^{1/2}$.

[c] Absolute difference ± standard error of the difference between the means. All differences were highly significant ($p < 0.001$), with the exception of that for defecation in the F_2 generation, which was significant at 0.005.

Table 5.3

Open-field behavioral scores of mice tested under different illumination

Illumination	Activity		Defecation		Number tested	
	Albino	Pigmented	Albino	Pigmented	Albino	Pigmented
White	8.8	12.9	2.10	1.95	39	37
Red	13.3	14.1	1.73	1.76	38	38

SOURCE: After DeFries, Hegmann, and Weir, "Open-field behavior in mice: evidence for a major gene effect mediated by the visual system," *Science,* 154, 1577–1579, 1966. Copyright © 1966 by the American Association for the Advancement of Science.

crossing over. However, the difference seems to be as large in the later generations as in F_2 and F_3. Thus, these data clearly indicate pleiotropic behavioral effects at the c locus.

Merrell (1965) has suggested that the study of gene substitutions, one or a few at a time, may provide information not only about the genetics of behavioral characters, but also about the characters themselves. The validity of Merrell's suggestion is evident from subsequent analyses of this character. It had previously been observed (McClearn, 1960) that the difference in open-field activity between a low-activity albino strain and a high-activity pigmented strain was somewhat less for tests run under red light. Under such illumination, visual stimulation should be low or absent for mice. Thus, if the difference in open-field behavior observed under white light were due to the albinos' having a greater photophobic reaction, the observed difference should be decreased or eliminated when subjects are tested under red illumination.

In order to test this hypothesis, the open-field behavior of albino and pigmented littermates was observed under either red or white illumination. The resulting data are summarized in Table 5.3. Differences attributable to illumination and to pigmentation were both significant ($p < 0.001$) with regard to activity. In addition, the interaction between illumination and pigmentation was significant ($p < 0.02$), indicating that albino and pigmented mice respond differently to changes in illumination. *Post hoc* comparisons among the four means indicated that albino mice tested under white light were significantly ($p < 0.001$) less active than each of the other three classes, which did not differ among themselves. Thus, the differences attributable to illumination and pigmentation, and their interaction were primarily due to the effect of white light upon albino subjects.

The pattern of defecation means corresponds closely to that expected of a character that is negatively correlated with open-field activity. However, only the difference attributable to illumination was significant with these data (p < 0.02).

As will be discussed in Chapter 9, it is clear that open-field behavior is influenced by genes at many loci. Nevertheless, the results of this study demonstrate that albino mice have lower activity and higher defecation scores than pigmented mice when tested under white light and that this difference largely disappears when these subjects are tested under red illumination. It may thus be concluded that this single-gene effect is mediated through the visual system and, if we accept the interpretation that a pattern of low activity and high defecation indicates heightened emotionality, albinos may be regarded as being more photophobic than pigmented animals.

In the study just described, effects of the c locus were assessed by crossing two highly inbred strains and obtaining their derived generations. This approach is limited to the extent that genes closely linked to the c locus may also influence the character under study. A more satisfactory approach is to utilize subjects from a *heterogeneous* population in which mating has been at random for many generations. With such a population, even genes on the same chromosome should be assorting independently. Such a population has recently been utilized to study the genetics of performance by mice in a learning situation offering positive rewards (Tyler, 1970).

The heterogeneous population (symbolized HS) that Tyler utilized was derived from an initial cross of eight inbred strains and was subsequently randomly mated. The formation of this HS population is discussed in some detail by McClearn, Wilson, and Meredith (1970). Subjects were deprived of food until their body weight reached about 85 percent of initial weight and then subsequently maintained at this weight during testing. At about 60 days of age, each mouse was given 5 trials per day for 8 consecutive days in a straight runway. A peanut pellet was present at the end of the runway on the first 5 days (acquisition period), but none was there during the last 3 days (extinction period). The raw data (time taken to run down the runway) were transformed (logarithmic transformation) prior to statistical analysis. The mean transformed running times of 56 albino and 212 pigmented subjects on each of the five days of acquisition and three days of extinction are plotted in Figure 5.1.

It may be seen in Figure 5.1 that initially albino and pigmented mice had similar running times. However, the mean running speed of albinos decreases more slowly than pigmented animals during acquisition and increases more slowly during extinction. It might be argued that this difference in running speed is due to the influence of albinism on activity, rather than on learning. However, in this experiment the difference was in the slope of the observed points, rather than overall mean running speed. In addition, although the runway was illuminated, it was constructed out of colored plastic that transmitted only red light, and it will be recalled that albino and pigmented sub-

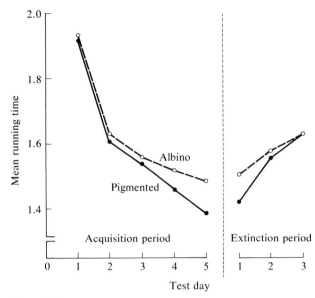

Figure 5.1
Mean running times of albino and pigmented mice. (From Tyler, 1970, p. 152.)

jects had similar activity scores when tested under red illumination. Thus, it seems likely that the single-locus effect observed in Tyler's experiment was not mediated by the visual system.

Coisogenic Strains. It might be argued that even after many generations of random mating, genes linked very closely to the locus under study may still not be segregating independently. Thus, the strongest evidence for behavioral pleiotropism is provided by comparing individuals with exactly the same genotype except for a newly arisen mutation. When a new mutation arises and is maintained within an inbred strain, mutant and nonmutant subjects within the strain are called *coisogenic* (Green, 1966).

Henry and Schlesinger (1967) compared coisogenic albino (mutant) and pigmented C57BL/6J mice for several behaviors: open-field behavior, avoidance conditioning, alcohol preference, and susceptibility to audiogenic seizures. The avoidance-conditioning apparatus consisted of a wooden box that contained a grid floor and a wooden escape shelf that went all the way around its inside walls. The conditioned stimulus (light and buzzer) was presented for three seconds and was then followed by the unconditioned stimulus (shock through grid floor), which was maintained until the subject jumped to the escape shelf. The subject was then placed back on the grid

Table 5.4

Behavior of albino and pigmented coisogenic C57BL/6J mice

Behavior	Albino Mean	N	Pigmented Mean	N	t	p <
Trials to acquisition criterion	22.8	20	12.3	23	3.3	0.01
Trials to extinction criterion	25.9	17	30.3	23	0.9	—
Alcohol preference ratio	0.46	19	0.71	19	10.3	0.001
Activity 1st 5 min.	196	20	228	23	2.2	0.05
2nd 5 min.	171	20	157	23	0.8	—

SOURCE: After Henry and Schlesinger, "Effects of the albino and dilute loci on mouse behavior. Journal of Comparative and Physiological Psychology, 63, 320–323, 1967. Copyright 1967 by the American Psychological Association and reproduced by permission.

floor for the next trial. The training schedule was continued until the subject avoided the shock in eight out of ten consecutive trials (acquisition criterion) or, if the subject failed to attain criterion, after 50 trials. An identical procedure was used for extinction, except that the unconditioned stimulus was never presented. The extinction criterion was *not* jumping on the shelf in eight out of ten consecutive trials.

The alcohol-preference test consisted of providing subjects with two cylinders from which the amount consumed could be measured. One cylinder contained a 10 percent solution of ethyl alcohol, and the other contained plain tap water. The amount of liquid consumed from each cylinder was recorded daily for a 14-day period. In order to avoid the bias of subjects' preferring to drink from a cylinder merely because of its location in the cage, the positions of the cylinders were changed every three days. Resulting data were expressed as preference ratios, i.e., the amount of fluid consumed from the alcohol cylinder divided by the total fluid consumed from both bottles.

No difference in incidence of audiogenic seizures (a character discussed in more detail in the next section) was observed between albino and pigmented mice. However, differences in the other behavioral tasks were obtained. The mean scores of albino and pigmented coisogenic mice for learning, alcohol preference, and open-field activity are presented in Table 5.4. As may be seen from this table, pigmented mice require fewer trials to achieve acquisition criterion, consume more alcohol, and have higher

Table 5.5

*Data observed and expected on the basis of
a single-locus autosomal-recessive model
for F_2 Nijmegen waltzer mice*

	Normal	*Mutant*
Observed	124	47
Expected	128.25	42.75

activity scores during the first five minutes in the open field than albinos. It should be noted that this confirms the evidence for a major-gene effect at the *c* locus that was obtained by using subjects in which the genetic background was randomized.

Mendelian Crosses

Discontinuous Characters. One of the earliest mutant behavioral conditions studied in mammals was that of "waltzing" in mice. In spite of the name, motor behavior of these animals is quite ungraceful and consists of head shaking, rapid circling, and hyperirritability. Waltzer mice were prized by mouse fanciers and imported to Europe and North America from Asia around 1890 (Gruneberg, 1952). The first scientific observations were by von Guaita (1898) prior to the rediscovery of Mendel's work.

Several different waltzer conditions are known, some of which have only recently been described. For example, van Abeelen and van der Kroon (1967) recently discovered the *Nijmegen waltzer,* a mutant condition characterized by running in tight circles in both directions with both horizontal and vertical head shaking. When mutant males were crossed with mutant females, 234 mutant and no normal offspring were obtained. However, when mutant mice were crossed with normal mice from other stocks, 254 normal F_1 offspring resulted. These results suggest that Nijmegen waltzing may be determined at a single locus by a recessive autosomal gene. In order to test this hypothesis, F_1 males were mated to F_1 females in order to produce a segregating F_2 generation. The resulting F_2 data are summarized in Table 5.5.

As shown in Chapter 3, when a chi-square test (one degree of freedom) is applied to these data, a chi-square value of 0.56 with an accompanying probability value greater than 0.40 is obtained. Thus, these results are compatible with a single-locus autosomal-recessive hypothesis.

In Chapter 4, the problem of ascertainment was illustrated with human family data. Another neurological mutant, "twirler," first described by Mary F. Lyon (1958), illustrates that this problem may also occur with animal material. Twirler mice are characterized by head shaking (more often in a horizontal than in a vertical plane) and circling. Although these mice are not deaf, otoliths are absent and canals of the inner ear are reduced in size and malformed.

Preliminary evidence indicated that this condition was probably due to a dominant gene. However, when twirler males were crossed with twirler females (both presumed to be heterozygotes), the following results were obtained:

$Tw+ \times Tw+ \longrightarrow$ 84 twirler and 58 normal offspring, where Tw symbolizes the dominant mutant gene and $+$ symbolizes the recessive normal allele. If twirler were due to a simple dominant gene, the expected numbers among the offspring would be 106.5 twirler and 35.5 normal. The departure from expectation is highly significant ($\chi^2 = 19.01$, df $= 1$, p < 0.0001).

Many dominant mutant genes are lethal when homozygous. If individuals having the genotype $TwTw$ die before investigators observe their behavior, the expected ratio among offspring of a $Tw+ \times Tw+$ cross would be 2 twirler:1 normal instead of 3:1. When this hypothesis is tested, a slight deficiency of twirler offspring is still noted, but the departure from expectation is not sufficiently great to be statistically significant ($\chi^2 = 3.6$, df $= 1$, p < 0.1). The slight deficiency of twirler mice could relate to the fact that some twirlers are smaller and thinner than normals: more twirlers than normals may die before investigators observe them. In addition the twirler phenotype is variable in expression; thus, some mice having the heterozygous genotype but only a low expression of the twirler condition may be misclassified as normals.

Subsequent work with twirler mice has indicated that about 25 percent of newborn offspring resulting from $Tw+ \times Tw+$ matings have a cleft palate or cleft lip and palate. These pups die within 24 hours of birth and consequently would not be observed for behavioral abnormalities. These mice are believed to be the missing $TwTw$ homozygotes, accounting for the 2:1, rather than the 3:1, ratio.

Many inherited neurological defects in the mouse are now known. Most appear to be due to single-locus autosomal-recessive genes, although some are caused by dominant genes and a few appear to be due to the combined effects of genes at several loci. A concise review has been provided by Fuller and Wimer (1966). Their classification system and a partial list of the known conditions are presented in Table 5.6. The highly descriptive names convey some of the diversity of behavioral anomalies that have been described. It is interesting to note that neurological mutations have also been described in other organisms, e.g., pigeons (Entrikin and Erway, 1972) and *Drosophila* (Kaplan and Trout, 1969).

Considerable effort has also been devoted to the study of sound-induced seizures in laboratory rodents. Some subjects respond to intense high-frequency sound with wild running, convulsions, and even death; other sub-

Table 5.6

Partial list of inherited neurological defects in mice

Class of syndrome	Name
Waltzer-shaker	shaker (1 and 2), pirouette, jerker, waltzer, varitint-waddler, fidget, twirler, zig-zag
Convulsive	trembler, spastic, tottering
Incoordination	quaking, jimpy, reeler, agitans, staggerer

jects are apparently unaffected by it. Interest in this phenotype has had a variety of sources. It is currently being used as a model system to study epilepsy in man, as well as an index of central-nervous-system excitability.

Witt and Hall (1949) first conducted a genetic analysis of the observed difference in seizure susceptibility between two inbred strains of mice and concluded that high susceptibility was determined by a single-locus auto-somal-dominant gene. Other investigations have subsequently been performed and more complex genetic models have been postulated. For example, Ginsburg and Miller (1963) proposed a two-locus model and Fuller, Easler, and Smith (1950) have hypothesized multiple-factor inheritance.

Much of the early work with sound-induced seizures utilized repeated testing of the same subjects. This method of assessing seizure risk would be appropriate if the probability of a seizure is independent of previous test experience. However, Henry (1967) has demonstrated that animals from a normally seizure-resistant strain become susceptible if exposed to a loud sound at an early age. This "acoustic priming" has since been found to be quite general in both inbred and outbred strains. Thus, repeated testing of subjects may have led to the confounding of two or more phenomena: response to initial presentation and response to later presentations.

When these responses are considered separately, the situation is greatly clarified (Collins and Fuller, 1968; Collins, 1970). Mice of the C57BL/6J strain, a strain whose members only rarely convulse upon initial exposure to a loud noise, were crossed to mice of the DBA/2J strain, whose members almost always convulse upon initial exposure. The resulting F_1 animals were backcrossed to the two parental lines (P_1 and P_2) and also mated among themselves, resulting in the production of B_1, B_2, and F_2 subjects. The backcross subjects were subsequently crossed to F_1 mice and the two backcross lines were mated, yielding generations symbolized B_1F_1, B_2F_1, and B_1B_2. Mice from these nine generations were individually tested for initial seizure susceptibility at about 21 days of age. Each subject was placed in a box and then exposed to an electric bell that was rung until the onset of a clonic convulsion or for a maximum of one minute. The number

Table 5.7

Summary of data and genetic analysis for the incidence of initial audiogenic seizures in mice

Generation	Number of subjects	Proportion observed to have seizures	Proportion expected to have seizures (single-locus autosomal model)	χ^2
P_1 (C57BL/6J)	45	0.000		
P_2 (DBA/2J)	58	.983		
F_1	89	.011		
B_1	115	.035	0.006	
B_2	119	.513	.497	0.12
F_2	105	.247	.251	0.01
B_1F_1	128	.070	.128	3.81
B_2F_1	96	.344	.374	0.37
B_1B_2	185	0.168	0.191	0.66

SOURCE: From Collins and Fuller, "Audiogenic seizure prone (asp): a gene affecting behavior in linkage group VIII of the mouse," *Science,* 162, 1137–1139, 1968. Copyright © 1968 by the American Association for the Advancement of Science.

of mice tested from each generation and the proportions of seizures observed and expected, assuming a single-locus autosomal model are presented in Table 5.7. As indicated in this table, C57BL mice had no seizures, almost all DBA mice had seizures, and almost no F_1 mice had seizures. These results suggest that susceptibility to audiogenic seizure upon initial exposure to the loud noise may be determined at a single locus by a recessive autosomal gene.

In order to test the adequacy of this single-locus autosomal-recessive model, it may be hypothesized that DBA mice are *asp asp,* where *asp* symbolizes a recessive gene, *audiogenic seizure prone.* C57BL mice would thus be ++, where + symbolizes the normal dominant allele, and F_1 mice would be *asp* +. The observed proportion of seizures by these genotypes may be utilized to predict the proportion having seizures in each of the six segregating generations. For example, in the F_2 generation, the genotypic proportions would be $\frac{1}{4}$ *asp asp,* $\frac{1}{2}$ *asp* +, and $\frac{1}{4}$ ++. Thus, the expected proportion having seizures in the F_2 generation may be calculated as follows: $\frac{1}{4}(0.000) + \frac{1}{2}(0.011) + \frac{1}{4}(0.983) = 0.251$. It may be seen from Table 5.7 that the results closely conform to those expected on the basis of a single-locus autosomal-recessive model, except in generation B_1F_1 where, however, the departure from expectation is not statistically significant (p > 0.05). In

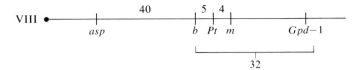

Figure 5.2

A portion of linkage group VIII of the mouse illustrating the position of the asp locus in relation to the four other genetic loci studied in these experiments. The centromeric end of the chromosome is located to the left. The numbers represent the percentage recombination values between pairs of loci. (From Collins, 1970, p. 106.)

generation B_1, the expected number was too small for application of the chi-square test to be appropriate.

Collins (1970) has subsequently conducted an extensive linkage analysis and has presented evidence which indicates that the *asp* locus is loosely linked to the *b* locus, which is part of a complex of linked loci referred to as linkage group VIII. This linkage relationship is illustrated in Figure 5.2.

If the investigator is sufficiently ingenious, a behavioral response may be assessed with a minimum of time and effort. For example, while a graduate student at the University of Minnesota, Glayde Whitney observed that mice from some strains almost never vocalize when lifted by their tails during cage changing, but those from another strain usually squeak when so lifted. This observation prompted Whitney (1969) to perform a systematic study of handler-induced vocalization in mice. Because of the ease of measurement, a very large number of subjects could be tested for this behavioral response. In his first experiment, subjects were tested by merely lifting them by the tail from a hardware cloth platform and placing them on an adjacent surface. If a subject vocalized in any manner audible to the experimenter during this procedure, it was assigned to the vocalizer class. Data from this experiment are summarized in Table 5.8.

C57 mice rarely vocalize when handled in this way, but about two-thirds of the JK mice vocalize. F_1 mice from crosses of the two strains vocalize somewhat less than JK mice, but considerably more than C57 mice; thus, the character manifests partial dominance. The expected proportions in Table 5.8 are calculated according to the method of Collins and Fuller (1968) and agree quite closely with observed results.

Whitney (1969) has pointed out, however, that although these results are consistent with a single-locus model, alternative explanations are possible. For example, it might be hypothesized that C57 and JK mice differ at two autosomal loci that influence vocalization; they are, say, of genotypes $A_1A_1B_1B_1$ and $A_2A_2B_2B_2$. If effects at one locus combine in a simple additive fashion with those at the other locus, i.e., there is no epistasis, the expected results would be exactly equal to those based upon a single-locus model.

Table 5.8

Genetic analysis of handler induced vocalization in mice

Generation	Number of subjects	Proportion observed to vocalize	Proportion expected to vocalize (single-locus model)
P$_1$ (C57)	70	0.03	
P$_2$ (JK)	71	.68	
F$_1$	99	.56	
B$_1$	47	.26	0.29
B$_2$	45	.62	.62
F$_2$	80	0.46	0.46

SOURCE: After Whitney, 1969, p. 338.

Thus, when effects are additive across loci, it is not possible, using this approach, to discriminate between a single-locus and an n-locus model, where n is two or more. On the other hand, it may be argued that an n-locus situation with no epistasis is unlikely to occur in nature.

Subsequent research by Whitney lends further support to the single-locus model. Using a similar handling procedure in a different laboratory, Whitney tested mice from seven different inbred strains. These data are presented in Table 5.9. If two or more loci influenced this character, it would seem likely that strains with intermediate levels of vocalization would be found. However, six of the seven strains vocalized to about the same extent as the C57 mice in the earlier study. I^s/Bi mice, on the other hand, vocalize almost as frequently as JK mice in the previous experiment. Thus, there appears to be a qualitative, rather than a quantitative, difference in incidence of vocalization upon being handled. When the origins of these strains were considered, it was found that several of the low-vocalizing strains are related. In addition, it was found that the two high-vocalizing strains are related to each other, but not to the low-vocalizing strains. Three inbred strains, I, J, and K, were produced many years ago in Strong's laboratory at Minnesota, the latter two of which were subsequently crossed to produce the JK strain. Thus, it is possible that the I^s/Bi strain, derived from the I strain, and the JK strain share by common descent a dominant gene that results in a high incidence of vocalization.

Whitney (1973) has subsequently utilized a gene-frequency-analysis approach similar to that discussed in Chapter 4 with human data. When 1,338 mice from a heterogeneous population were tested, the observed incidence of vocalization closely conformed to that based upon a single-locus model, but deviated significantly from expectation based upon various two-locus models (with no epistasis and with epistasis).

Table 5.9

Observed vocalization upon being handled in seven inbred mouse strains

Strain	Number of subjects	Proportion vocalizing when handled
A	134	0.00
BALB	179	.02
DBA	109	.04
C57BL	126	.02
C3H	32	.00
CBA	10	.00
I^s/Bi	27	0.48

SOURCE: After Whitney, 1969, p. 339.

Single-gene analysis has also been applied to nest-cleaning behavior in honeybees. Bee strains differ in their resistance to diseases that may strike the colony and kill larvae during development. Investigations of resistance to American foulbrood by Rothenbuhler (1967) indicated that bees differ in their response to the presence of dead larvae. Resistant strains were found to remove the foulbrood-killed larvae quickly from cells in the comb, i.e., displayed "hygienic" behavior, whereas susceptible strains did not.

As part of a genetic analysis of this behavior, Rothenbuhler crossed a resistant and a susceptible strain to produce an F_1 generation of worker and queen bees. Performance of F_1 workers indicated dominance of the allele(s) for susceptibility. Male bees (drones) are haploid, i.e., have only half the diploid number of chromosomes, because they develop from unfertilized eggs. Therefore, sons of F_1 queens carry only the genes that were present in the gamete produced by their mother. When such males are backcrossed to an inbred queen, a whole colony of like genotypes will result. The phenotype under study is thus the social behavior of an entire colony, rather than that of an individual.

Rothenbuhler obtained 29 such backcross colonies, of which 6 displayed hygienic behavior, 9 uncapped cells in which dead larvae were found but did not remove them, and 14 were nonhygienic. Since 6:23 is not close to a 1:1 ratio, it was hypothesized that two loci may be involved. It was further hypothesized that one locus determined uncapping of the cells and the other removal of dead larvae. This was tested experimentally by uncapping cells that contained dead larvae in colonies of the 14 nonhygienic strains. If the two-locus hypothesis were correct, half of these 14 colonies would be expected to be $u+rr$ and half should be $u+r+$, where u symbolizes a hypothesized recessive gene for uncapping and r a recessive gene for removing.

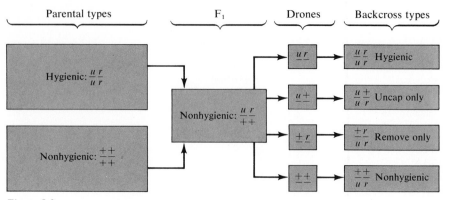

Figure 5.3

Two-locus model hypothesized to account for resistance to American foulbrood in honeybees. (After Rothenbuhler, "Genetic and evolutionary considerations of social behavior of honeybees and some related insects," In Hirsch (Ed.) Behavior-genetic Analysis, 1967, McGraw-Hill.)

When the uncapped cells were returned to the colonies, dead larvae were removed in 6 and not removed in 8, in close agreement with the expected 1:1 ratio. The hypothesized two-locus model is illustrated in Figure 5.3.

Continuous Characters. Each of the characters discussed in the previous section has an "either-or" sort of expression: waltzer or normal; twirler or normal; having audiogenic seizures or not having them; squeaking or not squeaking; and hygienic or nonhygienic. Such qualitative, or discontinuous, characters are especially amenable to Mendelian analysis. Although the approach is somewhat less direct, single-gene analysis may also be employed for continuously variable characters as will be illustrated using data pertaining to taste sensitivity to PTC in mice.

As discussed in Chapter 4, individual differences in ability to taste PTC and related compounds in human populations are largely due to genes at a single autosomal locus. Individual differences in taste sensitivity to PTC in subhuman primates and rodents have also been reported, but the mode of inheritance in infrahuman species has only recently been studied. As part of a genetic analysis of taste sensitivity to PTC in mice, Klein and DeFries (1970) assessed strain differences. Six males and six females from each of five inbred strains were offered tap water and PTC solutions in a two-choice situation similar to that used in alcohol-preference studies. The concentration of PTC was doubled every second day over a 26-day period, resulting in 13 concentrations ranging from 0.1 to 409.6 gm PTC per liter of tap water. A preference ratio was calculated for each subject at each concentration by dividing the amount of PTC solution consumed by the total liquid (PTC solution + tap water) ingested. A ratio of 0.5 would indicate no preference, and one lower than 0.5 would indicate avoidance of (and thus ability to taste) the PTC solution.

A threshold concentration for each subject was thus determined in a manner somewhat analogous to that used in human research (see Chapter 4). The threshold concentration was arbitrarily defined as the concentration at which a subject obtained a preference ratio of less than 0.2 together with its obtaining even lower ratios for all higher concentrations. A plot of the resulting threshold concentration distributions for the five strains is presented in Figure 5.4. It may be seen that the modal threshold concentration for BALB mice is lower than those of the other four strains, indicating that BALB mice taste PTC (or at least avoid PTC solutions) at lower concentrations. It should be noted that solution concentration increases with increasing concentration number in Figure 5.4, but decreases with increasing solution number in Figure 4.11.

Several experiments were then performed to identify a single concentration that maximally discriminates between "taster" and "nontaster" strains. The largest and most consistent strain difference was found for concentration 6.5 (4.5 mg PTC per liter of tap water) when subjects were maintained on a single concentration for 10 days. Mean preference on days 9 and 10 was used as the preference score, since it appeared that subjects required several days to learn to discriminate the solutions reliably. Such data were obtained on taster and nontaster parental strains, BALB/c and C57BL, and their derived F_1, plus backcross, F_2, and F_3 generations. A few subjects were found who apparently liked the PTC solution, i.e., more than 90 percent of the liquid they consumed was PTC solution. Although this behavior is of interest in its own right, the experiment was concerned only with avoidance versus having no preference. Therefore, data on the animals preferring PTC solution to tap water were excluded from subsequent analyses. The distribution of preference-ratio scores of the remaining subjects is presented in Figure 5.5.

The bimodality of the distributions in the B_2, F_2, and F_3 generations suggests single-locus determination with dominance for avoidance (ability to taste). The method of Collins and Fuller (1968) was applied to these data to test the adequacy of a single-locus hypothesis. The expected frequency within each preference-ratio-class interval for the segregating (B_1, B_2, F_2, and F_3) generations was calculated from the corresponding frequencies in the isogenic generations. None of the resulting chi-square values was significant, indicating no evidence for rejecting the single-locus hypothesis.

With continuously varying characters, other approaches may be used to obtain an estimate of the number of segregating gene pairs that influence a character. Such estimates are only very rough approximations (usually underestimates) and thus should be interpreted with caution. An estimate of about one is consistent with a single-locus model, but by no means proves that the hypothesis is correct.

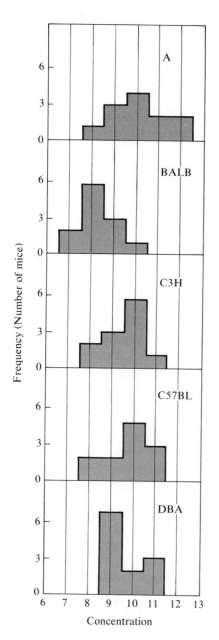

Figure 5.4

Frequency histograms of threshold concentrations for PTC tasting in each of five inbred strains of mice. (From Klein and DeFries, 1970, p. 556.)

Figure 5.5 (facing page)

Frequency histograms of mean PTC-preference ratios on days 9 and 10 at concentration 6.5 for each of two inbred strains and their derived generations. For each segregating generation the expected distribution based on a single-locus model is indicated by vertical lines. N = 39 BALB, 40 C57BL, 60 F_1, 62 B_1, 63 B_2, 62 F_2, and 97 F_3 subjects. (From Klein and DeFries, 1970, p. 557.)

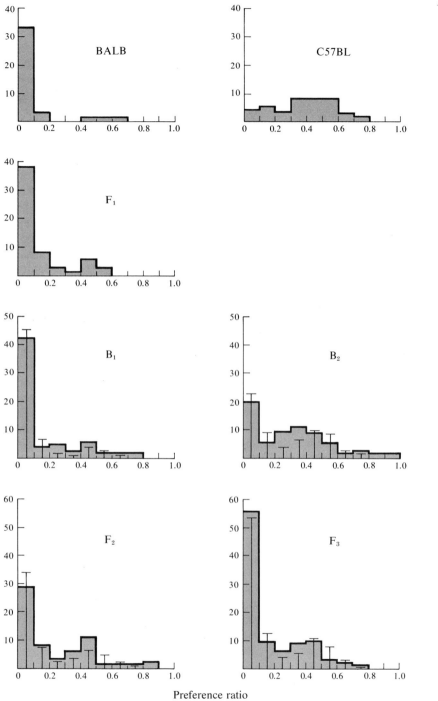

Preference ratio

The following is one of several expressions that have been derived for estimating the number of segregating gene pairs (cf. Wright, 1968):

$$n = \frac{3(\bar{P}_1 - \bar{P}_2)^2}{16(V_{F2} - V_E)},$$

where n is the number of independently segregating Mendelian units that influence the character under study, $\bar{P}$ is the mean value of the character in the indicated parental strain, V_{F2} is the observed variance in the F_2 generation, and V_E is the environmental variance. With highly inbred strains, the variation within both parental strains and the F_1 must be due to environmental causes. Thus, an average of the observed variance within these isogenic generations may be used as an estimate of V_E. Use of this formula requires the assumptions of complete dominance and of equal gene effects at each of the n loci. When this method was applied to the PTC data on mice just described, an estimate of $n = 1.05$ was obtained, again consistent with the single-locus hypothesis.

The genetics of continuously varying characters will be discussed in greater detail in Chapter 9, including a fairly detailed description of gene action in statistical terms. Before doing so, however, we shall consider the biochemical and physiological bases of gene action and behavior.

Chapter 6

The mechanisms of gene action

In 1902, Garrod discussed a rare human condition, called alkaptonuria, in which the affected have the remarkable symptom of excreting urine that turns black upon exposure to air. He and Bateson, whom he consulted on the matter, concluded that the condition was inherited, and, indeed, that it obeyed the newly rediscovered Mendelian laws. As important as it was at that time to have an example like this of Mendelian inheritance in man, of even greater importance was the conclusion that Garrod drew about the physiological basis of the disorder. Analysis revealed that the urine of the affected individual contained homogentisic acid rather than the normal urea. Garrod suggested (1908) that, somehow or other, the normal metabolic route whereby homogentisic acid is converted to urea had been blocked, and that alkaptonuria represented an "inborn error of metabolism." Further-more, he suggested that several other defects in man — albinism, cystinuria and porphyria — were due to similar metabolic blocks. Bateson (1909) pro-posed that these conditions might be due to the failure of the enzymes that control the normal reactions.

Other work distributed over the next twenty years on the inheritance of pigmentation in plants and animals (see Sturtevant, 1965, for review) pro-vided results that were interpretable by the hypothesis that genes are involved in the production of some sort of biochemical substance. Another line of investigation gave support to this general proposition; the existence of human blood groups had been described in 1900, their Mendelian inheritance had been shown in 1929, and thus a link between genetics and an immunological (and presumptively biochemical) property of an organism was established. These results all showed that genes *could* influence physiological functioning of an organism, but it was not clear whether these were typical or unusual situations.

In the 1930's a new research organism, the common bread mold *Neurospora,* was introduced into genetics. This organism has extremely simple nutritional requirements, and normally can survive on a simple medium containing certain inorganic salts, glucose, and an organic compound called biotin. From this simple diet, the organism is capable of metabolizing all of the complex chemicals required for its life. Beadle and Tatum (1941) irradiated spores of the fungus, and found that some of the organisms had undergone mutation and were no longer able to survive on the simple medium. By analyzing the nutritional requirements of these mutants they were then able to describe the normal metabolic sequence and showed that each step in the sequence is under the control of a single gene. These results gave rise to the "one-gene–one-enzyme" hypothesis, and it became increasingly reasonable to assume that the basic mechanism of gene action operates through the production — or the control of the production — of enzymes.

The Chemical Nature
of the Genetic Material

A great deal of research had also been done in the attempt to elucidate the chemical nature of the gene itself. For any chemical substance to qualify, of course, it would have to meet several requirements. It would have to be found in the nucleus of the cell, because it had been shown that the chromosomes are the carriers of the genes and they are found within nuclei. The substance would have to be capable of self-duplication, because the genes are. The chemical would have to be capable of existing in various forms, or to put it another way, to carry different genetic information, because it was known that there are a large number of genes and that they occur in different allelic forms. Actually, the correct answer had been guessed about the same time that Galton wrote *Hereditary Genius* but a great deal of work and many years were to be required before the supporting evidence was conclusive, and the detailed mechanism could be outlined.

The brilliantly successful synthesis of all of the data available was provided by Watson and Crick (1953a, 1953b). Basically, they hypothesized that deoxyribonucleic acid (DNA) is the fundamental component of the

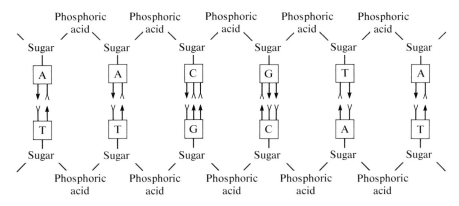

Figure 6.1

Flat representation of a DNA molecule. A = adenine; T = Thymine; C = cytosine; G = guanine. (From Lerner, Heredity, Evolution, and Society. W. H. Freeman and Company. Copyright © 1968, p. 76.)

hereditary material and proposed a molecular structure that could account for its biological properties. This structure was confirmed by subsequent research and the confirmation signalled the explosive growth and development of molecular biology.

Briefly, and in necessarily oversimplified form, the basic features of the molecular structure of the gene and of its action are as follows: A DNA molecule consists of two strands, each composed of phosphate and deoxyribose sugar groups, and the strands are held a fixed distance apart by pairs of bases (nitrogenous compounds). There are four bases involved: adenine, thymine, guanine, and cytosine. Due to structural properties of these bases, adenine always pairs with thymine and guanine always pairs with cytosine (Figure 6.1). The strands coil around each other to form a double helix (Figure 6.2).

The double nature of the helix and the restrictions on base pairing make possible the self-duplication of the DNA molecule. In the process of cell division, the helices of the DNA molecule unwind, the base pairs separate and one of each pair remains attached to each strand (Figure 6.3). Within the nucleus of the cell, the raw materials necessary for the construction of new DNA are found in the form of nucleotides, consisting of one of the four bases, a deoxyribose sugar, and a phosphate. Nucleotides come to pair with the exposed bases of the unwound strands, and ultimately a complementary strand is formed against each of the originals. It may be seen that by this process two molecules of DNA come to exist where there was previously but one.

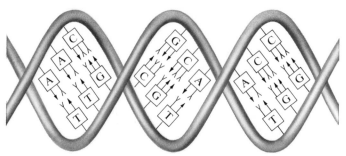

Figure 6.2
A three-dimensional view of a segment of DNA. (From Lerner,
Heredity, Evolution, and Society. W. H. Freeman and Company.
Copyright © 1968, p. 76.)

Genes and Protein Synthesis

It is known that much of the biochemical functioning of the cell takes place in the cytoplasm, yet the chromosomes with their DNA content are located within the nucleus of the cell. Therefore, the information in the DNA molecule has to be transmitted to the cytoplasm. This occurs in several steps, illustrated schematically in Figure 6.4. First, the information of the DNA molecule is transcribed onto a different sort of nucleic-acid molecule. This single-stranded molecule, ribonucleic acid (RNA), is composed of a ribose sugar, a phosphate, and the same bases as DNA with the exception that uracil substitutes for thymine. By a process of base pairing similar to that of the duplication of DNA, a complementary RNA strand can be formed against a DNA strand. (In Figure 6.4, the dark DNA strand is being transcribed.) This RNA molecule, called messenger RNA, enters the cytoplasm where it becomes associated with ribosomes, which are the site of protein synthesis. Within the cytoplasm is found another form of RNA, transfer RNA. This RNA, which has a helical structure, exists in a variety of forms and each of the forms corresponds to a specific amino acid. The transfer RNA's with their attached amino acids (indicated by numerals in Figure 6.4) line up on the messenger RNA in a sequence dictated by the limitations of base pairing. The amino acids join to form polypeptide chains, and polypeptide chains constitute proteins. Enzymes are proteins; thus, the genetic information of the DNA, through this series of steps, becomes expressed as specific enzymes.

The basic unit of the genetic code has been shown to be triplet sequences of bases, with each succeeding triplet specifying an amino acid. For example, three adenines in a row on the DNA molecule (AAA) will be transcribed in the messenger RNA as three uracils (UUU). When on the ribosome, this messenger RNA triplet will attract transfer RNA with the triplet sequence

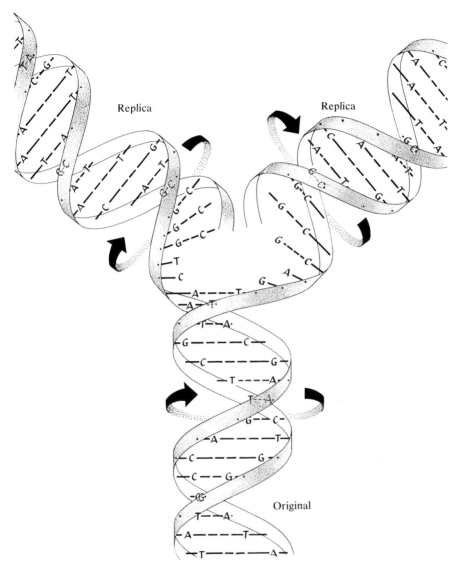

Figure 6.3
Replication of DNA. (After Stent, Molecular Biology of Bacterial Viruses, W. H. Freeman and Company. Copyright © 1963, p. 221.)

AAA. This particular transfer RNA is the one that "carries" the amino acid phenylalanine. Other triplets code for other amino acids, as shown in Table 6.1. The "breaking" of this code has been one of the great triumphs of molecular biology.

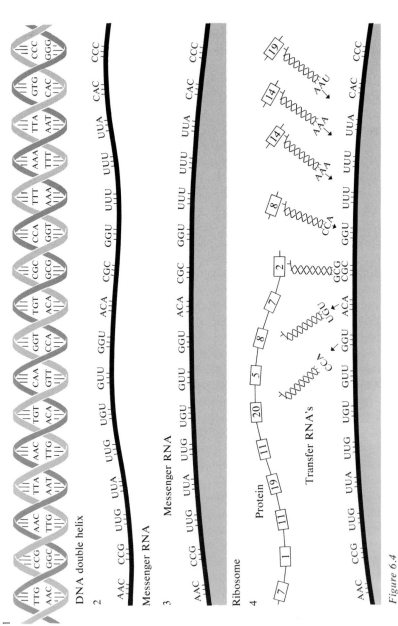

Figure 6.4

DNA, RNA, and protein synthesis. (From Nirenberg, "The genetic code: II." Copyright © 1963 by Scientific American, Inc. All rights reserved.)

Table 6.1

The genetic code

		Second letter			
		A	G	T	C	
A		Phe	Ser	Tyr	Cys	A
		Phe	Ser	Tyr	Cys	G
		Leu	Ser	chain end	chain end	T
		Leu	Ser	chain end	Try	C
G		Leu	Pro	His	Arg	A
		Leu	Pro	His	Arg	G
		Leu	Pro	Gln	Arg	T
		Leu	Pro	Gln	Arg	C
T		Ile	Thr	Asn	Ser	A
		Ile	Thr	Asn	Ser	G
		Ile	Thr	Lys	Arg	T
		Met	Thr	Lys	Arg	C
C		Val	Ala	Asp	Gly	A
		Val	Ala	Asp	Gly	G
		Val	Ala	Glu	Gly	T
		Val	Ala	Glu	Gly	C

(First letter at left; Third letter at right)

SOURCE: After Cavalli-Sforza and Bodmer, *The Genetics of Human Populations.* W. H. Freeman and Company. Copyright © 1971, p. 6. Data from Crick, 1966.

NOTE: Each amino acid is coded by a triplet of three bases, as shown in the table, which is a compact way of setting out the sixty-four possible triplets.

The four bases are denoted by the letters A, G, T, and C. In DNA the four bases are:

A = Adenine   T = Thymine
G = Guanine   C = Cytosine

The twenty amino acids are identified as follows:

Ala = Alanine   Lys = Lysine
Arg = Arginine   Met = Methionine
Asn = Asparagine   Phe = Phenylalanine
Asp = Aspartic acid   Pro = Proline
Cys = Cysteine   Ser = Serine
Glu = Glutamic acid   Thr = Threonine
Gln = Glutamine   Try = Tryptophan
Gly = Glycine   Tyr = Tyrosine
His = Histidine   Val = Valine
Ile = Isoleucine   Chain End.
Leu = Leucine

These developments have provided a dramatically new view of Mendel's hypothetical "elements." As Lerner (1968) has succinctly put it,

> *the gene is a stretch of DNA coding for a particular polypeptide.* The average number of nucleotides in a gene is about 1,500; the average polypeptide of a protein has, perhaps a sequence of 300 amino acids. The number of genes in different organisms varies from no more than three in small viruses, . . . to several thousand or enough to code for 2,000–3,000 different proteins in the bacterium *E. coli,* and, perhaps, to some hundreds of thousands in man. (p. 79)

The enzymes, whose molecular configurations are determined by the genetic information in the DNA, are organic catalysts. They permit biochemical reactions to occur rapidly that would otherwise be sluggish or would not occur at all under the conditions of temperature and pressure prevailing within an organism's body. These reactions are fundamental to the development of all of the organ systems of an organism (as will be discussed later) and to the functioning of these organs. The influence of genes, therefore, is not to be regarded as being exerted through some extracorporeal mechanism; the pathways from genes to behavior run through the skeletal system, the muscles, the endocrine glands, the digestive, respiratory, and excretory systems, and the autonomic, peripheral, and central nervous systems. Investigations of these pathways, therefore, involve the domains of molecular biology and biopsychology.

Although this schema can be understood as a general theoretical proposition, the specific details have been elucidated in only a limited number of examples. This area of research is becoming increasingly popular, however, and it can be confidently predicted that growth of knowledge concerning mechanisms of genetic influence on behavior will be very rapid in the near future.

Heritable Defects in Man

Phenylketonuria. The largest accumulation of evidence relating biochemistry to behavior is that pertinent to conditions of mental retardation in man. The earliest and still best described condition is that of phenylketonuria, the Mendelian analysis of which has been previously discussed. The trail of research that has led to our current understanding of this condition had its beginning in Norway in 1934. A dentist with two feeble-minded children was distressed because they exuded a peculiar odor that so aggravated his asthmatic condition that he was unable to stay with them in a closed room. When he mentioned his problem to a colleague, the colleague suggested that an acquaintance, a Dr. Asbjörn Fölling, was interested in such matters and should be contacted (Centerwall and Centerwall, 1963). Fölling, who was Professor of Nutritional Research at the University of Oslo, began his search for the cause of the peculiarity of these children by use of urinalysis. Suspecting the presence of diacetic acid, he added a ferric chloride solution to the urine in the expectation that the mixture would turn reddish brown. To his surprise, the mixture turned green. After about ten weeks, Fölling had purified the compound causing this color change and had identified it as phenylpyruvic acid. He postulated that the disease was inherited, that the urinary symptom was the consequence of a disturbance in the metabolism of phenylalanine, an essential amino acid, and that the biochemical anomaly in some way caused the mental retardation.

Originally called "imbecilitas phenylpyruvica," the disease came to be known as Fölling's disease, or phenylketonuria (PKU), and became the subject of active research in a number of laboratories around the world. This research has revealed that the metabolic derangement is attributable to the absence or inactivity of the enzyme phenylalanine hydroxylase in phenylketonuric patients. This enzyme catalyzes the conversion of phenylalanine to tyrosine (see Figure 6.5). If this conversion is blocked, phenylpyruvic acid accumulates, which, in turn, generates abnormal quantities of O-hydroxyphenylacetic acid, phenyllactic acid, and phenylacetic acid. Evidently, the presence of one or more of these metabolic products is toxic in some fashion to the developing nervous system. Other metabolic blocks that have been identified in the same general pathway are also shown in Figure 6.5. One block results in alkaptonuria, demonstrating the essential correctness of Garrod's view. Another block results in a diminished production of the pigment melanin, leading to albinism, already described as having some behavioral consequences in Chapter 5.

This knowledge concerning the location of the metabolic lesion of phenylketonuria made possible a rational search for means of identifying heterozygotes. Hsia et al. (1956) described phenylalanine tolerance tests in which samples of blood were assayed for phenylalanine at different times after an oral dose of phenylalanine was given. A number of variations on the original scheme have been developed since that time. It is Hsia's (1970) judgment that these techniques do distinguish a population of heterozygotes from a population of normals, but the overlap between distributions is sufficiently great that the sensitivity of classification of specific individuals is less than might be desired.

This knowledge of the metabolic consequences of an absence of phenylalanine hydroxylase has also made possible the search for a rational therapy for the condition. Phenylalanine is an essential amino acid found in a wide variety of foods, particularly meats. Bickel and colleagues (1953) prepared a special diet that was very low in phenylalanine. Administration of this diet to a phenylketonuric child results in a normalization of his blood biochemistry, but Hsia et al. (1958) found no improvement in intelligence when the

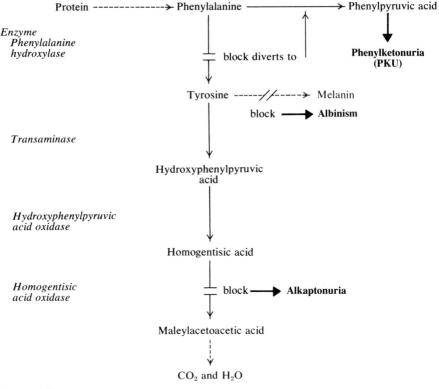

Figure 6.5

Metabolic pathways of phenylalanine metabolism, showing blocks in phenylketonuria, alkaptonuria, and albinism. Intermediary steps are indicated by dashed arrows. (After Lerner, Heredity, Evolution, and Society. W. H. Freeman and Company. Copyright © 1968, p. 91.)

diet was administered to older phenylketonuric children. When the diet was administered to very young phenylketonuric patients, however, the results seemed to be quite favorable, and on this basis routine screening programs for the purpose of identifying phenylketonurics at birth were established.

Through these programs, and through related programs of testing of relatives of affected persons, it became possible to assay the intelligence of individuals identified on the basis of the chemical defect. Previously, most research had been conducted on individuals biochemically identified as phenylketonurics from a population of individuals already determined to be mentally retarded. The assumption had been that all phenylketonurics were probably institutionalized for mental retardation, and, on this assumption, calculations had been made that the mean IQ of phenylketonurics was approximately 30. With the new screening procedure, a surprising number of individuals were discovered who were biochemically phenylketonurics, but whose intelligence was in the normal range. This discovery necessitated a re-evaluation of the results on the efficacy of dietary treatment. If some of the individuals treated with the special diet would have developed normal intelligence in any case, then the report of average IQ's of treated subjects in the range of 80 to 90 could hardly be taken as evidence that the diet prevented retardation.

A great deal of research has subsequently been devoted to the problem of the efficacy of the low phenylalanine diet, and it remains somewhat controversial. A straightforward experiment that would compare the outcome in a treated and an untreated group, both identified at birth, would provide a critical test, of course. This would require the ethically dubious withholding of a *potentially* effective treatment from a group of patients, however, and is unlikely ever to be done. An approach to this type of comparison can be made, however, by comparing the IQ's of treated patients with those of older siblings untreated because the dietary therapy had not yet been invented. The results of such a comparison by Hsia (1970) are given in Figure 6.6. With late treatment or no treatment at all, the distribution of IQ's (or developmental quotients, for those subjects too retarded for accurate IQ assessment) ranges from 10 to 110; siblings treated from an early age have IQ's ranging from 65 to 120. Furthermore, the distribution of IQ's in this latter group is negatively skewed with a clustering of cases at the higher end of the distribution. These results constitute reasonable evidence that the diet is in fact a useful therapy.

The discovery of normal IQ's in untreated phenylketonurics also raised the issue of the possible heterogeneity of the condition and a number of variant forms have now been described. Hsia (1970) recommends as a working definition that patients with persistent plasma phenylalanine levels in excess of 25 milligrams percent be diagnosed as "classical" phenylketonurics. Of these, approximately one-fourth may achieve normal intellectual functioning without dietary treatment. It is difficult to be sure of the diagnosis of patients with levels between 25 milligrams percent and 15 milligrams percent, but those with levels below 15 milligrams percent probably exhibit one of the variants of phenylketonuria.

Variants of Phenylketonuria. Most cases of hyperphenylalanemia (but not PKU) during early infancy are due to hypertyrosinemia. Deficiencies in tyrosine transaminase and hydroxyphenylpyruvic acid oxidase (see Figure 6.5) result in a marked accumulation of tyrosine and a moderate increase in phenylalanine. The phenylalanine levels in the blood due to hypertyrosinemia seldom exceed 10 milligrams percent and do not appear to be harmful (Hsia, 1970). Hyperphenylalanemia may also occur due to a deficiency of phenylalanine transaminase, resulting in a level of phenylalanine in the range of 5–10 milligrams percent, but is not accompanied by ketones in the urine. No mental deficiency has been found in hyperphenylalanemic patients.

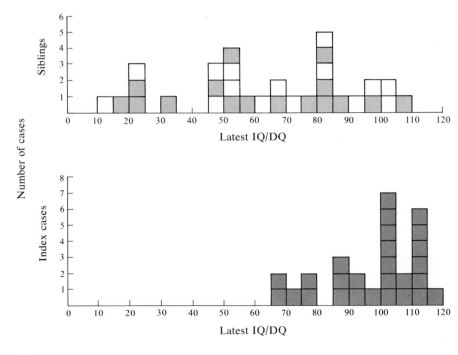

Figure 6.6

Frequency histograms of IQ or DQ scores of early treated index cases versus late-treated or untreated siblings. (From Hsia, "Phenylketonuria and its variants." In Progress in Medical Genetics, Steinberg and Bearn (Eds.) Grune & Stratton, Inc. 1970. Used with permission.)

There have also been several reports of transient hyperphenylalanemia. In some instances, high levels of phenylalanine were found when a child was a few weeks or a few months old, but did not persist at later ages even if the child was fed a normal diet. The converse has also been observed, i.e., a child who first appeared to have a normal level of phenylalanine, but later developed into a classical PKU patient.

Persistent hyperphenylalanemia of unknown cause also occurs. In fact, recent data from the newborn screening programs indicate that there is about one case of hyperphenylalanemia for every two cases of PKU in the United States and Western Europe. Most patients with persistent hyperphenylalanemia apparently develop normal intelligence without dietary treatment. This condition may also be due to an autosomal recessive gene

Liver biopsies for patients with hyperphenylalanemia have shown a decrease in activity, but not a complete absence, of phenylalanine hydroxylase. Thus, genes at two loci or several alleles at the same locus may be responsible for the coding of this enzyme.

Other Aminoacidurias.   The very considerable success in elucidating the biochemical mechanism of phenylketonuria has inspired an intensive research effort directed toward identifying and analyzing other conditions of mental retardation associated with abnormal metabolism of amino acids. To date, a number have been identified: arginosuccinicaciduria, citrullinuria, histidinemia, cystathioninuria, Hartnup's disease, homocystinuria, Lowe's oculocerebro-renal syndrome, maple-syrup-urine disease, microcephaly with spastic dyplegia, Oasthouse urine disease, and Wilson's disease. To date none of these conditions is as well understood as phenylketonuria, and some are in fact represented by only one or two cases. Some are characterized by excessive amounts of a single amino acid in urine, while others display an excess of several amino acids. Mental retardation is a common feature. The intensity of the research effort in this area would seem to assure that great strides in understanding the etiology and in devising rational therapies for these conditions will occur in the relatively near future.

Lesch-Nyhan Syndrome.   Types of metabolic abnormality other than those of amino acid metabolism have been related to specific behavioral anomalies. Lesch-Nyhan syndrome, for example, is a familial neurological disease characterized by cerebral palsy, mental retardation, choreoathetosis (irregular, spasmodic, involuntary movements of the facial muscles, limbs, and extremities), aggressive behavior, and compulsive biting that results in self-mutilation of the lips and fingers. The condition is also characterized by an overproduction of uric acid, the degree of which greatly exceeds that of adults with clinical gouty arthritis. This overproduction of uric acid in subjects with Lesch-Nyhan syndrome is due to the greatly reduced activity of an enzyme involved in the control of purine synthesis, hypoxanthine-guanine phosphoribosyltransferase (Seegmiller, Rosenbloom, and Kelley, 1967).

The limitation of the condition to males, as well as its familial distribution, was indicative of sex-linked recessive inheritance. Subsequent research utilizing autoradiography of cultured cells has yielded evidence to support this hypothesis. Unlike those from normal control subjects, cells from subjects with Lesch-Nyhan syndrome do not incorporate radioactive hypoxanthine from the medium in which they are grown and thus appear unlabeled. In contrast, biopsies of skin from women presumably heterozygous for the condition (mothers of sons with Lesch-Nyhan syndrome) reveal that each of the women carries two classes of cells: deficient (unlabeled) cells, in which the X chromosome bearing the normal allele was inactive; and normal cells, in which the X chromosome bearing the normal allele was active. This mosaicism of cells from heterozygous females conforms with expectation based upon Lyon's hypothesis of random inactivation of maternal X chromosomes, a subject discussed in Chapter 7.

More recently, DeMars and co-workers (1969) have shown that this syndrome may be diagnosed prenatally by culturing cells from the amniotic fluid surrounding a fetus. The utility of fetal diagnosis, as well as heterozygote detection, for genetic counseling will be discussed in Chapter 11.

Carbohydrate and Lipid Metabolism. The condition galactosemia illustrates the class of carbohydrate metabolism defects. Persons homozygous for the autosomal recessive allele lack the enzyme galactose-1-phosphate uridil transferase, and are unable to convert galactose to glucose. An intermediate metabolite, galactose-1-phosphate, accumulates and, in some as yet unknown manner, causes a variety of severe symptoms that may include early death. Affected individuals who do survive are severely mentally retarded. The early identification of homozygous recessive infants and a dietary replacement of milk by galactose-free substitutes is quite successful. Heterozygotes have half the normal enzyme activity, which is apparently sufficient to prevent appearance of the behavioral symptoms.

An example of lipid metabolism derangement is provided by the autosomal recessive condition, infantile amaurotic idiocy, also known as Tay-Sachs disease. Homozygous individuals have a generalized absence of a component of the enzyme beta-D-N-acetylhexosaminidase (Okada and O'Brien, 1969). Such individuals are apparently normal at birth but begin to show symptoms of nystagmus (spasmodic movement of the eyes) and paralysis when a few months old. The condition steadily worsens to a state of profound idiocy, paralysis, and blindness. Death usually intervenes before two years of age. Autopsy has shown that nerve cells of the brains of affected individuals contain abnormal amounts of a lipoid substance and the neurons show degenerative changes. Knowledge of the biochemical lesion responsible for this disease has facilitated identification of heterozygotes (O'Brien et al., 1970) and diagnosis of affected fetuses by sampling amniotic fluid.

A related condition, juvenile amaurotic idiocy, also known as Spielmeyer-Vogt disease, is also inherited as an autosomal recessive; however, it is not known whether the locus is different from that for Tay-Sachs disease or whether the condition is determined by a different allele at the same locus. Juvenile amaurotic idiocy also involves storage of abnormal lipoid substances and degenerative changes in the neurons of the brain, but differs from the infantile form in having a later age of onset. Typically, the degenerative neurological signs appear at about the age of six years and proceed through profound idiocy to death at an average age of about sixteen years.

In addition to the types of mental retardation just described, there are others associated with chromosomal anomalies that will be discussed in

Table 6.2

*Frequencies of alleles at five loci detected in
an enzyme survey in an English population*

Locus	Alleles		
	1	*2*	*3*
Red-cell acid phosphatase	0.60	0.36	0.04
Phosphoglucomutase			
Locus *PGM₁*	0.76	0.24	[a]
Locus *PGM₂*	1.00	—	[a]
Locus *PGM₃*	0.73	0.27	—
Adenylate kinase	0.95	0.05	—

SOURCE: From Harris, 1967, p. 210.
[a]There are several different rare alleles.

Chapter 7. For most of these, although the relationship to the abnormal chromosomal constitution is clearly established, details of the biochemical route through which behavior is affected are not yet known.

It should be pointed out that even when all types of mental retardation due to defective metabolism or chromosome anomalies are considered, there remains unaccounted for a very large proportion of the total population of retardates. Some of these are undoubtedly suffering from conditions similar to those just described but for which the mechanism has not yet been discovered. Others are attributable to environmental causation, and still others to genetic segregation at two or more loci; for these latter groups, it will no doubt be much more difficult to delineate the metabolic mechanisms.

Enzyme Polymorphisms in Man. It is obvious that most of our knowledge concerning the mechanism of gene action in human behavioral genetics is derived from the study of abnormal conditions. Although the current dogma of the mode of gene action, outlined briefly above, suggests that enzymatic differences underlie quantitative variation, explicit evidence on this point has been hard to obtain. Harris and his collaborators have approached the problem directly by seeking in populations of normal individuals enzyme varieties for which no *a priori* evidence exists. (It is of interest that this work is being conducted in the Galton Laboratory at University College, London.) In one series of observations twelve enzymes were examined. Three were found to exhibit polymorphism, that is, several alleles exist. The results are shown in Table 6.2. It may be seen that three alleles have been found that determine red-cell acid phosphatase, that three different loci are involved in phosphoglucomutase, and that there are two alleles for adenylate kinase. The three alleles for red-cell acid phosphatase give rise to six different enzyme-structure phenotypes. If enzyme activity is measured quantitatively, a continuous distribution is obtained. Combining results from his own laboratory and those from the work of others, Harris (1967) stated:

> Since each of these polymorphisms appears to occur independently of the others a large number of different combinations of enzyme phenotypes may be found among individuals in the general population. Indeed the commonest of these occurs in no more than 2 per cent of the population, and the probability that two randomly selected people would have the same combination of these phenotypes is less than 1 in 140. Thus quite a high degree of individual differentiation in enzymic make-up can already be demonstrated from this limited series of examples, and it is of interest that most of this is attributable to variation in molecular structure of the enzymes. This must surely be only the tip of the iceberg, and one may plausibly imagine that in the last analysis every individual will be found to have a unique enzymic constitution. (p. 211)

Infrahuman Animal Phenotypes

For obvious reasons many investigators have worked with mouse and rat strains in the effort to identify mechanisms through which genes influence behavior. A frequently used approach has been to choose strains known to differ behaviorally and then to compare them for some anatomical or physiological trait hypothesized to be part of the causal network. There are, unfortunately, some rather serious limitations to this approach. Consider two strains, A and B, that do in fact exhibit a striking difference in behavioral trait X. Now, in the process of selective breeding or inbreeding that gave rise to these strains, stochastic processes would have been at work on all traits that are completely unrelated to the behavioral trait in question. For some of these unrelated traits, these processes would have brought about the accumulation of genes making for high level of the trait in strain A and lower level in strain B; for others of these unrelated traits, there will be low levels of expression in strain A and high in strain B; and for others still, there would be no average difference between the strains. Thus, even if a trait hypothesized to be part of the mechanism of the behavioral difference is in fact totally unrelated to it, the likelihood that the strains differing in the behavioral trait also differ with respect to the hypothesized trait is quite appreciable. From this type of comparison, therefore, the strongest information that can be obtained is the demonstration that the strains do *not* differ in the hypothesized trait. This is fairly conclusive evidence that the initial hypothesis was incorrect. If the strains do in fact differ in the hypothesized direction, it can be taken only as very modest support of the hypothesis, justifying more intensive research into the issue.

Some of the shortcomings of strain comparisons can be overcome by use of more than two comparison groups. In selective breeding experiments, if replicate selected lines are available and the differences are in the expected direction in both replicates, the evidence is very much stronger than if the observation were made only on a single replicate. Similarly, with inbreds, a large number of strains can be employed, and if the relationship holds across the entire series, it may be regarded with considerable confidence. In many ways, a more powerful approach to the question of correlated characters is to examine the relationship in a segregating population, such as an F_2 or a more highly heterogeneous stock.

The widespread use of inbred strains for nonbehavioral research has provided a large reservoir of hypothesis-generating information and has shown that there is no dearth of differences in biological intermediates to be explored in relation to behavioral differences. An example might be taken from Amin et al. (1957) who studied differences in thyroid activity of a number of mouse strains and their F_1 hybrids. The order of the strains for thyroid-secretion rate, from highest to lowest, was C57BL/6, C57BR/cd, BALB/c, and A. Comparison of these results with the results on activity described in earlier chapters immediately suggests the relationship of thyroid activity to activity in open-field and arena situations. Less directly relatable to established behavioral differences, but suggestive of the ubiquity of genetic control of endocrine systems, are the results showing strain differences in mouse adrenal structure (Meckler and Collins, 1965; Shire and Spickett, 1968) and in anterior pituitary growth hormone (Yanai and Nagasawa, 1968). An even more general indication of genetic control of bodily structure is provided by Dawson (1970), who investigated body composition of inbred mice. Strain differences were found for body weight, fat-free body weight, water composition, fat-free combustible matter, and fat and ash constituents of the bodies of these animals. A particularly noteworthy study is that of Wimer and collaborators (1969), who assessed volume of total brain, relative and absolute volume of neocortex and hippocampus, and cross-sectional areas of these structures in nine inbred strains. The aim was to obtain information on the degree of genetic determination of these characteristics in preparation for selective breeding studies designed to manipulate them systematically.

Endocrine Systems.   Considerable attention has been given to the thyroid gland. Yeakel and Rhoades (1941) examined the thyroid glands of the strains of rats that Hall (1938) had selected for "emotionality" in the open-field situation. It was found that the weight of the thyroid glands was higher in the emotional than in the nonemotional strains. This observation was repeated by Feuer and Broadhurst (1962) who made a similar study on the Maudsley reactive and nonreactive selected strains. The finding of greater thyroid weight in the reactives was repeated in this experiment, but it was shown that the amount of thyroid hormone and the rate of secretion were lower in the reactive than in the nonreactive animals.

The adrenal glands have also been the focus of attention. Thiessen and Nealey (1962) showed strain differences in eosinopenia, used as a measure of adrenal response to stress, in inbred mice. Measuring plasma steroid levels directly, Levine and Treiman (1964) found significant differences in the temporal pattern of corticosterone response to stress in C57BL/10, A/J, DBA/2, and AKR mice. Vale and colleagues (1971) have shown an interesting interaction of genotype with environment in response to crowding. As population density increased, the BALB/c mice showed a striking rise in the average number of attacks of one mouse upon another; the C57BL/6, C3H, DBA/2, and A strains were unaffected over the same range of population densities. The BALB/c animals and the C3H animals showed initial decline in adrenal weight followed by a rapid rise as population density increased; this relationship was not found for the A, C57BL/6, or DBA/2 animals, whose adrenal weights were relatively unaffected by the crowding. The Maudsley reactive rats have also been shown to have larger adrenal glands than the nonreactive (Feuer and Broadhurst, 1962). The female reactive animals have smaller ovaries than the nonreactive females, but no differences were found in the testis size of the males or in the pituitary size of either sex.

Riss and colleagues (1955) have provided a particularly instructive example. Significant differences in male mating behavior have been shown to exist among three strains of guinea pig. In all strains, castration eliminated mating behavior and exogenous replacement of testosterone restored it, but no amount of overdosing could bring the behavior of animals of a given strain to a level exceeding that of the uncastrated controls of the same strain. In this case, although it is clear that the hormone must be necessary for the behavior to occur, the genetically influenced behavioral differences appear to be mediated by differences in the sensitivity of the target tissue rather than in amounts of the endocrine substance.

Central Nervous System.   Naturally, the central nervous system has been a favorite site of investigation. That genes can influence both the structure and function of the nervous system (within "normal" ranges) has been demonstrated by, for example, the previously described work of Wimer and associates (1969) concerning strain differences in forebrain structures, and the research of Hegmann (1972), who showed substantial strain differences in peripheral-nerve-conduction velocity in mice.

The neurological mutants in mice discussed in Chapter 5 provide clear examples of correlates of nervous system abnormalities and gross behavioral anomalies due to single loci. For example, the conditions reeler, agitans, and staggerer are the expressions of defects of the cerebellum. Quaking mice, on the other hand, suffer from a general reduction in myelin throughout their central nervous system. Other research has focused upon nervous system attributes and genetically mediated differences in normal behavior. Rosenzweig, Krech, and Bennett (see Rosenzweig, 1964), for example, have utilized descendants of rats selectively bred in a study of maze performance in rats (that of Tryon; see Chapter 9) in a systematic series of explorations of brain biochemistry and learning behavior. In their initial observations they found that descendants of the "maze-bright" rats differed from descendants of the "maze-dulls" in activity of the enzyme acetylcholinesterase (AChE), which accelerates breakdown of the neurotransmitter substance acetylcholine (ACh). The initial hypothesis was that the strains might differ with respect to this neurotransmitter substance. However, AChE was much more readily measurable than ACh, so the former was used as an index of the latter. In an attempt to confirm the relationship, Roderick (1960) selectively bred in a replicate experiment rats high and low in the enzyme measure. Behavioral testing of the resulting strains gave results contrary to expectation: the animals selectively bred to have higher AChE levels actually performed more poorly in the maze-learning situations than did those selected for low AChE levels. This outcome illustrates well the point made earlier about the hazards of two-group comparisons. Subsequent results suggested that the critical factor is the ratio of ACh to AChE, with higher ratios characteristic of the "brighter" animals.

A rather different approach to the nervous system has been taken by a number of investigators who have employed susceptibility to audiogenic seizure as an index of central-nervous-system excitability. Representative of this work is that of Schlesinger and colleagues (1970) who have concentrated on measurements of neurochemistry of the seizure-resistant C57BL/6 mice and of the seizure-susceptible DBA/2 mice. The neurotransmitter substances serotonin (5-HT) and norepinephrine (NE) have been found to be significantly lower in the DBA/2 mice than in C57BL/6 mice. This difference is particularly pronounced at 21 days of age, an age that coincides with the peak of susceptibility of the DBA mice. Further evidence that these neurochemicals are involved in seizure susceptibility has been obtained by various drug treatments that either deplete or increase the levels of 5-HT and NE. Depletion increases susceptibility and increased amounts protect against seizures.

The work of Ginsburg and colleagues (1967) has explored neurochemical differences in very localized areas of the brain. Seizure-prone animals have been shown to differ from seizure-resistant ones by possession of a high level of activity of the enzyme adenosine triphosphatase in the granular cell layer of the dentate fascia of the hippocampus.

Liver Enzymes and Alcohol-related Behavior.   The differences among mouse strains in preference for alcohol and in susceptibility to the effects of alcohol have provided yet another model system for the investigation of biochemical parameters. Alcohol is metabolized principally in the liver, and the enzyme mediating the first step in the process is alcohol dehydrogenase (ADH). Several investigators (Wilson et al., 1961; Rodgers et al., 1963; Eriksson and Pikkarainen, 1968) have shown that the liver ADH activity of C57BL mice is higher than that of various mouse strains that have a lower preference for alcohol. A subsequent exploration of the magnitude of the correlation between ADH activity and alcohol preference in HS mice, a segregating population derived from crosses of eight inbred strains (McClearn, 1968), showed a significant but low correlation ($r \simeq 0.30$). The magnitude of this correlation suggests that we can account for approximately 10 percent of the variability in alcohol preference in this population in terms of individual differences in ADH activity.

The second step in the metabolism of alcohol (also occurring in the liver) is mediated by the enzyme aldehyde dehydrogenase (ALDH). Sheppard et al. (1970) showed C57BL/6 mice to have higher ALDH activity than DBA/2 mice. Belknap (1971) performed a correlational study between ALDH activity and pharmacological sleeping time (duration of "sleep" induced by an injection of a standard dose of alcohol) in HS mice. The correlation obtained was approximately 0.6. Thus, variation in ALDH activity can account for approximately one-third of the variability in this index of sensitivity to the effects of alcohol.

Another approach to mechanisms in alcohol-related behavior has been taken by Kakihana et al. (1966), who examined the rate of elimination of alcohol from blood and brain tissue of BALB/c ("long sleeping") and C57BL ("short sleeping") mice. Since the rates of elimination of alcohol were found to be almost identical in the two strains, it appears that their striking differences in sleeping time, as defined above, are due to differences in sensitivity of the brain rather than to differences in the rate of metabolism of alcohol.

Genetic Dissection of the Nervous System

As discussed earlier in this chapter, Beadle and Tatum elucidated various biosynthetic pathways by the induction of mutations in the common bread mold *Neurospora*. A number of *Neurospora* mutants were found that were unable to convert the simple substances present in a minimal medium into the more complicated compounds necessary for their growth. An analysis of the nutritional requirements of these mutants revealed the stepwise sequence of normal metabolism and demonstrated that each step in the sequence is under the control of a single gene. Benzer (1967) has suggested that a similar approach may be fruitful when applied to the complex structures and events underlying behavior, and has applied it to *Drosophila* behavior. *Drosophila melanogaster* was chosen for this research because of

Table 6.3

Offspring from mating of normal male and attached-X female Drosophila

		Male gametes	
		X	Y
Female gametes	X̂X̂	X̂X̂X (nonviable)	X̂X̂Y (attached-X female)
	Y	XY (normal male)	YY (nonviable)

its high fecundity and short generation interval and because of the great accumulated store of information currently available concerning its genetics.

In order to induce mutations, Benzer (1967) fed male *Drosophila* from a highly inbred wild-type strain a potent mutagen and then mated them with virgin females that carried X chromosomes that were attached to each other. The offspring produced by such a mating are indicated in Table 6.3. The chromosomal basis of sex determination in *Drosophila* differs from that in mammalian species. Although XX and XY individuals are normal females and males, respectively, in both *Drosophila* and man, XXY individuals are abnormal males in our species (see Chapter 7), but are fertile females in *Drosophila*. Therefore, the mating of a mutagen-treated male and an attached-X female *Drosophila* yields two types of viable progeny: (1) males that carry X chromosomes from their treated fathers and that will thus express any induced sex-linked recessive mutations; and (2) attached-X females that received their X chromosomes from their untreated mothers and, hence, would only express induced autosomal dominant mutations.

Benzer (1967) subjected offspring of such matings to a countercurrent-distribution-sorting technique that fractionates the population according to phototactic response. The apparatus consists of two test tubes joined by a celluloid sleeve and laid horizontally in a black rack. A 15-watt fluorescent lamp, located perpendicular to the tubes, is utilized as the light source. As illustrated in Figure 6.7, flies are placed at one end of a double tube (tube B) at the start and then allowed to move freely for one minute. The tubes (A and B) are then separated and joined by new tubes, resulting in two double tubes. In order to begin cycle 2, the flies in both double tubes are brought back to the same end. After one minute, the two double tubes are separated and new tubes are added to A and B. The remaining tubes are then joined since each contains flies that have made one positive response. This sequence may be continued for as many transfers as the investigator desires. In his studies, Benzer utilized 15 transfers, yielding a total of 16 fractions (zero to 15 positive responses).

The distributions of wild-type flies when moving toward light and away from light are shown in Figure 6.8. Data presented in the top figure were

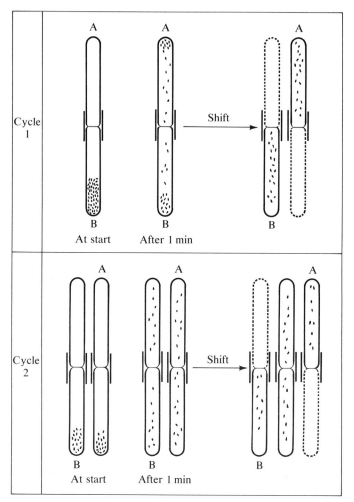

Figure 6.7

Countercurrent distribution procedure for fractionating a Drosophila
*population. In each cycle, the flies are divided according to their
phototactic response. Dotted lines indicate new tubes introduced at the
end of each cycle. Only the first two transfer cycles are shown.
(After Benzer, 1967, p. 1113.)*

obtained when the light source was placed at the tube end distal to the
starting point, whereas those resulting in the bottom distribution were
obtained when the lamp was located proximal to the starting tube. These
distributions demonstrate that wild-type *Drosophila* of this particular strain
are positively phototactic, i.e., they move toward light. Benzer has found
that several hundred flies may be tested simultaneously in this apparatus
without seriously affecting the distribution.

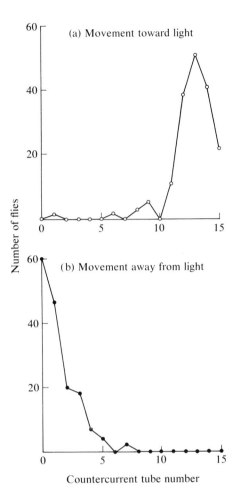

Number of flies

(a) Movement toward light

(b) Movement away from light

Countercurrent tube number

Figure 6.8
Countercurrent distribution of flies
showing the number in each tube after
15 transfers. The upper curve shows
high frequency of movement toward
light. The lower curve shows low
tendency to move away from light.
(After Benzer, 1967, p. 1115.)

In his initial report, Benzer (1967) described two male offspring of mutagen-treated males and attached-X females that were nonphototactic and that when mated to virgin attached-X females produced male progeny that were also nonphototactic. The distribution of the progeny of one of these "strange behavior" mutants is shown in Figure 6.9. When all flies are tested at once, two peaks are observed. However, when the data are plotted separately according to sex, it becomes clear that only the male offspring are nonphototactic. Since male offspring carry the X chromosome of their non-phototactic father, whereas female offspring carry the attached-X chromosomes of their mother, these results demonstrate that the nonphototactic response of the father was due to a sex-linked gene mutation.

In a subsequent report, Hotta and Benzer (1970) described a series of nonphototactic mutants with defects detectable by the electroretinogram

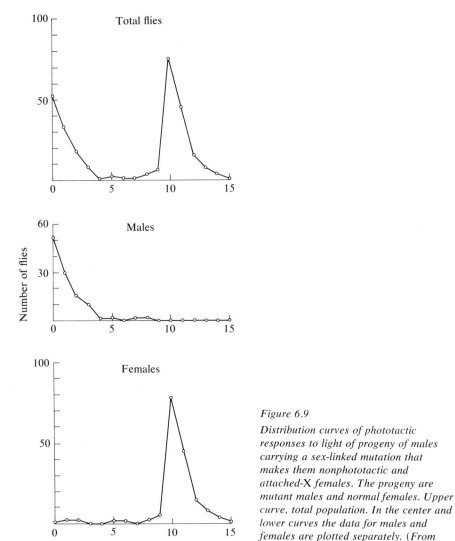

Figure 6.9

Distribution curves of phototactic responses to light of progeny of males carrying a sex-linked mutation that makes them nonphototactic and attached-X females. The progeny are mutant males and normal females. Upper curve, total population. In the center and lower curves the data for males and females are plotted separately. (From Benzer, 1967, p. 1117.)

(ERG), all of which were induced by mutagenesis and isolated by the countercurrent-distribution technique. All mutations were found to map at five loci on the X chromosome and all involved defects within the eye or in very closely associated tissue. In order to demonstrate eye involvement, the behavior of mosaic flies was studied. Mutant nonphototactic males carrying various sex-linked recessive morphological markers (such as mutations for body color) were crossed to females from a strain in which one of the X chromosomes is frequently lost during the first mitotic division after fertili-

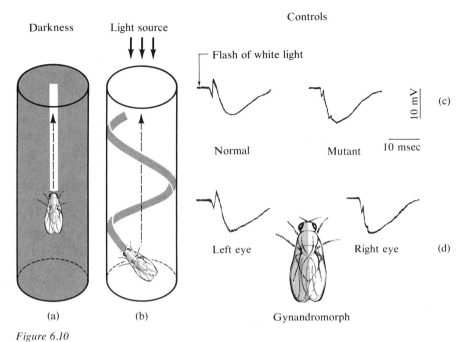

Figure 6.10

(a) *Behavior in darkness of a mosaic fly having normal vision in one eye and defective in the other. The fly, being negatively geotactic, climbs straight up.* (b) *When light is shining from above, the same fly turns its defective eye toward the light and climbs a helical path.* (c) *Electroretinograms of normal and mutant flies. Stimulus is a 20-microsecond strobe flash of white light.* (d) *Schematic drawing shows a gynandromorph in which the shaded left half is* XX *female and the right half is* XO *male, as indicated by markers. The male eye gives a mutant ERG, while the ERG of the female eye is normal.* (*From Hotta and Benzer, 1970, p. 1159.*)

zation. Among female progeny, elimination of an X chromosome during this stage of development results in the individual's being an XO/XX mosaic. The XO part (in *Drosophila*, XO individuals are phenotypically male) is indicated by expression of the sex-linked recessive morphological characters.

In order to assess the behavior of mosaics, flies were placed in a vertical tube. Since wild-type (normal) flies are both positively phototactic and negatively geotactic (i.e., they move away from gravity), they climb straight up the tube both in darkness and when a light source is located at its top. In the dark, a mosaic fly having one eye normal and the other with deficient visual function due to a mutation also climbs straight up. However, when a light source is located at the top of the tube, the mosaic fly climbs in a helical path in an apparent attempt to equalize the light sensations on the two sides (see Figure 6.10(a) and (b)). Since the defective eye is always turned toward the light, the fly traces a right-handed helix if the right eye is abnormal, and a left-handed helix if the left is.

The ERG of a normal *Drosophila* has two main components when the fly is responding to a short flash of light (see Figure 6.10(c)). The first is recorded as a negative wave and is apparently due to the depolarization of the photoreceptor cells. The second, due to a neural discharge, is seen as a positive peak. In mosaic flies with one normal eye and one mutant eye, a different ERG may be obtained for each eye. The eye with normal structure yields a normal ERG, whereas the mutant eye gives a defective ERG (see Figure 6.10(d)). Hotta and Benzer have shown that the same result is obtained regardless of the amount of normal female tissue present, as long as one eye is normal and the other is mutant. Thus, the mutation is autonomous and the defect must occur within the eye itself or in closely associated tissue.

Of the nonphototactic mutants studied by Hotta and Benzer (1970), some showed a near absence of the ERG receptor potential, indicating a lack of response of the photoreceptor cells to light. In other mutants, the positive spike of the ERG was considerably reduced. In still others, the rhabdomeres of the receptor cells degenerated with age, resulting in a small and delayed ERG. Since these mutants were chosen for study because of their ERG abnormalities, it is not surprising that the primary effects were found to be located in the eye. When other criteria are employed in choosing subjects for study, other mutations are likely to be found that affect the nervous system more centrally.

As Benzer (1967) has clearly indicated, phototaxis is a complex behavioral response: "Light is absorbed by a pigment in the receptor cell, producing neural excitation, transmission at synaptic junctions, integration in the central nervous system involving comparison with other inputs, and generation of appropriate motor signals such that the fly walks in a particular direction" (p. 1118). A mutation resulting in a defect in any of these structures or processes could lead to a change in phototaxis. If many nonphototactic mutants were isolated by the countercurrent-distribution technique, analogous to the mutant strains of *Neurospora* that do not grow on minimal medium, defects at various steps in the sequence might be uncovered. By comparing the nonphototactic flies with normal wild types, information concerning the normal sequence of events that takes place in the nervous system might be obtained. Potential application to stimuli other than light is obvious.

Chapter 7

Chromosomes and behavior

Chromosome mechanics, including the processes of mitosis and meiosis, were briefly discussed in Chapter 2. In this chapter, the human chromosome complement and associated anomalies will be considered. This will be followed by a discussion of chromosomal analysis in *Drosophila*, an organism whose chromosome map is known in much greater detail than is man's.

The Chromosomes of Man

Although the chromosomes of *Drosophila* and other organisms were being subjected to detailed analyses as early as the 1930's, human cytogenetics lagged far behind. When the authors of this volume were students, not *that* many years ago, we were taught that the number of chromosomes in man was 48 (24 pairs). However, after using improved techniques, Tjio and Levan in 1956 reported that the normal diploid chromosome number in man is 46, not 48! Since that date, important developments in human cytogenetics have occurred with great rapidity.

In order to study the chromosome complement, or *karyotype*, of an individual, a sample of white blood cells (leukocytes) is usually obtained and cultured in the laboratory for two or three days. A chemical (phytohemagglutinin) is added to the culture to stimulate growth and cell division.

Dividing cells are then exposed to colchicine, a chemical that inhibits the separation of doubled chromosomes. This results in the accumulation of cells in the metaphase stage of mitosis, illustrated in Figure 2.5(c), a stage in which the doubled chromosomes have shortened, thickened, and are still attached at the centromere, or kinetochore. The cells are then washed with a saline solution, resulting in the swelling of the cell and dispersal of the chromosomes. When these cells are squashed or air dried, the chromosomes tend to lie in the same optical plane. The cells are then stained and photographed under high-power magnification. The chromosomes in the photograph may then be cut out and rearranged according to their size and location of the centromere. The karyotype of a normal male is shown in Figure 7.1.

An international conference was held in Denver, Colorado, in 1960 for the purpose of standardizing the classification of human chromosomes. The resulting "Denver classification" is based upon both chromosome length and location of the centromere. If the centromere divides a chromosome into arms of approximately equal length, the chromosome is said to be *metacentric*. If the centromere is very close to one end of the chromosome, the chromosome is referred to as being *acrocentric*. If the centromere is located somewhere between the middle and one end, the chromosome is described as being *submetacentric*.

As indicated in Figure 7.1, the 23 pairs of human chromosomes are classified into seven distinct groups. Group A includes chromosomes 1, 2, and 3. These are large metacentric chromosomes that may be distinguished from each other on the basis of size and of location of the centromere. Group B includes chromosomes 4 and 5, large submetacentric chromosomes. Group C is the largest group, including chromosomes 6 through 12 and the X chromosome, all of which are medium sized and submetacentric. Group D includes the medium-sized acrocentric chromosomes, 13, 14, and 15. Group E chromosomes (16, 17, and 18) are relatively short and metacentric or submetacentric; those in Group F (19 and 20) are shorter and metacentric. Group G chromosomes are very short and acrocentric. This group includes chromosomes 21 and 22, as well as the Y chromosome.

A system for describing the chromosome complement of an individual has also been devised. The total chromosome number is indicated first, followed by the sex-chromosome constitution and any autosomal abnormality. A plus or minus sign after a chromosome number or letter indicates that an entire autosome is represented an extra time or is missing and a question mark indicates uncertainty. This system of nomenclature is illustrated in Table 7.1.

In 1968, another significant advance was made in the technology of chromosome identification. In that year, Caspersson and co-workers reported that metaphase chromosomes, after being specially prepared, may be stained by fluorescent DNA-binding agents, such as quinacrine mustard, to yield a clear pattern of light and dark cross striations when viewed in the fluorescence microscope. The complex mechanism by which these compounds

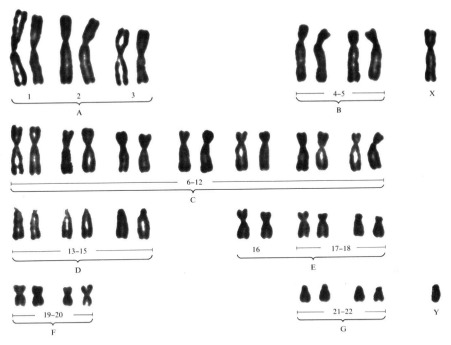

Figure 7.1

Male karyotype. (From Stern, Principles of Human Genetics, 3rd Ed. W. H. Freeman and Company. Copyright © 1973, p. 27. Original photomicrograph by Dr. Margery Shaw.)

Table 7.1

Nomenclature for human chromosome complements, including aberrant ones

Abbreviation	Description
46, XY	Normal Male
46, XX	Normal Female
45, X	22 pairs of autosomes, one X chromosome; one sex chromosome missing
47, XXY	22 pairs of autosomes; one extra sex chromosome
45, XY, C−	Male; one chromosome missing in group C
47, XX, 21+	Female; one extra chromosome number 21
45, XX, ?C−	Female; one autosome missing, probably in group C
45, X/46, XX	A mosaic, some cells like those of a normal female and some missing an X chromosome

SOURCE: After Hsia, *Human Developmental Genetics*. Year Book Medical Publishers. Copyright © 1968, p. 176. Adapted from Chicago Conference: Standardization in Human Cytogenetics: Birth defects: Original Article Series II: 2. New York: The National Foundation, 1966.

bind DNA appears to be influenced by the distribution of protein components along the intact chromosome, resulting in a characteristic pattern of cross banding for each chromosome. Thus, by observing the banding pattern, it is possible to identify unambiguously each chromosome and each of its arms. This is clearly an improvement over previous techniques that often resulted in no more than the assignment of individual chromosomes to groups on the basis of their size and location of the centromere. With special pretreatment of the chromosomes, Giemsa staining, one of the classical staining techniques of cytology, yields similar banding patterns (see Figure 7.2). As will be seen later in this chapter, these improved techniques have greatly facilitated the identification of various chromosomal anomalies.

Chromosomal Anomalies of Man

Although meiosis is usually a very orderly process, irregularities occasionally occur. For example, two breaks may occur in the same chromosome. The breaks may rejoin normally or the broken segment may become inverted. If the two breaks are within one arm of the chromosome, the *inversion* is referred to as *paracentric*. If both arms are involved, the location of the centromere within the chromosome will be changed, resulting in a *pericentric* inversion. The formation of paracentric and pericentric inversions is illustrated in Figure 7.3 (a).

When two breaks occur within a chromosome, the segment not containing the centromere may become lost, resulting in a *deletion* as is shown in Figure 7.3 (b). If a third break occurs in a replicated chromosome (sister chromatid) or in a homologous chromosome, rather than becoming lost, the segment may become incorporated into the gap provided by the third break. As seen in Figure 7.3 (c), this results in two abnormal chromosomes, one with a *duplication* and one with a deletion.

A *translocation* between two nonhomologous chromosomes is illustrated in Figure 7.3 (d). When breaks occur near the centromere in the long arm of one chromosome and in the short arm of the other, two acrocentric chromosomes may be converted into metacentric chromosomes, a large one and a very small one that may eventually be lost. This would not only change chromosomal morphology; if the very small chromosome is lost, the chromosome number will be changed.

If during cell division a centromere divides in a transverse rather than in a longitudinal plane, one daughter chromosome will possess both long arms and the other both short arms of the replicated chromosome. This *isochromosome* formation is illustrated in Figure 7.3 (e).

During meiosis, sister chromatids occasionally migrate to the same daughter cell. This process of *nondisjunction* results in the formation of two aberrant daughter cells, one with an extra chromosome and one missing a

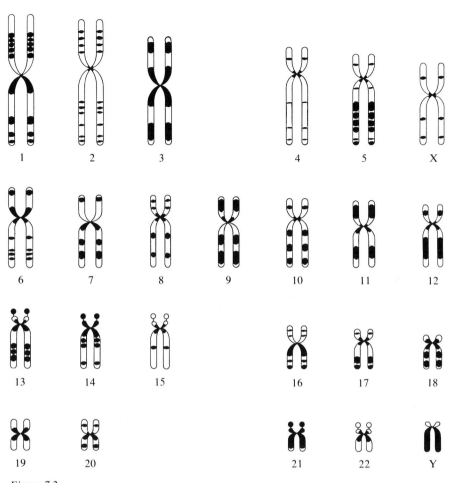

Figure 7.2

Diagram of banding patterns for each chromosome, seen after use of a special staining method. (From Stern, Principles of Human Genetics, 3rd Ed. W. H. Freeman and Company. Copyright © 1973, p. 29. Modified from Fig. 3, Drets and Shaw, Proc. Nat. Acad. Sci., 68, 1971.)

chromosome (see upper portion of Figure 7.4). During meiosis, homologous chromosomes also sometimes fail to segregate. This form of nondisjunction is contrasted to normal gamete formation with regard to the sex chromosomes in the bottom part of Figure 7.4.

Sex-chromosome Anomalies. In 1949, Barr and Bertram observed that nondividing cells of normal males and females differ in the morphology of their nuclei. In normal females, a small distinct mass, or body, that stains

(a) Inversions

Paracentric   Pericentric

(b) Deletion   (c) Duplication

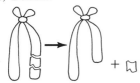   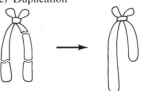

(d) Translocation (centric fusion) between nonhomologous chromosomes

Acrocentrics   Metacentrics

(e) Isochromosome formation

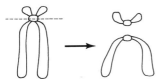

Figure 7.3

Common types of chromosomal rearrangements. (After Nadler and Borges, "Chromosomal structure and behavior" in Hsia (Ed.) Lectures in Medical Genetics. Copyright © 1966 by Yearbook Medical Publishers. Used by permission.)

positively for DNA was found to adhere to the inner surface of the nuclear membrane. In normal males, however, there is no such chromatin mass. Subsequent research has shown that the number of chromatin masses found in a cell is one less than the number of X chromosomes. This correspondence

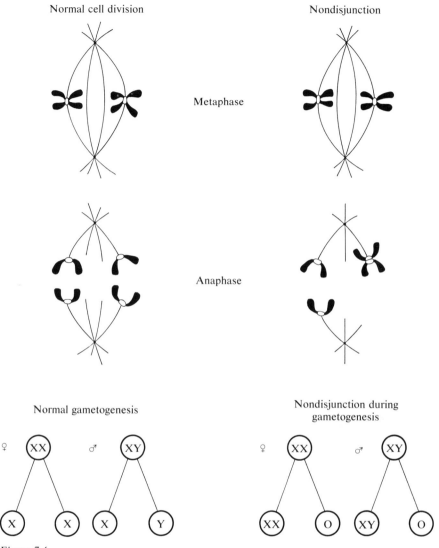

Figure 7.4
Diagrammatic representation of nondisjunction. (After Nadler and Borges, 1966, p. 37.)

between number of chromatin masses and sex-chromosome complement is illustrated in Figure 7.5.

Subsequent research has also shown that the chromatin mass is a single, condensed X chromosome. In 1961, Mary Lyon hypothesized that this condensed X chromosome was genetically inactive. She also hypothesized that

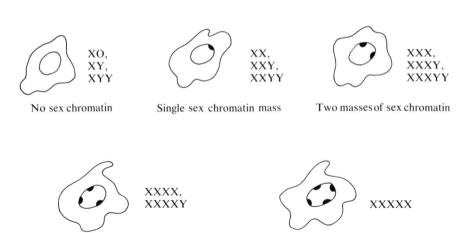

Three masses of sex chromatin   Four masses of sex chromatin

Figure 7.5

Correspondence between sex-chromatin patterns and sex-chromosome complements.
(After Hsia, 1968, p. 184.)

the inactive X chromosome could be either of maternal or paternal origin, even in different cells within the same individual, and that inactivation occurred early during embryonic development. Considerable evidence has subsequently been accumulated in support of Lyon's hypothesis. This inactivation of an X chromosome is apparently a mechanism that compensates, at least to some extent, for the disparity between the amounts of genetic material in the sex chromosomes of normal males and those of females.

The sex-chromatin test has greatly facilitated the discovery of individuals with sex-chromosome anomalies. This test is much simpler and less expensive than karyotype analysis. Cells may be easily obtained for examination by lightly scraping the inside of the cheek. These cells are then spread on a slide, stained, and examined microscopically for the presence of sex-chromatin masses. Because of the economy of this test, large scale surveys have been undertaken. For example, among 8,621 mentally defective, institutionalized males and mentally handicapped schoolboys, about 0.8 percent were found to have sex-chromatin anomalies (Hsia, 1968). This is approximately twice the incidence observed in the general population.

Several distinct syndromes have been reported that are due to an abnormal number of sex chromosomes. Individuals with *Klinefelter's syndrome* are phenotypic males, but sexual development is abnormal. An almost invariable clinical feature is the presence of abnormally small testes together with otherwise apparently normal external genitalia. About half of the patients with Klinefelter's syndrome are mentally retarded and a variety of personality and psychiatric problems has been reported. Although individuals with this condition are males, they usually test positively for the presence of a sex-chromatin mass in the nuclei of their cells. In about two-thirds of the cases, the karyotype is 47, XXY. However, 48, XXXY; 49, XXXXY; 48, XXYY; 49, XXXYY; and various other arrangements including mosaicisms have been described. This condition is generally considered to be due to nondisjunction during meiosis, resulting in a gamete that has an extra sex chromosome (see bottom portion of Figure 7.4). Fertilization either of an XX-bearing egg by a Y-bearing sperm or of an X-bearing egg by an XY-bearing sperm would result in an offspring with Klinefelter's syndrome. It is possible, however, that errors during early mitotic cell division in a normal zygote may also occasionally be the cause of an individual's having this syndrome. There is an increased risk of Klinefelter's syndrome among children of older mothers, although this association is less marked than that with Down's syndrome, which is discussed in the next section. The incidence of Klinefelter's syndrome is believed to be about 0.2 percent among newborn males.

Within the last several years, another sex-chromosome anomaly in phenotypic males has received considerable publicity. In 1965, Jacobs and co-workers reported that the incidence of chromosomal anomalies among individuals institutionalized because of "dangerous, violent, or criminal propensities" was higher than that in the population at large. Of 197 institutionalized volunteers who were karyotyped, 12 were found to have a chromosomal anomaly of some kind. One was a 46, XY/47, XXY mosaic, one was 48, XXYY, and seven were 47, XYY. Three had no sex-chromosome anomalies, but only minor autosomal defects. The average height of the 47, XYY males was 73 inches, in contrast to an average height of 67 inches for the males of normal karyotypes in the institution.

To ascertain that an individual is 47, XYY, a karyotype analysis is necessary, i.e., a sex-chromatin test will not reveal the presence of an extra Y chromosome. Thus, the prevalence of this condition in the general population is not yet well known; however, in a heterogeneous sample of 9,327 "normal" adult males, nine 47, XYY karyotypes were identified (Price and Jacobs, 1970), indicating a prevalence rate of approximately 0.1 percent. The incidence may be slightly higher among newborn males, 0.1–0.2 percent (Shah, 1970; Hook, 1973), but it is considerably lower than the 3.5 percent found in the institutionalized sample of Jacobs and colleagues (1965).

Because of the importance of this initial discovery of a possible association between the presence of an extra Y chromosome and violent aggressive behavior, a number of related studies have been subsequently undertaken. These surveys have usually been of tall prisoners confined to special security sections because of their violent behavior. Thus, the possibility of sampling biases is clear. Nevertheless, a fairly consistent pattern of results has been obtained. Data summarized by Shah (1970) are presented in Table 7.2. The total number of individuals surveyed in these studies is 5,342. Of these (excluding the two 46, XY/47, XYY mosaics), a total of 103, i.e., 1.9 percent,

Table 7.2

Summary of institutional surveys pertaining to the 47, XYY karyotype

Investigator and year	Reference group size	Classification	Minimum height criterion (cm.)	No. XYY	Approx. XYY %
Casey et al. (1966)	100	Maximum Security Hospitals, England	183	16	16.0
Jacobs et al. (1968)	315	Mentally Disordered Offenders, Scotland	None	9	2.8
Akesson et al. (1968)	86	Two Mental Hospitals, Sweden	183	4[b]	2.3
	86	Juvenile Detention Center (14–16 yrs), Pa.	183	1	1.2
	42	Juvenile Detention Center (15–17 yrs), Pa.	183	0	0
	74	State Prisons, Pa.	183	2	2.7
	72	City Prisons, Pa.	183	1	1.3
Baker et al. (1970)	58	City Prison (penal & mentally ill), Pa.	183	1	1.7
	82	State Prison (penal & mentally retarded), Pa.	183	0	0
	102	State Hosp. for Criminally Insane, Pa.	183	3	3.0
	230	State & V.A. Hospitals, Pa.	183	0	0
	130	School for Mentally Retarded, Pa.	183	1	0.8
Welch et al. (1967)	20	"Defective Delinq." aggress. offenders, Md.	188[a]	0	0
	35	"Defective Delinquent" Offenders (various criteria, e.g., particular aggressivity.)	183[a]	1	3.0
Goodman et al. (1967)	52	Caucasian Prison Inmates, Ohio	185.4[a]	2	4.0
	100	Caucasian & Negro Prison Inmates, Ohio	185.4[a]	2	2.0
Griffiths & Zaremba (1967)	34	Wandsworth Prison, England	183	2	6.0
Wiener et al. (1968)	34	Adult Prisoners, Australia	175.3[a]	4	12.0
Hunter et al. (1968)	29	Boys in Approved Schools, England	"tall"	3	10.0
Court Brown (1968)	71	Epileptic Colony, England	None	1	1.4
	605	Hospital for Mentally Subnormal, England	None	0	0

	N	Institution	Height	Number	%
Cited by: Price & Jacobs (1970)	607	New Entrants (1 yr), Scottish Borstals	None	1	0.2
	302	Allocation Center, Soughton Prison, Edinburgh	None	0	0
	204	Recidivist Criminals, Grendon Prison, England	None	2	1.0
	419	Inmates, all Scottish Prisons	178	1	0.3
	74	Young Offenders' Institution, Scotland	178	1	1.4
	17	Scottish Detention Center	178	0	0
	34	Wandsworth Prison, England	183	2	6.0
	24	Nottingham Prison, England	183	2	8.0
	40	Pentridge Prison, Australia	175	5	12.5
	19	Hospital for Mentally Subnormal, England	183	2	10.0
	11	Hospital for Mentally Subnormal, England	183	2	18.2
	30	Scottish Mental Subnormality Hospital	183	2	6.6
	183	Scottish Mental Disease Hospitals	183	2[c]	1.6
	40	English Mental Disease Hospitals	183	0	0
Melnyck (1969)	200	Mentally Disordered Sex Offenders & Criminally Insane, Calif.	183	9	4.5
Nielsen (1968)	41	State Hospital, Denmark	180	0	0
Nielsen et al. (1968)	37	Institution for Criminal Psychopaths, Denmark	180.3	2	5.4
Sergovich (1969)	230	Hospital for Criminally Insane, Canada	None	4[b]	1.7
Daly (1969)	210	Maximum Security Hospitals, U.S.A.	183	10	5.0
Marinello et al. (1969)	86	Attica State Prison, N.Y.	183	2	2.3
	76	State Mental Hospital, N.Y.	183	1	1.3
	57	Juvenile Offenders (Detention Home & Court Psychiatric Referrals), N.Y.	None	1	1.8
Abdullah et al. (1969)	18	Criminal Psychiatric Patients (known for violent behavior), N.Y.	183	1	5.5
	26	Psychiatric Patients (known for "violent, destructive behavior"), N.Y.	183	0	0

SOURCE: After Shah, 1970, p. 8.

NOTE: This list is not meant to be complete or exhaustive.

[a] Height converted to centimeters from inches.

[b] Includes one mosaic—46, XY/47, XYY.

[c] One 48, XXYY is not included in this figure.

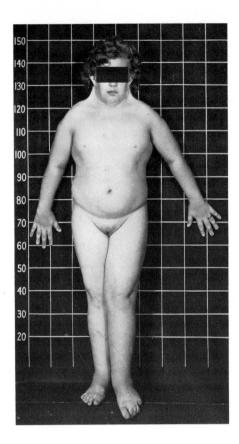

Figure 7.6
Patient affected with Turner's syndrome.
(Courtesy of John Money.)

possessed an extra Y chromosome, 10–20 times the incidence found among newborn males.

Kessler and Moos (1970) have also reviewed the evidence for an association between XYY constitution and criminality and conclude that the facts do not appear necessarily to corroborate the conclusion that XYY males are particularly aggressive. That greater-than-average height might so alter the psychological environment that a tall individual is predisposed to aggressive behavior, however, is not discounted. Regardless of whether the mechanism is social or physiological, the evidence for an association between antisocial behavior and XYY constitution cannot be rejected outright and is too important to ignore.

Sex-chromosome anomalies among phenotypic females have also been reported. *Turner's syndrome* is characterized by gonadal dysgenesis, infantile sexual development, and various physical stigmata including short stature and webbed neck. About 80 percent of Turner's patients are chromatin-*negative* females whose karyotype is 45, X or who are mosaics, having some 45, X cells. A photograph of a Turner's patient is included as Figure 7.6.

Table 7.3

Average IQ's of persons (numbers in parentheses) having various combinations and numbers of sex chromosomes

	0	Y	YY	YYY
0		unlikely to occur		
X	100 (60)	100	76 (6)	80 (1)
XX	100	84 (43)	58 (19)	n.r.[a]
XXX	51 (28)	52 (12)	48 (1)	n.r.
XXXX	40 (3)	35 (22)	n.r.[a]	n.r.
XXXXX	very low (2)	n.r.[a]	n.r.	n.r.

SOURCE: Moor, 1967.
[a]*n.r.* = not reported

Although it was once thought that Turner's patients were below the average in general intelligence, it now appears that the cognitive defect is highly specific. Shaffer (1962) first reported that these patients had low perceptual organization scores, but were nearly normal with respect to full-scale IQ. Subsequent research by Money (1964, 1968) has shown that the most serious deficiency in Turner's patients is in spatial abilities, although a deficiency in arithmetical operations has also been reported.

The incidence of chromatin-negative Turner's syndrome is about 0.03 percent of female births. In contrast to Klinefelter's syndrome, frequencies of this syndrome are not higher than those in the population at large either among individuals institutionalized because of mental defect or among children of older mothers.

Another sex-chromosome anomaly among phenotypic females is that of *triple-X* or *early menopause syndrome*. Although triple-X females are sexually normal and usually fertile, menstrual irregularities and early menopause may occur. The most striking clinical feature, however, is mental retardation. Most triple-X females have a 47, XXX karyotype, although 48, XXXX; 49, XXXXX; and mosaics have also been reported. The incidence is about 0.14 percent in the general newborn population, but 0.39 percent among institutionalized mental defectives. There is little or no evidence that the frequency is higher among children of older mothers.

Vandenberg (1971) has recently published a convenient summary, based on an earlier review by Moor (1967), of the average IQ of individuals with sex-chromosome anomalies. These data are presented in Table 7.3. It should be recognized that problems of ascertainment exist in these data. Most of the individuals with abnormal karyotypes were located in institutions.

Nevertheless, a striking pattern of relationship between average IQ and number of extra X chromosomes is observed, even within the abnormal categories. The first column is based on data from phenotypic females. The first entry in it is the average IQ of Turner's patients and the second, of normal females. Considering only the lower three entries in this column (all abnormal and probably institutionalized), we see that the average IQ clearly decreases with increasing number of extra X chromosomes. The same pattern may be observed in the other three columns, all of which are based upon data from phenotypic males (one Y chromosome is sufficient to result in phenotypic maleness, regardless of the number of X chromosomes in the karyotype). A similar, but less marked, trend across rows may be noted, indicating that extra Y chromosomes result in less IQ deficit than extra X chromosomes.

The finding of an association between IQ and the number of extra X chromosomes indicates that the dosage-compensation mechanism that results in inactivation of extra X chromosomes does not completely neutralize the excess. It is possible that a genetic imbalance during very early embryonic development, prior to inactivation of extra X chromosomes, is sufficient to result in retardation. It is clear, however, that the inactivation of extra X chromosomes has some neutralizing effect on the resulting genic imbalance. As will be shown in the next section, relatively few autosomal trisomies have been found and all of these have been chromosomes from smaller groups. Thus, it seems likely that having extra chromosomes from the larger groups, other than the X chromosome, is lethal. Inactivation of extra X chromosomes, although not sufficient to result in normal development, nonetheless permits survival of the affected zygote.

Autosomal Anomalies.   The first autosomal anomaly in man was discovered in 1959 by Lejeune, Gautier, and Turpin. They reported that patients with *Down's syndrome*, or "mongolism," now sometimes referred to as *trisomy-21*, have 47 chromosomes instead of the normal 46. One of the small chromosomes of group G is present in triplicate, rather than duplicate, yielding the following karyotype: 47, XY or XX, 21+.

Down's syndrome is so common (an incidence of about 0.15 percent of newborns) that its general features should be familiar to everyone. Infants affected with Down's syndrome are usually quiet and uncrying during the early weeks of life; various physical stigmata that they bear (see Figure 7.7) are well known: the presence of an upward and outward slant of the eyelid fissures and epicanthal folds (small folds of skin over the inner corners of the eyes typical of members of the Mongoloid race and thus suggesting the earlier name of the condition); and characteristic features of hand and finger prints. There is a higher-than-average incidence of respiratory infections, heart malformations, and leukemia among Down's patients, resulting in a high mortality rate during the first few months of life. One of the most

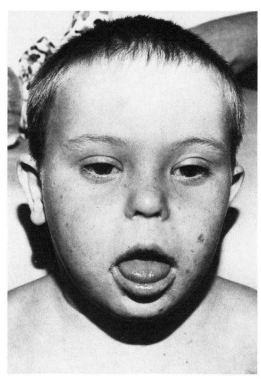

Figure 7.7
Typical patient affected with Down's syndrome.
(Courtesy of George F. Smith, M.D.)

striking features of Down's syndrome, however, is the severe mental defect. Although problems of ascertainment exist, the average IQ among institutionalized patients is only 23; that of mosaics having some Down's-type cells is apparently considerably higher.

Down's syndrome, first described by Langdon Down in 1866, was one of those baffling conditions that defied explanation for many years. Although it was occasionally found to be familial, it was clearly not due to a simple dominant or recessive gene. Its higher incidence among children of older mothers gave rise to many environmental explanations (reproductive exhaustion, and so on), which, since the discovery of the trisomy-21 condition, have now been laid to rest. Standard trisomy-21 is believed to be nondisjunctional in origin (Figure 7.8).

The degree of mental defect in individuals having an extra chromosome 21, a very small chromosome belonging to group G, is more severe than that of individuals with an extra X chromosome (group C). This greater deficit

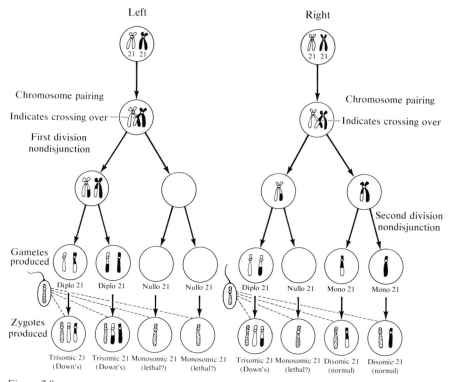

Figure 7.8
Schematic representation of nondisjunction occurring during (left) *the first division and* (right) *the second division of meiosis in standard trisomy-21. (After Penrose and Smith, 1966, p. 121.)*

is apparently due to the at-least-partial inactivation of extra X chromosomes. Thus, although individuals with trisomy-21 carry less extra genetic material, the genic imbalance is actually greater.

The incidence of standard trisomy-21 increases greatly as a function of increasing maternal age. In Table 7.4, the percentage of Down's infants (9,441 cases) born to mothers of various ages is compared to the percentage of normal babies born to women in the same age groups. The ratio of the percentage of Down's births to the percentage of normal births to women in each age group is also indicated (column on the far right). It may be seen that the ratio changes relatively little from group to group until the age of 30–34, when a definite increase is evident; it then rises abruptly in the older categories. This ratio increases more than 36-fold when mothers 19 or younger are compared to those 45 and older.

The percentages in Table 7.4 may also be thought of as probabilities. For example, the probability that the mother of a child with Down's syndrome

Table 7.4

Distribution of Down's syndrome by mothers' ages

Mothers' age	% Down's infants born	% of normal births	% Down's births / % normal births
–19	1.9	4.9	0.39
20–24	10.5	26.1	0.40
25–29	14.5	30.9	0.47
30–34	16.6	22.1	0.75
35–39	27.0	12.0	2.25
40–44	25.2	3.7	6.81
45–	4.3	0.3	14.33
Total	100.0	100.0	1.00
Mean Age	34.43	28.17	—

SOURCE: After Penrose and Smith, 1966, p. 157.

was between the ages of 35 and 39 when this child was born is 0.27. The probability that the mother of this child was 35 or older is equal to $0.270 + 0.252 + 0.043 = 0.565$. In contrast, the probability that the mother of a normal child was 35 or older at the time of birth is only $0.120 + 0.037 + 0.003 = 0.160$. Thus, more than 56 percent of Down's infants are born to mothers 35 or older, although only 16 percent of normal children are born to mothers in this age range. This indicates that the number of Down's infants born would be reduced by more than half if all women completed their childbearing before the age of 35.

About 90 percent of all Down's cases are standard trisomy-21. However, in a small proportion of cases the cause is a translocation. This translocation, usually the reciprocal exchange of a large part of the long arm of chromosome 21 with a small part of the short arm of a chromosome from group D or G, may occur spontaneously or it may be transmitted from one generation to the next. Unlike standard trisomy-21, the incidence of this form of Down's syndrome is independent of maternal age. The origin and consequences of a translocation between chromosomes 21 and 15 are illustrated in Figure 7.9.

Individuals with standard trisomy-21 have a total of 47 chromosomes. However, in translocational Down's syndrome, the number of chromosomes is 46. Although the chromosome number is apparently normal, such individuals have a pair of intact number 21 chromosomes plus a large segment of a third chromosome 21 that has become attached to a nonhomologous chromosome (see zygote 4, Figure 7.9). Zygote 1 is probably inviable due to the absence of a large segment of chromosome 21, resulting in considerable

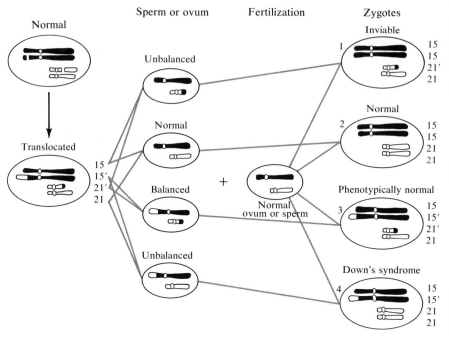

Figure 7.9

Diagram showing suggested origin of 15/21 translocation and its genetic consequences.
(After Polani et al., 1960, p. 723.)

genic imbalance. Zygote 2 is completely normal, whereas zygote 3 is phenotypically normal, but carries the translocation. Since this phenotypically
normal sib of an individual with translocational Down's syndrome has a
greatly increased risk of producing a Down's child, the practice of karyo-
typing all normal sibs in such families would be most useful for the purposes
of genetic counseling.

Evidence (Johnson and Abelson, 1969) obtained from institutionalized
cases indicates that the average IQ of subjects with translocational Down's
syndrome may be somewhat higher than that of standard trisomy-21 subjects. Translocational subjects were also found to be more active, more
aggressive and to have a higher incidence of "problem behavior."

Within a year after the discovery of trisomy-21 in Down's subjects, two
other autosomal trisomies were reported. Edwards et al. (1960) described a
female child with multiple congenital abnormalities whose karyotype was
47, XX, 18+. Patients with *trisomy-18 syndrome*, sometimes referred to now
as *Edwards' syndrome*, usually have a generalized hypertonicity of the
skeletal muscles, resulting in a characteristic flexing of the fingers with the
index finger overlapping the third finger (see Figure 7.10). Other clinical

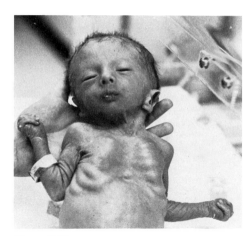

Figure 7.10
Patient affected with trisomy-18 syndrome.
(Courtesy of George F. Smith, M.D.)

Figure 7.11
Patient affected with D-trisomy
(trisomy-13) syndrome. (Courtesy of
George F. Smith, M.D.)

features include micrognathia (small jaw), ear deformities, and severe retardation. The mean survival time of patients with this syndrome is only a little more than three months.

Although the overwhelming majority of patients with Edward's syndrome have primary trisomy, presumably due to nondisjunction, translocational cases may also occur. The probable frequency of the syndrome is about 0.02 percent among newborns. The incidence of this syndrome is somewhat higher among children of older mothers, although the age effect is not as great as for Down's syndrome or Klinefelter's syndrome.

In 1960, Patau and colleagues described a female child with multiple congenital abnormalities whose karyotype was 47, XX, D+. The most striking clinical features of individuals with this *D-trisomy syndrome* are cleft palate and lip and various anomalies of hands and fingers (see Figure 7.11). All patients with this syndrome are severely retarded and the mean life span is only about four months. Recent evidence obtained using quinacrine-mustard fluorescence analysis (Caspersson and co-workers, 1971) suggests that D-trisomy syndrome is due to trisomy for chromosome 13; thus, in the more recent literature, it is sometimes referred to as *trisomy-13 syndrome*.

The majority of patients with D-trisomy syndrome are indeed trisomic. However, a few translocational cases have been reported. The incidence among newborns is about 0.02 percent, and there appears to be a moderate maternal-age effect.

A different kind of chromosomal anomaly in man, a deletion, was reported by Lejeune and co-workers in 1963. Individuals with *cri du chat syndrome* were observed to be missing a part of the short arm of a chromosome in the B group. More recent studies employing banding pattern techniques have revealed that about one-half of the short arm of chromosome 5 is deleted in these individuals. A few families have been described in which a phenotypically normal parent has been found to carry a balanced translocation. Offspring who inherited the deleted chromosome, but not the balanced recipient chromosome, have the syndrome.

One of the most unusual features of *cri du chat* syndrome is the peculiar weak cry that sounds like the mewing of a cat. Other clinical features include microcephaly, downward and outward slant of the eyelid tissue, and widely spaced eyes, resulting in the so-called "moon face" (Figure 7.12). All patients with this syndrome have severe mental retardation, but life span is variable. The incidence is apparently less than that of the three trisomy syndromes previously discussed and apparently there is little maternal-age effect.

The discovery of these anomalies, particularly the finding that Down's syndrome is due to the presence of an extra chromosome, must be regarded as a most important breakthrough in the genetic analysis of behavior. Although several other chromosome anomalies have since been described, it would seem that chromosome analysis in man is still in its infancy. Application of the new banding-pattern techniques will almost certainly result in the identification of new syndromes due to small deletions and translocations. In addition, these methods are particularly well suited for application to automatic machine recognition and classification of chromosomes (Caspersson, Lomakka, and Møller, 1971). The banding patterns may be measured photoelectrically and subjected to computer analysis, which may perhaps some day make karyotype analysis as routine as blood typing. Finally, increased resolution of the morphology of individual chromosomes may reveal sufficient variation that it may be possible to identify the paternal, maternal, and even grandparental origin of each chromosome of an individual. In addition to greatly facilitating linkage studies, such a capability would have enormous potential for application in genetic counseling.

Chromosomal Analysis in Drosophila

In contrast to those of man, the chromosomes of *Drosophila* have been studied in great detail for many years. This is principally due to the presence in the salivary glands of *Drosophila* larvae of giant chromosomes whose structure is thus easy to observe in relatively simple preparations.

Hirsch and his students, pioneers in the application of chromosomal analysis in *Drosophila* to the study of behavior, have investigated positive and negative geotaxis (movement towards and away from gravity) in *Drosophila melanogaster*. A mass screening maze (Figure 7.13) has been developed that efficiently, automatically, and reliably sorts flies on the basis

(a)

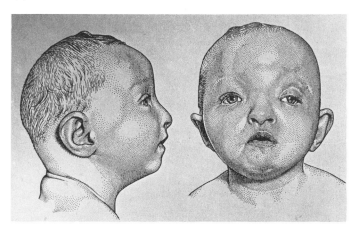

(b)

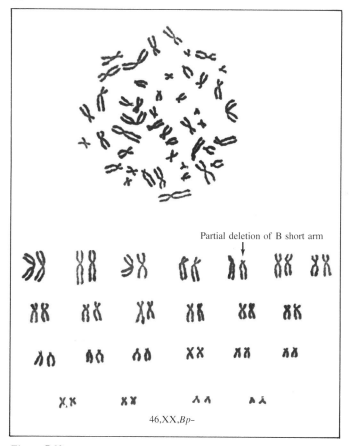

Partial deletion of B short arm

46,XX,Bp–

Figure 7.12
Cri du chat. Top: *Patient.* (*Photograph courtesy of George F. Smith, M.D.*) Bottom: *Karyotype of a patient, showing partial deletion of the short arm of number 5.* (*Courtesy of Arthur Robinson, M.D.*)

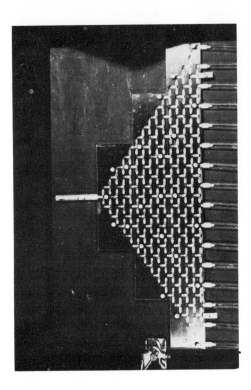

Figure 7.13

Photograph of a 15-unit maze. (From Hirsch and Erlenmeyer-Kimling, "Sign of taxis as a property of the genotype," Science, 134, 835–836, 1961. Copyright © 1961 by the American Association for the Advancement of Science.)

of their behavior. About 200 flies are introduced into the vial at the left and allowed to migrate toward food vials on the right. In each tube, the flies can go up or go down. With a 15-unit maze (16 collection tubes), resulting scores range from −7.5 to +7.5. Subjects that obtain a score of −7.5 climbed upward at each of the 15 choice points, whereas those with a score of +7.5 made 15 positive responses toward gravity. Cone-shaped funnels (Figure 7.14) at each choice point apparently discourage backward movement in the maze and entrance into adjacent rather than forward units.

In one experiment (Hirsch and Erlenmeyer-Kimling, 1962), three populations of *Drosophila* were assayed: a strain selectively bred for positive geotaxis; a strain selectively bred for negative geotaxis; and an unselected control population. These strains were each crossed to a special stock that carried several marker genes and chromosomal inversions. The markers permitted identification of each chromosome in subsequent generations and the inversions were used to maintain the integrity of these chromosomes by effectively suppressing crossing over within the inverted segments.

Drosophila melanogaster have four pairs of chromosomes, three relatively large pairs and a small fourth pair. Only effects of the three large

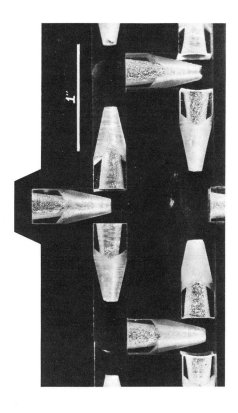

Figure 7.14

Photograph of cones at choice points in maze. (From Hirsch, "Studies in experimental behavior genetics," Journal of Comparative and Physiological Psychology, 1959, 52, 304–308. Copyright 1959 by the American Psychological Association and reproduced by permission.)

chromosomes were assayed in this experiment. The mating design was such that females were produced that were either homozygous or heterozygous for the chromosomes to be assayed. The difference in a geotactic response between flies homozygous and heterozygous for a particular chromosome was then used to infer the cumulative effect of the genes on this chromosome.

The mating system is illustrated in Figure 7.15. Tester females bearing the indicated markers on chromosomes X, II, and III were crossed to males from one of the stocks to be assayed. Since tester females were heterozygous for loci on two intact chromosomes, four (2^2) combinations of F_1 females resulted. Only one type of F_1 female was used for subsequent breeding. F_1 females heterozygous for the dominant markers Bar eyes (B), Curly wings (Cy) and Stubble bristles (Sb) were backcrossed to the tested male. Since these F_1 females were heterozygous for loci on three chromosomes, eight (2^3) chromosomal types were observed among the backcross progeny. These eight combinations are indicated in Table 7.5.

The four genotypes listed in the upper part of Table 7.5 are each heterozygous ($B+$) for the X chromosome to be tested, whereas those below are

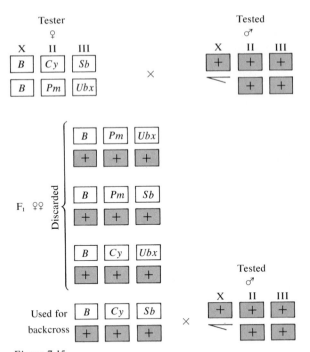

Figure 7.15

*Mating design employed in chromosomal analysis of
geotaxis in* Drosophila. *(After Hirsch and Erlenmeyer-
Kimling, "Studies in experimental behavior genetics,"*
Journal of Comparative and Physiological Psychology, 1962,
55, 732–739. *Copyright 1962 by the American Psychological
Association and reproduced by permission.)*

homozygous (++). An estimate of the average difference between $B+$ and
++ flies may be obtained by multiplying the coefficients listed in the column
designated X by the mean score of females of the indicated genotypes and
dividing the sum of these products by four. In a similar manner, the average
difference between flies heterozygous and homozygous for chromosomes II
and III may be assessed by utilizing the coefficients in columns designated
II and III.

Using these procedures Hirsch and Erlenmeyer-Kimling (1962) estimated the cumulative effects of genes on each of the three major chromosomes of three strains. Ten replications were utilized in the analysis of each
strain, yielding a total of 9,752 subjects in this experiment. The results are
indicated in Table 7.6. In the unselected population, chromosomes X and II
contribute to positive geotaxis, and chromosome III contributes to negative
geotaxis. In the strain selected for positive geotaxis, little change is observed

Table 7.5

Coefficients applied to backcross progeny means to assess effects of individual chromosomes

Backcross progeny	Chromosome		
	X	II	III
X B+ II Cy+ III Sb+	−1	−1	−1
X B+ II Cy+ III ++	−1	−1	+1
X B+ II ++ III Sb+	−1	+1	−1
X B+ II ++ III ++	−1	+1	+1
X ++ II Cy+ III Sb+	+1	−1	−1
X ++ II Cy+ III ++	+1	−1	+1
X ++ II ++ III Sb+	+1	+1	−1
X ++ II ++ III ++	+1	+1	+1

with regard to the effects of chromosomes X and II; however, the sign of III changes from negative to positive. In the strain selected for negative geotaxis, the positive effect of chromosomes X and II is considerably reduced and the negative effect of chromosome III is greatly increased.

Table 7.6

*Estimates of cumulative effects on geotaxis
of genes located on the three major chromosomes
of* Drosophila melanogaster

Population	Chromosomes		
	X	II	III
Positive	1.39	1.81	0.12
Unselected	1.03	1.74	−0.29
Negative	0.47	0.33	−1.08

SOURCE: After Hirsch and Erlenmeyer-Kimling,
"Studies in experimental behavior genetics: IV.
Chromosome analyses for geotaxis," *Journal of
Comparative and Physiological Psychology*, 55,
732–739, 1962. Copyright 1962 by the American
Psychological Association and reproduced by
permission.

The results of this study provide direct evidence for the polygenic determination of this behavior, i.e., genes affecting geotaxis were shown to be
located on each of the three major chromosomes of *Drosophila*. Other
analyses of polygenic characters will be considered in Chapter 9. Before
embarking on this more statistically oriented level of analysis, however, we
will discuss some research in developmental behavioral genetics.

Chapter 8

Developmental behavioral genetics

Among the most intriguing and persistent problems of biology are those associated with the development of multicellular organisms. How does a single cell, the fertilized egg, become elaborated into a highly complex and differentiated organism such as a fruit fly, a mouse, or a man? How do the cells of the various organ systems, all of which contain the same basic genetic information, come to have highly distinctive structures and functions? Part of the explanation for differentiation is attributable to the fact that the biochemical constituents of the cytoplasm of the fertilized egg are arranged in gradients. Thus, after the initial mitotic divisions, the various daughter cells may have quite different concentrations of a variety of substances. Given these early differences, it is easy, in principle, to see that cells can become progressively more chemically differentiated over time. In some cells the absence of a substance will mean that any reactions requiring that substance as a substrate will not be able to proceed, the products of that reaction will be unavailable for subsequent reactions, and so on.

As the number of cells of the developing organism increases, different physical forces come into play on its different parts. For example, cells on the surface of the embryo can participate more directly in gaseous exchange with the environment than can those buried more deeply, providing another basis for chemical differentiation of the various cells. Environmental forces are thus obviously of great significance from the very moment of conception.

Developmental Genetics

That genes are important in directing and regulating developmental processes has been demonstrated. Although the situation is very complex, the data now in hand are sufficient to illustrate the ways in which genes and environment interact and coact in guiding development.

Chromosome Puffs.   In the tissues of the larvae of certain insects, particularly in salivary glands, there exist exceptionally large chromosomes, thought to be many replicated chromosome strands remaining in close association. The visibility of these large chromosomes has made them a favorite material for research, and much basic information about genetics has emerged from their study. One particularly interesting phenomenon that has been observed is the "puffing" of different regions of the chromosomes. These swellings of particular chromosome regions are transient, being present at one time and absent at another. Furthermore, the particular chromosome bands that puff differ from tissue to tissue. Evidence that this puffing is functionally significant is provided by the fact that the puffs show indications of intensive RNA production.

Chromosomal puffing provides an example of the timing of gene action in multicellular animals. Developmental control is revealed in observations on *Chironomus* (a small Dipteran fly) in which the normal molt is found to be accompanied by a particular pattern of puffing on one chromosome followed 30 minutes later by puffing on a different chromosome. The former lasts throughout the molting process; the second persists for two days and then recedes. Insights into the regulation of gene timing are provided by the observation that the insect hormone ecdysone is also found in the larvae at the time of the molt. Injection of ecdysone into a larva not yet old enough to undergo spontaneous molting initiates the same series of chromosomal puffing events as those associated with normal molting.

The Operon Model.   Another line of research that promises to illuminate genetic control of development is the work of Jacob and Monod (1961) on the regulation of protein synthesis. The theory of Jacob and Monod, which has come to be known as the operon model, was based on research on microorganisms, and the extent to which it applies to more complex organisms remains to be established. Nevertheless, it would seem to be a reasonable working hypothesis that this or similar mechanisms function in higher organisms.

The operon model distinguishes several types of genes (see Figure 8.1). *"Structural"* genes are those that specify polypeptides, as discussed in Chapter 6. *"Regulator"* genes produce a substance, called *repressor*, that may bind chemically with certain metabolic products or with a third type of

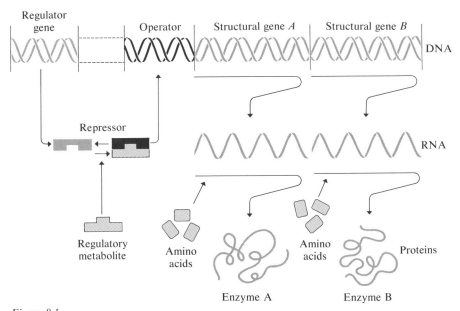

Figure 8.1

The operon model. (From Changeux, "The control of biochemical reactions." Copyright
© *1965 by Scientific American, Inc. All rights reserved.)*

gene, an *"operator"* gene. Operator genes are contiguous to one or more
structural genes, and together, the operator and the structural genes associated with it are termed the *"operon."* If there is none or not enough
regulatory metabolite for the repressor produced by a particular regulator
gene, then the repressor will bind with operator. The operator appears to be
a critical starting point for transcription of RNA from the DNA of the
structural genes, so that binding of repressor to operator prevents the
synthesis of any of the enzymes coded for in the DNA of the structural
genes of that particular operon. If the appropriate regulatory metabolite
becomes available, either through exogenous sources or as an end product
of the functioning of some other operon, then repressor will bind preferentially to this regulatory metabolite, releasing operator and permitting the
synthesis of all of the enzymes of the operon to proceed. In effect, presence
of the regulatory metabolite turns an operon on, and its absence turns an
operon off.

Interactions among operons can provide mechanisms for relatively permanent changes in gene functioning. Figure 8.2 shows two operons, in each of
which one of the structural genes is in fact a regulatory gene (RG) for the
other operon. Thus, if operon 1 is active, the product of RG-1 will turn

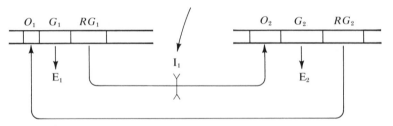

Figure 8.2

*An operon circuit that would switch the production of one enzyme off and
another on. (From Srb, Owen, and Edgar, General Genetics, 2nd Ed.
W. H. Freeman and Company. Copyright © 1965, p. 364.)*

operon 2 off. If some inducer from outside the system, (I_1), a regulatory
metabolite for the repressor of RG-1, appears even briefly, operon 2 can
begin to produce protein, including the product of RG-2, which switches off
operon 1. Operon 2 is now locked on and operon 1 is locked off until such
time as an inducer for operon 2 might appear. Far more complex types of
interactions could be postulated, but the foregoing should make it clear that
operon-type systems could accomplish much of the turning on and off of
genes in developmental sequence and the selective turning on of genes in
different tissues that must accompany developmental processes.

Quantitative Models. The example of chromosome puffing clearly suggests that genes may be turned on and off in a particular tissue or may be
active in one tissue but not in others, and the operon model shows how the
turning on or off might be accomplished. However, for complicated situations that must exist when many genes are involved, sequentially or simultaneously, in guiding development, a quantitative approach, parallel to the
quantitative formulations necessary for dealing with polygenic inheritance
discussed in Chapter 9, seems to be required. The analysis is complicated by
the fact that the relevant phenotype is not static, measured only once in the
organism's life, but dynamic. Waddington, who has made particularly
important contributions to this area, regards the basic elements of development to be stabilized or buffered pathways of change, which he calls
"creodes," a term derived from Greek words meaning necessary path. The
development of a polygenically determined system may be regarded as the
progression of a point through n-dimensional space. The creode is the trajectory of this point toward some predetermined endpoint. A creode is so
constituted that when deviations from the trajectory result from some exter-
nally applied disturbance, there is a tendency for the system not to adopt a
new path to the old endpoint but to return to the original pathway leading
toward the endpoint. Empirical evidence for this sort of "catch up" process

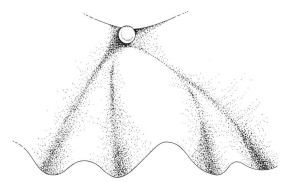

Figure 8.3

Waddington's "epigenetic landscape" — a hypothetical model describing gene-environment interaction in development. For interpretation, see text. (After Waddington, The Strategy of the Genes. Copyright © 1957 by Macmillan Publishing Company. p. 29.)

from children whose growth has been interrupted by illness of one kind or another has been provided by Tanner (1963).

Complex mathematical formulations have been constructed in attempts to analyze the great variety and complexity of feedback loops that must be involved in the buffering of a creode (see Waddington, 1962). Further discussion of these formulations is beyond the scope of the present book, but a three-dimensional analogy provided by Waddington does serve to illustrate some of the essential points. Figure 8.3 portrays an "epigenetic landscape," which Waddington has utilized to illustrate the development of organ systems. For our present purposes, we may understand the contour of the landscape to be determined by the individual's genotype, and the location of the ball to represent the value of the phenotype in question. During development, which is represented in this analogical model by the ball rolling forward, environmental forces may act to displace the ball laterally from its path. Depending upon the steepness of the walls of the valley in which the ball is located, it will return more or less quickly after such a deviation to its former path. There exist certain critical periods when a lateral excursion will direct the ball into one of two or more alternative pathways which represent distinct pathways of development. Genetically determined individual differences in susceptibility to environmental forces could be represented in this model by varying widths of the valley floor and steepness of its sides, and by different points of bifurcation of the valleys. Waddington (1957) warns that this model is meant to be suggestive only and is not to be interpreted in any literal sense. However, it obviously serves as a convenient heuristic model for developmental processes, and makes explicit the interaction between genes and environment during development.

Genetics of Behavioral Development: Animal Research

Two general approaches have been taken in exploring the genetics of the development of behavioral patterns in experimental animals. The first of these is essentially descriptive. The developmental sequence of some particular behavior is described in two or more groups of differing genotype reared under standard environmental conditions. The second approach is experimental. Within each of two or more genetically distinct groups, one subgroup is administered an environmental treatment of some type and another subgroup is reared normally. Group differences in response to the environmental treatment provide evidence for genotypic variation in developmental susceptibility. This is equivalent to testing for the presence of genotype-environment interaction (a concept discussed in some detail in Chapter 9), in which a time delay between administration of the treatment and measurement of the behavior makes it possible to implicate developmental processes occurring during the interval.

Pattern of Development as a Function of the Genotype.   An example of the descriptive approach is provided by Thiessen (1965) who compared the development of several behaviors in mice homozygous for the recessive gene wabbler-lethal to that of heterozygous siblings. The condition of the homozygous recessive animals involves degeneration of myelin of the central nervous system. At about 14 days of age, incoordination of the hind limbs and paralysis are displayed, and the animals die before reaching the age of reproduction. Not surprisingly, the affected animals are less active in an arena-type situation, and are less able to rise on the hind legs or to climb. Interestingly, the depression in arena activity is noticeable as early as day 10, prior to the average onset of signs of gross motor incoordination and of degenerating myelin. Thiessen proposes that this behavioral effect perhaps reflects some biochemical processes that later participate in demyelination.

Other studies have concentrated on specific behaviors for which the genetic basis is not so well known. Schlesinger et al. (1965), for example, have established that a very sharp peak in seizure susceptibility to audiogenic stimulation of DBA/2 mice occurs at 21 days of age. That the brain serotonin and norepinephrine of DBA/2 animals differs from C57BL mice only at this age strongly suggests a causal relationship. In investigation of other inbred strains, Fuller and Sjursen (1967) have described a variety of different patterns of development of convulsive risk and of death risk shown by animals exposed to audiogenic stimulation.

Learning performance has also been examined from a developmental genetic perspective. Meier and Foshee (1963) and Meier (1964) showed not only that mouse strains differed in performance at a particular age, but also that there were widely divergent developmental functions. Other results on learning were obtained in rats by McGaugh and Cole (1965), who demonstrated that both maze-bright and maze-dull rats performed better under distributed practice than under massed practice when tested at 142 to 164 days of age. However, although the maze-bright animals also showed this improvement at 29 to 33 days of age, distribution of practice had no effect upon maze-dull animals of that age.

These studies illustrate well that genotype can influence behavioral development, but descriptive studies are limited in the type of information they provide. Approaches to understanding the details of the genotype-environment interaction and the mechanisms involved require the use of experimentation.

Differential Effects of Environmental Variables on Later Behavior: Postnatal. In the experimental approach, both the duration of the environmental probe and its timing during the ontogeny of the individual have varied. We may first consider those procedures administered after the subjects have been born.

In some of these experiments, general circumstances are changed in which the specific environmental features that mediate an effect, if one is found, are not specifiable. In foster rearing, for example, the fostered young are presented with a wide array of circumstances different from the control animals, ranging from quality and quantity of milk to olfactory stimuli to the complex behavior of the foster mother or foster parents. In a number of investigations, it has been shown that cross-fostering has no effect on a genetically influenced difference in behavior. For example, Ginsburg and Allee (1942) found cross-fostering to have no effect upon aggression in inbred mice and Broadhurst (1961) found maternal effect on open-field emotionality of inbred rats to be negligible. Other studies, however, have revealed such an effect. Ressler (1963), for example, conducted an experiment on the effects of cross-fostering on subsequent behavior of C57BL/10 and BALB/c mice. All animals were reared by foster parents; half were reared by foster parents of the same strain, and the other half by foster parents of the other strain. The behavioral testing apparatus was a small compartment in which the animals were individually placed at 60 days of age. A panel at one end of this compartment activated a switch that, at the experimenter's discretion, could turn on a light in the compartment. Each mouse was tested for one-half hour. During the first 15 minutes pressing the panel had no effect, but during the second 15-minute period a panel press resulted in one second of illumination. Thus, both a "manipulation" and a "visual exploration" score were obtained. There was no relation between the manipulation scores obtained and strain of foster mother; BALB/c mice received higher scores than C57BL/10 mice regardless of maternal environment. For visual exploration scores, however, pups reared by BALB/c mothers, regardless of whether they themselves were C57BL/10 or BALB/c, received higher scores than those reared by C57BL/10 mothers.

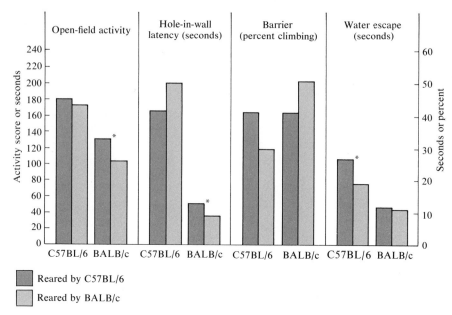

Figure 8.4

Scores on activity measures of C57BL/6 and BALB/c mice foster-reared by mothers of their own or the other strain. Open-field-activity and hole-in-wall-latency results should be read from the left scale. Barrier and water-escape results should be read from the right scale. Asterisk indicates significant difference (p < 0.05) between groups of animals of same strain under different conditions of rearing. (After Reading, "Effect of maternal environment on the behavior of inbred mice," Journal of Comparative and Physiological Psychology, 1966, 62, 437–440. Copyright © 1966 by the American Psychological Association and reproduced by permission.)

Reading (1966) compared C57BL/6 and BALB/c animals in a fostering experiment similar to that just described. Between 7 and 10 weeks of age, the pups were tested on a battery of activity measuring devices, including an open-field test, a hole-in-wall emergence test, a water-escape task, and a barrier-climbing task. Strain differences are apparent in all measurements except that for climbing the barrier (see Figure 8.4). The C57BL/6 mice were more active in the open field, took longer to emerge from the hole-in-wall and required longer to escape from the water trough than did BALB/c mice, regardless of mode of rearing. However, significant effects of fostering were apparent for the BALB/c mice in open-field activity and hole-in-wall performance. In each case, those BALB/c pups who were reared by C57BL/6 mothers differed from their controls in the direction of the C57BL/6 strain's typical performance. Likewise, the water-escape performance of C57BL/6 mice was affected by fostering.

These results, in addition to showing that some traits in some strains may be susceptible to the general influence of foster rearing, and others not, raise an interesting point about the subtleties of separating genetic and environmental influences. Under standard rearing conditions, differences between inbred strains are attributed to their genetic differences. If cross-fostering reduces or eliminates such a difference, then part of the initially observed difference between the strains is ascribable to maternal behavior. This does not reduce the importance of genetics in determining the difference between strains; it simply shows that some of those genetic differences are mediated through the maternal behavior. From the viewpoint of the individual young, maternal behavior is an environmental effect, but that effect is a consequence of the mother's genotype.

More specific and delineated environmental differences have been employed in other investigations. Freedman (1958) established "indulgent" and "disciplinary" rearing regimes for dogs of four breeds (shetland sheep dogs, basenjis, beagles, and wire-haired fox terriers). Under the indulgent regime, pups were never punished and were encouraged to undertake a variety of activities. Under the disciplinary regime, pups were restrained and were taught to obey a series of commands. These conditions persisted from the third to the seventh week of age, at which time each animal was observed in a standardized test situation. Food was placed in a bowl in the center of a room. For three minutes the animal was prevented from eating by the experimenter's swatting it on the hind quarters and shouting "no." Thereupon, the experimenter left the room and the time until the animal began to eat was recorded. Both beagles and terriers showed a marked effect of rearing mode, with the disciplined animals beginning to eat much sooner than the indulged animals. For both basenjis and shetlands the mode of rearing had no effect, but for quite different reasons. The basenjis ate soon after the man left the room regardless of mode of rearing; the shetlands of both groups, on the other hand, were so affected by punishment in the test situation that they never did approach the food.

Another study on dogs has been reported by Fuller (1967) who studied the effects of experiential deprivation on later behavior of wire-haired terriers and beagles. Responses to a human handler, to toys, and to another puppy were compared for animals that had been reared in isolation and those reared as pets. The results showed that genotype can be an important determinant of the direction, the duration, and the intensity of the influence of the isolation period. All of these effects are to be seen in the data of Figure 8.5, which shows results for an index of activity displayed by the animals in the arena testing situation.

Another approach to assessment of the developmental effects of experiential diversity has been made by Henderson (1970) who compared mice of various inbred strains and their F_1's for brain-weight increases resulting from environmental enrichment. Control animals were reared in standard

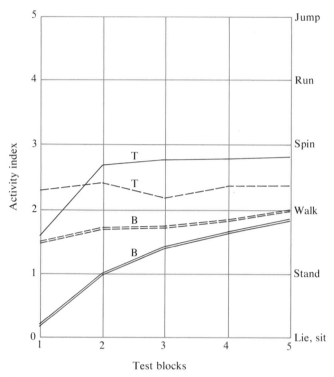

Figure 8.5

Average activity scores for terriers (T) and beagles (B) reared in isolation (solid lines) and as pets (dashed lines). Each block includes four tests given during one week (at ages 16–20 weeks). The descriptive terms to the right provide a guide to the actual observations. However, a mean value of two indicates a mixture of standing, walking, jumping, etc., and not just continuous walking. (From Fuller, "Experiential deprivation and later behavior," Science, 158, 1645–1653, 1967. Copyright © 1967 by the American Association for the Advancement of Science.)

laboratory cages, while the experimental animals lived in cages which provided an enriched environment in the form of objects for climbing and exploring. In brief, the results were that the hybrid animals, on the average, showed a greater percentage increment in brain weight as a result of the enriched environment during development than did the animals of the inbred strains. Henderson interprets these results as indicating that this character, the capacity to respond to environmental enrichment by increasing brain size, has been subjected to selection pressure, and is therefore closely related to fitness. (See Chapters 9 and 10 for a discussion of heterosis, inbreeding depression, and fitness.)

Another examination of the effects of enriched environment was conducted by Cooper and Zubek (1958) who worked with rats selectively bred for maze brightness and maze dullness in a Hebb-Williams graded difficulty task. Enrichment of environment had no effect upon maze-bright rats, but maze-dull animals reared under enriched conditions were substantially superior to maze-dull rats reared under standard conditions. On the other hand, an impoverished environment was extremely detrimental to maze-bright rats, but had little effect on the maze-dull ones.

Experimental interventions of shorter duration have been employed by Lindzey and his colleagues (Lindzey, Lykken, and Winston, 1960; Lindzey, Winston, and Manosevitz, 1963) in studies of the effects on mice of the trauma of a loud, high-frequency sound administered for two minutes daily from the fourth to the seventh day of age. Beginning at 30 days of age, the mice were tested in open-field and emergence-test situations, designed to assess emotionality, activity, and timidity. The open-field test was repeated at 100 days of age. The results showed the C57BL mice to be particularly sensitive to noxious infantile stimulation of this type, relative to the other strains (C3H, DBA/8, and JK) tested.

The C57BL strain also showed itself to be labile in another study by Lindzey and Winston (1962) in which the effects of systematic handling ("gentling") on maze performance were investigated. The handled C57BL mice were superior to their controls, but the treatment had almost no effect on C3H animals.

Differential Effects of Environmental Variables: Prenatal. It has been clear for some time that various hormones, drugs, and other substances can have profound effects on the physiology and morphology of a developing fetus. In addition, some evidence has suggested that severe emotional stress undergone by females during pregnancy may affect the behavior of their offspring. In order to test this hypothesis, Thompson (1957) mated female rats that had been trained to avoid electrical shocks in a double-compartment shuttlebox and then exposed them three times each day during pregnancy to the shock side of the shuttlebox, but with the door to the escape area locked. The offspring of these females were found to be significantly less active than controls in open-field tests administered at both 30 and 130 days of age and to have higher latency scores. Low activity and high latency were assumed to be indices of high emotionality; thus, it was concluded that the maternal anxiety generated in the shuttlebox situation had increased the emotionality of the offspring.

Other environmental variables manipulated during prenatal development have also been found to influence adult behavior. In addition to handling (Ader and Conklin, 1963) and crowding (Keeley, 1962), maternal adrenalin injections have been found to modify offspring behavior (Thompson, Watson and Charlesworth, 1962). In 1961, Thompson and Olian reported the

Table 8.1

Effects of prenatal maternal stress on offspring open-field behavior in mice

Strain	Mean daily activity in feet		Numbers	
	Experimental	Control	Experimental	Control
BALB/cJ	4.8	3.6	39	54
C57BL/6J	30.0	36.5	66	65

SOURCE: After Weir and DeFries, "Prenatal maternal influence on behavior in mice: Evidence of a genetic basis," *Journal of Comparative and Physiological Psychology,* 58, 412–417, 1964. Copyright 1964 by the American Psychological Association and reproduced by permission.

results of the first study of the differential effects of prenatal maternal stress on offspring behavior of genetically distinct groups. Maternal adrenalin injections were found to increase open-field activity of a low-activity strain of mice (A/J) and to decrease activity of a high-activity strain (C57BL/6). However, no significant effect was found with a strain of intermediate activity (BALB/Ci).

DeFries and colleagues (1964, 1967) have conducted a series of researches in this area. The primary objective of the first study was to test the hypothesis that the response to prenatal behavioral stress is a function of genotype. Female BALB/c and C57BL/6 mice were the mothers; half were subjected daily to stress during about the latter half of pregnancy and the other half served as controls. The treated mice were subjected to stress by being forced to swim for three minutes, by being placed for three minutes in a brightly lighted tilt box in which a loud tone occurred at random intervals for about one-half of the period, and by being placed for two minutes in a brightly lighted open field. Upon littering, treatment was terminated. Beginning at 40 days of age, the offspring were subjected to open-field tests on each of five successive nights. The resulting data, pooled across sex and day of test, are summarized in Table 8.1.

Of major interest to the original purpose of this study is the interaction between strain and treatment, significant at the 0.001 level of probability. As may be seen from this table, offspring of treated BALB/c females were somewhat more active than controls, whereas offspring of treated C57BL/6 females were less active than controls. Thus, the differential response is not just in terms of the magnitude of the effect; the response of the two strains is in opposite directions.

The results of this study support the hypothesis that the response to prenatal stress is a function of genotype—but genotype of what? The primary objective of the next study was to assess the role of the genotype of

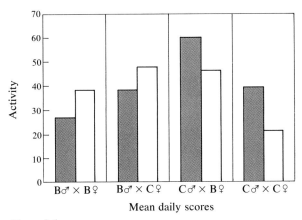

Figure 8.6

Mean daily open-field activity scores of treated (shaded bars) and control (unshaded bars) offspring of two inbred strains of mice and of reciprocal crosses between them. The parental strains employed were BALB/cCrgl *and* C57BL/Crgl, *denoted* B *and* C, *respectively. (After DeFries, 1964, p. 291.)*

the fetus versus that of the mother in the differential response in offspring to prenatal stress. If the response to prenatal stress were solely a function of the genotype of the mother, then both hybrid offspring and inbred offspring of mothers of the same genotype would be expected to respond alike. However, if the response were solely a function of the genotype of the fetus, all hybrid progeny having the same genotype should respond alike, regardless of the strain of the female parent. Stress was applied daily to pregnant females as in the earlier study, beginning 10 days after mating, and continuing throughout the remainder of the gestation period, which is approximately 20 days in the mouse. Testing was conducted in a different laboratory, however. Different substrains of the BALB/c and C57BL strains were employed, and a modified version of the open-field test was administered. The offspring were administered open-field tests on two consecutive days, beginning at 48 days of age. The results of this study, pooled across sex and day of test, are summarized in Figure 8.6.

It may be seen from this illustration that the strains employed behaved quite differently from those in the previous study. In fact, in this study, BALB/c offspring were somewhat more active than C57BL offspring, at least under control conditions. This difference is perhaps due to the different substrains and conditions employed. Nevertheless, a differential response is again apparent; prenatal stress reduced open-field activity of one strain and increased that of the other.

It may also be seen from Figure 8.6 that hybrids manifest considerable heterosis and that their response to treatment is most interesting. Hybrids do not respond like inbreds carried by mothers of the same strain and those carried by mothers of different strains respond in opposite directions; thus, the results of this study suggest that response to prenatal stress is neither solely a function of fetal genotype nor of maternal genotype. Instead, it appears that females of the different strains respond differently to being subjected to stress and that the direction of the resulting effect on the developing fetus depends upon whether it is inbred or hybrid.

Ovary Transplantation.   The technique of transplanting ovaries may be used to assess the relative importance of prenatal and postnatal maternal effects. Ovaries from inbred donors may be transplanted into hybrid recipients and, by appropriate matings, it is possible to obtain inbred offspring of a particular strain that have been carried by females of different genotypes (inbred versus hybrid) and inbred offspring of different genotypes that have been carried by females of the same strain (hybrid).

As an example of this approach, ovaries from donors of each of two inbred strains of mice (BALB/c and C57BL/6) were transplanted bilaterally into F_1 hybrid females. The mating system was such that inbred progeny were carried by either hybrid or inbred mothers. Ten days after surgery, hybrid females carrying transplanted ovaries from BALB/c donors were mated to inbred BALB males and hybrid females bearing C57BL/6 ovaries were mated to inbred C57BL/6 males. Offspring of these matings were designated B/H and C/H, respectively. Within-strain matings of inbred males to unoperated inbred females produced control offspring, designated B/B and C/C. At 40 ± 5 days of age, all offspring from first litters were tested in an open field. Body weight in grams was also recorded. The resulting mean transformed open-field activity and defecation scores and body-weight data from 29 litters produced by hybrid mothers and 32 litters from inbred mothers are summarized in Table 8.2.

Comparison of B/B versus B/H and C/C versus C/H indicates little or no evidence for the presence of maternal effects on open-field behavior, but does demonstrate the effect of the maternal environment on body weight, a character previously known to be susceptible to such effects. Comparison of B/H versus C/H indicates the importance of the genotype of the offspring for open-field behavior. Since no cross-fostering was employed in this study, it was not possible to determine the relative influence of prenatal versus postnatal effects on body weight.

Genetics of Behavioral Development: Human Research

It has already been noted that dietary treatment can ameliorate some of the effects of phenylketonuria, if initiated promptly enough. This necessity of early treatment has obvious developmental implications. Further insights into the nature of the developmental process have been gained from the study of children of phenylketonuric and hyperphenylalanemic mothers.

Table 8.2

*Mean open-field behavioral scores and body weight of
inbred progeny carried by inbred BALB/c (B),
C57BL/6 (C), or hybrid (H) mothers; B/H symbolizes
inbred BALB/c offspring carried by hybrid mothers, etc.*

Measurement	B/B	B/H	C/H	C/C
Activity	4.08	4.30	16.02	15.86
Defecation	2.71	2.95	1.09	1.17
Body Weight	16.22	18.24	18.93	16.56
N	72	79	39	71

SOURCE: From DeFries et al., 1967, p. 208.

The findings are likely to be of practical import in the near future, because in a few years many women homozygous for the recessive allele for phenylketonuria who have been successfully treated will be entering reproductive age. If their dietary treatment has at that time been terminated, as will be the case for many or most of them, the hazards to their children of developing in biochemically abnormal uterine conditions will become a matter of issue.

Hsia (1970) reviewed the cases that had been reported through 1969. Ninety-four children were born to 28 identified untreated phenylketonuric mothers. Of these 94 offspring, 8 were themselves phenylketonuric, 79 were nonphenylketonuric, and there were 7 for whom classification was uncertain. Of particular concern is the fate of the nonphenylketonuric children, all of whom must be heterozygotes for the phenylketonuria allele and would thus be normal in normal uterine circumstances. Briefly, these nonphenylketonuric children were characterized by retardation of growth, an increased incidence of congenital physical malformations and a greatly increased risk of mental retardation. Of the 79, 7 died too young for evaluation or before data on them were collected. Of the remaining 72 children, 61 are mentally retarded. Of the 11 presumably normal, evidence is rather incomplete on 9. Hsia (1970) concludes, therefore, that practically all heterozygote offspring of phenylketonuric mothers become mentally retarded. These results are particularly meaningful when compared with the results of offspring of hyperphenylalanemic mothers. Of a total of 19 such children, 2 had phenylketonuria, 1 had hyperphenylalanemia, and 16 had neither of these biochemical anomalies. Fifteen of these 16 children are mentally normal. Hsia (1970) proposes that the fetal brain is harmed by concentrations above a certain level of phenylalanine or its metabolites. The biochemical milieu of the fetus of an untreated phenylketonuric mother is presumably more severe than that of the fetus of a hyperphenylalanemic mother. The former is evidently above the threshold of critical damage and the latter condition below that threshold. The period of susceptibility evidently ends sometime beyond the

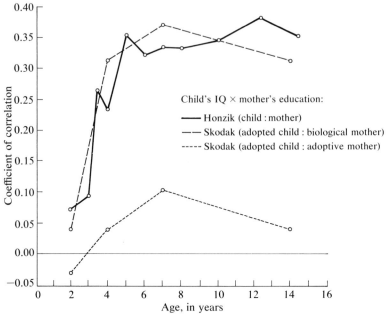

Figure 8.7

Coefficients of correlation at different ages between child's IQ and educational level of own or of foster mother. (After Honzik, "Developmental studies of parent-child resemblances in intelligence." Child Development, 28, 215–228, 1957. The Society for Research in Child Development, Inc.)

age of 3 years, thus making it possible to terminate the special dietary treatment.

In addition to these examples of abnormal conditions, it is possible to cite data concerning normal range of variation in intelligence. Particularly relevant are results of some long-term studies on adoption. Skodak and Skeels (1949) found that, although the mean IQ of adopted children was considerably higher than that of their biological parents, a substantial biological-parent–child correlation was nonetheless present, exceeding the correlation of adopted children with adoptive parents. The magnitude of this correlation was very low at two years of age (perhaps due to lack of test reliability at young ages), but increased rapidly to about 0.35 by six years of age. These correlations on adopted children were compared by Honzik (1957) to those she had obtained using children reared normally by their own mothers (Figure 8.7), where educational level is used as an index of mental ability of the mothers. It may be seen from Figure 8.7 that the IQ's of adopted children correlate as highly with educational levels of their biological mothers as IQ's of "own" children correlate with educational levels of their mothers.

Of special developmental interest is the gradual increase in resemblance between offspring and biological parent. The maximum correlation attained (about 0.35) is lower than that empirically obtained in studies of parent-offspring resemblance in intelligence, but it should be noted that, in this case, mothers' IQ scores were not utilized. The use of educational level as an indirect index of maternal IQ could easily account for this apparent reduction in degree of resemblance of parent and offspring. The resemblance of adopted child and foster mother also rises with age of child, but only a very modest degree of resemblance is ultimately attained. Hindley (1961) found the correlation between a child's IQ and his biological mother's education to rise from 0.26 at 6 months of age to 0.43 at 5 years of age; the correlation with father's educational level rose from 0.18 to 0.40. Interpretation of these results is clouded somewhat by the difficulties associated with questions of reliability and validity of the tests for intelligence employable with very young children. Yet it seems plausible that the increasing similarity between parent and child may be the result of progressive switching on of more and more genes that are involved in cognitive functioning.

An approach to the genetics of behavioral development using the twin method has recently been reported by Wilson (1972). Twins were participants in a longitudinal study of growth and development conducted by the Louisville Twin Study. At 3, 6, 9, 12, 18, and 24 months of age, members of each twin pair were tested on a research version of the Bayley scales of mental and motor development. The sample size and intraclass correlation for monozygotic (MZ) and dizygotic (DZ) pairs on the mental scale scores are presented in Table 8.3. At each age, the MZ correlation is higher than the DZ correlation, as well as being almost as high as the estimated reliability of the test (0.88 to 0.94). In contrast, the correlations between test scores obtained by the same individuals at different ages are much lower, ranging from 0.53 to 0.08. Thus, the score of one member of a twin pair at a given age is a better predictor of his co-twin's score than an earlier score of the co-twin itself. This was even true for DZ twins. It seems that developmental processes during infancy produce greater age-to-age changes for a given child than do the combined effects of heredity and environment within the average DZ twin pair at a given age. These results suggest that twin pairs follow the same pattern of mental development across ages as reflected by similar score profiles. Sample profiles for MZ twins are presented in Figure 8.8.

The score profile may be described in terms of contour and over-all elevation. The contour is a function of age-to-age changes in precocity (spurt-lag factor), whereas the overall elevation is more a reflection of enduring developmental maturity. The similarity for members of MZ and DZ twin pairs with respect to over-all level and profile contour were described by intraclass correlations (repeated-measures analysis adapted for use with twin data). As indicated in Table 8.4, MZ correlations based upon

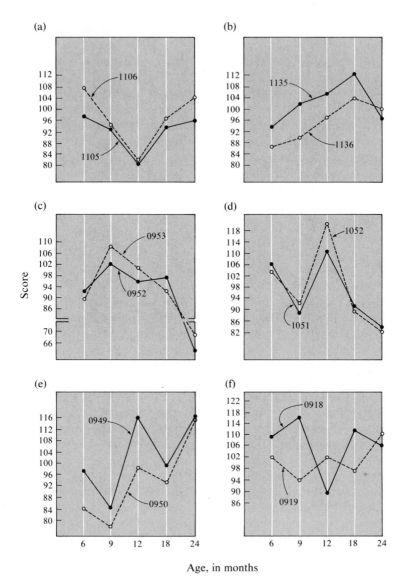

Figure 8.8

*Profiles of mental development scores for MZ twins at ages 6 through 24
months. The pairs in (a) through (e) exhibit moderate to high profile
congruence; the pair in (f) is obviously noncongruent. (From Wilson, "Twins:
early mental development," Science, 175, 914–917, 1972. Copyright © 1972
by the American Association for the Advancement of Science.)*

Table 8.3

Within-pair correlations of MZ and DZ twins on Bayley mental scale scores

Age in months	Number of pairs		Intraclass correlation	
	MZ	DZ	MZ	DZ
3	71	79	0.84[a]	0.67
6	85	98	.82	.74
9	82	101	.81[a]	.69
12	86	104	.82[a]	.61
18	88	91	.76	.72
24	57	77	0.87[a]	0.75

SOURCE: From Wilson, "Twins: early mental development," *Science,* 175, 914–917, 1972. Copyright © 1972 by the American Association for the Advancement of Science.
[a]MZ correlation significantly higher ($p < 0.05$) than DZ correlation.

Table 8.4

MZ and DZ correlations for aspects of Bayley mental-score profiles

Age in months	Source	Intraclass Correlation	Test for MZ > DZ
			$p <$
3, 6, 9, and 12	Overall Level:		
	MZ Pairs	0.90	0.01
	DZ Pairs	.75	
	Profile Contour:		
	MZ Pairs	.75	.01
	DZ Pairs	.50	
12, 18, and 24	Overall Level:		
	MZ Pairs	.89	.05
	DZ Pairs	.79	
	Profile Contour:		
	MZ Pairs	.67	0.05
	DZ Pairs	0.52	

SOURCE: From Wilson, "Twins: early mental development," *Science,* 175, 914–917, 1972. Copyright © 1972 by the American Association for the Advancement of Science.

over-all level indicate a very high level of concordance within each year. Combining scores across ages yielded an intraclass correlation equal to the estimated test reliability. This in fact might represent another way of obtaining reliability measures, i.e., correlating scores of individuals with identical genotypes and similar environments (raised in the same home and tested at the same age). Concordance for over-all level of DZ pairs is also high, but significantly less than that for MZ pairs. A similar pattern emerges when profile contour is considered. The significantly higher concordance of MZ than DZ pairs indicates that these spurts and lags during development are at least to some degree under genetic control.

In a very genuine sense, any demonstration of genetic influence on a behavioral trait implies genetic influence over the developmental processes that are pertinent to the trait. To acknowledge this fact, however, does not contribute materially to understanding the dynamics of genotype-environment interaction in behavioral development. In this chapter, it has been shown that suitable theoretical perspectives are available and that a respectable body of empirical data is already at hand. These data demonstrate the potential of research in the area, although it is fair to say that developmental behavioral genetics is in its early stages. Evidence of increasing interest in the area makes it likely that rapid advances will be made in the future.

Chapter 9

Population genetics and behavior

Population genetics is concerned with the transmission of characters within groups of breeding organisms, with the genetic structure of these populations, and with the forces that influence this structure. The characters studied by population geneticists may be either qualitative, i.e., readily classifiable into discrete categories, or quantitative. Qualitative characters may be subjected to the analytical methods outlined in Chapters 4 and 5. Quantitative characters, however, are influenced by genes at many loci and by environmental factors. Since effects of individual genes in *polygenic* systems are not readily discernible, alternative methods are required for the analysis of these more complex characters.

The concepts and methods introduced in this chapter are largely mathematical in nature and, for the sake of completeness, a number of derivations have been included. It is recognized, however, that these derivations are not essential for introductory-level comprehension of the basic concepts. Thus, as indicated later in this chapter, several sections may be conveniently skipped by the beginning student without undue loss of continuity.

The Hardy-Weinberg-Castle equilibrium was introduced in Chapter 4 and it was demonstrated that gene and genotype frequencies in populations are stable in the absence of opposing forces. In this chapter, the consequences of relaxing the assumptions of no migration, no mutation, and no selection will be examined, followed by discussions of quantitative genetic analysis and systems of mating in populations.

Forces That Change Gene Frequency

The frequency of genes in populations may be changed by migration, muta-tion, selection, and even by chance. If a population is small, the random sampling of gametes may lead, purely by chance, to changes in gene fre-quency. A clear exposition of the complexities of *random drift* in gene frequency from one generation to the next has been provided by Wallace (1968). Only the systematic processes that change gene frequency will be discussed here.

Migration.   Assume that in a population the frequency of some gene is q_0. How will the frequency of this gene change if a number of individuals from another population immigrate and become incorporated into this popu-lation? If the population then consists of a proportion m of immigrants and $1 - m$ natives, the new gene frequency (q_1) will be:

$$q_1 = mq_m + (1 - m)q_0 = m(q_m - q_0) + q_0,$$

where q_m is the frequency of the same gene among the immigrants. Thus, the change in gene frequency (Δq) as a function of migration is:

$$\Delta q = q_1 - q_0 = m(q_m - q_0),$$

i.e., change in gene frequency is a function of the rate of migration and the difference between the gene frequencies of the immigrant and native popu-lations. For example, assume a rate of migration of 10 percent, where the frequency of some gene is 0.20 in the natives and 0.30 in the immigrants. Thus,

$$\Delta q = 0.10(0.30 - 0.20) = 0.01.$$

Although this change in gene frequency may appear small, sustained immi-gration of this magnitude over many generations could have a substantial effect on gene frequency.

In a similar manner, it may be shown that the change in gene frequency as a result of one generation of selective emigration is as follows:

$$\Delta q = n(q_r - q_n),$$

where n is the proportion of the population which leaves, i.e., the rate of emigration, q_r is the frequency of some gene in those individuals that remain in the population, and q_n is the frequency of the same gene in those individ-uals that emigrate.

Mutation.   Let us now consider the change in gene frequency that may be brought about by mutation. Assume that there are two alleles at an auto-somal locus and that A_1 mutates to A_2 with frequency u per generation and

Table 9.1

Relative fitness of genotypes for three different cases of selection

Item	Genotype		
	A_1A_1	A_1A_2	A_2A_2
Frequency	p^2	2pq	q^2
Relative fitness			
A_1 completely dominant, selection against A_2A_2	1	1	$1 - s$
A_1 completely dominant, selection against A_1A_1 and A_1A_2	$1 - s$	$1 - s$	1
Overdominance, selection against A_1A_1 and A_2A_2	$1 - s_1$	1	$1 - s_2$

A_2 mutates back to A_1 at rate v per generation. If the initial frequencies of A_1 and A_2 are p_0 and q_0, respectively, the change in gene frequency as a function of mutation in one generation will be as follows:

$$\Delta q = up_0 - vq_0.$$

In general, the spontaneous mutation rate is usually very low, somewhere between 10^{-5} and 10^{-7}. For example, assume $p_0 = 0.9$, $q_0 = 0.1$, $u = 10^{-6}$ and $v = 10^{-7}$. Then, $\Delta q = (0.000001)(0.9) - (0.0000001)(0.1) = 0.00000089$. Thus, except when considered on an evolutionary time scale, mutation is not a powerful force in changing gene frequency.

Selection. A difference in reproductive rate among individuals or groups, i.e., *selection,* can be a powerful force in changing gene frequencies. When dealing with single-locus characters, the coefficient of selection, s, may be used to measure the intensity of selection. Consider the first case (complete dominance, selection against A_2A_2) listed in Table 9.1. When s = 0, all genotypes contribute equally to the next generation. However, when s = 1, there is complete selection against A_2A_2, i.e., the fitness of A_2A_2 is zero. Of course, s is not restricted to the extreme values of zero or one and may assume any value in between. It is important to note that the fitness values in Table 9.1 are expressed in *relative* terms, i.e., relative to the genotype with highest fitness. Thus, if A_1A_1 and A_1A_2 genotypes produce 20 offspring on the average, whereas A_2A_2 genotypes produce only 5, the relative fitness of these three genotypes would be 1, 1, and $\frac{1}{4}$, respectively, in which case s = 0.75.

The relative contribution of each genotype to the gene pool of the next generation is given by the *gametic contribution.* Let us symbolize the gametic contribution of A_1A_1, A_1A_2, and A_2A_2 genotypes by P_s, H_s, and Q_s, respectively. These gametic contributions are easily obtained from the product of the genotypic frequency and the relative fitness value. Thus, for complete dominance and selection against A_2A_2, $P_s = p^2$, $H_s = 2pq$, and $Q_s = q^2(1-s)$. The gene frequency after one generation of selection may be obtained as follows:

$$q_1 = \frac{Q_s + \frac{1}{2}H_s}{P_s + H_s + Q_s};$$

thus, for complete dominance and selection against A_2A_2,

$$q_1 = \frac{q^2(1-s) + pq}{1 - sq^2}.$$

The change in gene frequency that would accompany one generation of selection against A_2A_2 is given by the difference between q_1 and q:

$$\Delta q = q_1 - q$$

$$= \frac{q^2(1-s) + pq}{1 - sq^2} - q$$

$$= \frac{q^2(1-s) + pq}{1 - sq^2} - \frac{q(1 - sq^2)}{1 - sq^2}$$

$$= \frac{q^2 - sq^2 + (1-q)q - q + sq^3}{1 - sq^2}$$

$$= \frac{q^2 - sq^2 + q - q^2 - q + sq^3}{1 - sq^2}$$

$$= \frac{-sq^2 + sq^3}{1 - sq^2}$$

$$= \frac{-sq^2(1-q)}{1 - sq^2}.$$

After n generations of complete selection ($s = 1$) against A_2A_2, it may be shown that the frequency of A_2 will be:

$$q_n = \frac{q_0}{1 + nq_0},$$

where q_0 is the gene frequency in generation zero. As an example, consider some deleterious condition determined by an autosomal recessive gene, A_2, that we may wish to eliminate from a human population. How much could the frequency of this recessive gene be lowered if A_2A_2 individuals did not reproduce for a number of generations? Genes with detrimental effects tend to be in relatively low frequency; thus, let us assume that $q_0 = 0.02$. If no A_2A_2 individuals reproduced for 50 generations, the frequency of this undesirable gene would be:

$$q_{50} = \frac{0.02}{1 + 50(0.02)} = 0.01.$$

In other words, after some 1500 years of intense selection against A_2A_2 (assuming a generation interval of about 30 years), the frequency of this gene would only change from 0.02 to 0.01. This demonstrates the relative ineffectiveness of this form of selection when the frequency of a recessive gene is initially low.

With complete dominance and selection against A_1A_1 and A_1A_2:

$$\Delta q = \frac{sq^2(1-q)}{1 - s(1-q^2)}.$$

In this case, when $s = 1$, $\Delta q = 1 - q$. Thus, after one generation of such selection,

$$q_1 = q + \Delta q = 1.$$

When $s = 1$ and selection is against the dominant form, only A_2A_2 genotypes will reproduce. Thus, it is understandable that the frequency of the A_2 allele should become one after only one generation of such selection. In a similar manner, one generation of complete selection against A_1A_2 and A_2A_2 could result in the elimination of an undesirable recessive gene from the population, thereby indicating the utility of detecting carriers for recessive conditions. As seen above, the frequency of a recessive allele is changed very slowly when the recessive homozygote is selected against. When the frequency of A_2 is low, most of the A_2 alleles will remain undetected in heterozygous carriers. However, if carriers could be detected and if they refrained from reproducing, it would be possible to eliminate the undesirable gene in one generation. The feasibility of such eugenic measures is discussed in more detail in Chapter 11.

When overdominance exists at a locus, the relative fitnesses of A_1A_1 and A_2A_2 homozygotes may not be equal. Thus, two coefficients of selection are employed for this form of selection in Table 9.1. Change in gene frequency due to overdominance selection is as follows:

$$\Delta q = \frac{pq(s_1p - s_2q)}{1 - s_1p^2 - s_2q^2}.$$

Selection against both homozygotes results in a form of *balanced polymorphism*, which serves to maintain both alleles in the population. After many generations of such selection, an equilibrium will be achieved in which

$\Delta q = 0$. At that point, the frequency of the A_2 allele is a simple function of the selection coefficients:

$$q = \frac{s_1}{s_1 + s_2}.$$

The classic example of overdominance is that of sickle-cell anemia in man. Although relatively few individuals afflicted with this most serious disease survive to reproduce, the gene is nonetheless maintained in relatively high frequency in some African populations and among Afro-Americans. This high frequency of an essentially lethal recessive gene is apparently due to the relatively high fitness of heterozygotes. Individuals who carry the gene heterozygously seem to be more resistant than homozygotes for the normal gene to a form of malaria prevalent in certain parts of Africa.

Within the last decade it has become evident that there is a very great store of genetic variation being maintained in natural populations, including man. Considerable disagreement currently exists among population geneticists as to the relative importance of various mechanisms that may be responsible for the maintenance of this genetic variation. Spontaneous mutations occur at very low frequencies and most mutations are at least mildly deleterious in the homozygous state. Thus, one form or another of selectional balance may play a major role in maintaining this great store of genetic variation.

Huxley et al. (1964) have suggested that balancing selection may be responsible for maintaining schizophrenia at its current level of about one percent. Until the advent of drug therapy, schizophrenics were almost certainly at a considerable reproductive disadvantage. Although the relative fitness of schizophrenics was likely lower than that of normals, Erlenmeyer-Kimling and Paradowski (1966) have presented evidence to suggest that sibs of schizophrenics may actually have larger families than normal controls. Additional evidence concerning the inheritance of schizophrenia will be presented later in this chapter and in Chapter 11.

Another form of selectional balance is that due to frequency-dependent selection. A behavioral example involving sexual selection will be presented in Chapter 10.

Quantitative Genetic Analysis

The Model.   In statistics, a mathematical model is assumed that may be used to describe each observation:

$$Y = u + T + e,$$

where Y symbolizes the datum obtained from an individual exposed to a given treatment, u the population mean, T the effect due to that treatment, and e the "random error" (plus or minus) peculiar to each observation.

Table 9.2

*Mean phenotypic values of two
genotypes reared in two environments*

		Genotypes	
		G_1	G_2
Environments	E_1	$\overline{P}_{G_1E_1}$	$\overline{P}_{G_2E_1}$
	E_2	$\overline{P}_{G_1E_2}$	$\overline{P}_{G_2E_2}$

An analogous model is assumed in quantitative genetic theory:

$$P = G + E,$$

where P is the phenotypic value (the measured value for some character of an individual), G is the genotypic value (value conferred upon the individual by its genotype, also expressed in units of measurement), and E is an environmental deviation due to all nongenetic causes. Although P is analogous to Y and G is analogous to T, E includes both real environmental effects and random error.

The population mean is not included in the quantitative genetic model since it is usually most meaningful to think of P as a deviation from the population mean. Thus, hereditary and environmental variation result in positive or negative deviations of scores about the population mean.

If the joint effects of G and E combine in a nonlinear manner, another term should be added to the simple linear model: $P = G + E + (G \times E)$, where $(G \times E)$ symbolizes the deviation due to genotype-environment interactions. *Genotype-environment interaction* may be defined as the differential response of different genotypes to different environments. Consider two genotypes (e.g., members of two inbred strains) reared in two different environments (Table 9.2). Let the subclass symbols shown in the body of Table 9.2 refer to the mean values of a large number of individuals of the indicated genotype reared in the indicated environment. If no nonlinear interaction exists between G and E, the difference between the two genotypes should be the same in the two environments. Symbolically expressed,

$$\overline{P}_{G_1E_1} - \overline{P}_{G_2E_1} = \overline{P}_{G_1E_2} - \overline{P}_{G_2E_2}$$

if no genotype-environment interaction is present. Consider for example the cases indicated in Table 9.3. Case I is an example of no genotype-environment interaction, i.e., $12 - 10 = 8 - 6$. Cases II and III, however, illustrate situations in which the genotypes respond differently to the difference in environments. Case III is more extreme in that a more favorable environment for G_1 is less favorable for G_2.

Table 9.3

Hypothetical examples to illustrate the absence and presence of genotype-environment interactions

| Environments | Genotypes | | | | | |
| | Case I | | Case II | | Case III | |
	G_1	G_2	G_1	G_2	G_1	G_2
E_1	12	10	12	10	12	6
E_2	8	6	8	4	8	10

Numerous examples of genotype-environment interactions for behavioral characters have been observed with laboratory animals. As discussed in Chapter 8, this interaction is frequently manifest in studies of behavioral development. However, in such research, this interaction may have been amplified due to the use of inbred subjects. Although the magnitude of genotype-environment interactions in genetically segregating populations, a difficult subject to investigate, is little known, Jinks and Fulker (1970) have recently tested for the presence of such interactions in various human personality and cognitive measures and found them to be relatively unimportant.

In the several sections that follow, the concept of the genotypic value and its component parts is considered in some detail. Those readers not concerned with this subdivision and the consequent partitioning of the genotypic variance may proceed directly to the section on heritability (page 201).

Genotypic Value. Although quantitative genetics was developed for application to characters influenced by genes at many loci, the basic underlying model is based upon segregation at only a single locus. Once we have considered the statistical description of gene action at a single locus, it will be possible to generalize to the polygenic case.

With regard to a single locus, values may be arbitrarily assigned to the different genotypes as indicated in Figure 9.1. The homozygote with the higher value (homozygous for the "increasing" allele) is symbolized A_1A_1. The point that bisects the distance between the homozygotes into two equal parts may be defined as the origin so that all genotypic values may be expressed as deviations from this mid-homozygote point. Since A_1A_1 and A_2A_2 deviate from this point by the same amount, we may assign them the genotypic values +a and −a, respectively. The genotypic value of the heterozygote, symbolized by d, will depend upon the gene action at the locus under consideration. For example, if A_1 is dominant to A_2, the heterozygote will

Figure 9.1

Assigned genotypic values. (After Falconer, Introduction to Quantitative Genetics. Copyright © 1960, p. 113. The Ronald Press Company, New York.)

fall to the right of the mid-homozygote point and d will be positive. If A_2 is dominant to A_1, d will be negative. If there is complete dominance, d will equal either +a or −a. If there is overdominance, d will be greater than +a or less than −a. Finally, if there is no dominance, d will equal zero.

As an example, recall the pleiotropic effect of albinism on open-field behavior of mice. In Table 5.2, the mean open-field activity scores of 485 pigmented and 152 albino mice were shown to be 12.7 and 10.7, respectively, in generation F_2. Positive and negative environmental deviations tend to be cancelled out when large samples are averaged. Thus, we may assume that the difference between these observed values is largely a function of a difference in genotypic values. The genotypic value for animals of genotype *CC* may be calculated as follows:

$$a = \frac{12.7 - 10.7}{2} = 1.$$

Assuming complete dominance at the *c* locus for its effect on activity, d = a = 1. These genotypic values are illustrated in graphic form in Figure 9.2.

Mean Genotypic Value. The mean genotypic value for a character in a population is obviously a function of the genotypic frequency. When the system of mating is specified, the mean genotypic value may be expressed as a function of the gene frequency. Although the quantitative genetic model is quite general and may be applied for any system of mating, the derivations are much simpler when mating is at random. Thus, random mating will be assumed for the derivations in this section. Extensions to other systems of mating will be indicated later in this chapter.

The genotypic values and corresponding frequencies under random mating are presented in Table 9.4. Since $p^2 + 2pq + q^2 = 1$, the mean genotypic value may be obtained by multiplying the genotypic values by their frequencies and summing across genotypes:

$$\overline{G} = p^2a + 2pqd - q^2a = a(p^2 - q^2) + 2pqd$$

$$= a(p + q)(p - q) + 2pqd = a(p - q) + 2pqd,$$

where $\overline{G}$ is expressed as a deviation from the mid-homozygote point.

Figure 9.2

Genotypic values corresponding to open-field activity scores of mice.

For example, consider again albinism and open-field activity in mice. In the F_2 generation, the expected genotypic distribution was $\frac{1}{4}CC + \frac{1}{2}Cc + \frac{1}{4}cc$, corresponding to a Hardy-Weinberg-Castle equilibrium with $p = q = \frac{1}{2}$. Thus, the mean genotypic value should have been:

$$\overline{G} = a(p - q) + 2dpq = 1(\tfrac{1}{2} - \tfrac{1}{2}) + (2)(1)(\tfrac{1}{2})(\tfrac{1}{2}) = 0.5.$$

The location of this mean on the scale of genotypic values is illustrated in Figure 9.3. This measure of central tendency is located exactly half way between the mid-homozygote point and the genotypic value of $C-$ since the genotypic distribution is $3C-:1cc$ in this population. If the frequency of cc were higher, i.e., if q were higher than p, $\overline{G}$ would be lower. In addition, if d were less than a, $\overline{G}$ would be reduced. Thus, the mean genotypic value is a function of both gene frequency and level of dominance.

Average Effect. Sperm and eggs contain genes, not genotypes. Thus, in order to analyze the resemblance of relatives, it is necessary to assess the *average effect of genes* in the population, not just genotypic values.

The average effect of a gene does not depend upon whether the gene was transmitted by a sperm or an egg. Thus, for the sake of expediency, we may arbitrarily define the average effect of a gene, say A_1, as the mean genotypic value (expressed as a deviation from the population mean) of individuals resulting from the random fertilization of eggs by A_1-bearing sperm. In order to calculate the average effects of A_1 and A_2 genes, symbolized α_1 and α_2, respectively, it is necessary to determine the frequency of the various off-

Figure 9.3

Location of the mean on a scale of genotypic values.

Table 9.4 189

Genotypic values and frequencies

Genotype	Frequency	Value
A_1A_1	p^2	$+a$
A_1A_2	$2pq$	d
A_2A_2	q^2	$-a$

Table 9.5

Frequency of genotypes resulting from the random fertilization of eggs by A_1-bearing and A_2-bearing sperm

Genotype	Genotypic value	Sperm bearing	
		A_1	A_2
A_1A_1	$+a$	p	0
A_1A_2	d	q	p
A_2A_2	$-a$	0	q

spring genotypes that result from the random fertilization of eggs by A_1-bearing and A_2-bearing sperm (see Table 9.5). The first entry in the A_1 column of Table 9.5, p, represents the probability (or frequency) that A_1-bearing sperm will fertilize an egg drawn at random with the result that an A_1A_1 zygote is conceived. The mean genotypic value of offspring resulting from A_1-bearing sperm may be determined from the product of the entries in the A_1 column and the corresponding genotypic values, summed across genotypes. Thus, α_1 may be obtained as follows:

$$\alpha_1 = pa + qd + (0)(-a) - \overline{G}$$
$$= pa + qd - [a(p - q) + 2dpq]$$
$$= qd + qa - 2dpq$$
$$= q[a + d(1 - 2p)]$$
$$= q[a + d(p + q - 2p)]$$
$$= q[a + d(q - p)].$$

In a similar manner it may be shown that $\alpha_2 = -p[a + d(q - p)]$.

It is also useful in quantitative genetic theory to define a parameter that describes the average change in the genotypic value due to the substitution of an A_1 allele for an A_2 allele in the population. Consider the values and frequencies presented in Table 9.6. A new term is introduced in this table,

Table 9.6

Gene dosage (N^{A1}) *and genotypic*
value (G)

Genotype	N^{A1}	G	Frequency
A_1A_1	2	+a	p^2
A_1A_2	1	d	$2pq$
A_2A_2	0	−a	q^2

gene dosage, symbolized N^{A1}, which corresponds to the number of A_1 alleles present in a genotype. The average effect of a gene substitution in such a population is thus merely the average change in genotypic value due to a unit change in gene dosage. Recall that in Chapter 3 the regression of Y on X measured the average change in Y corresponding to a unit change in X. Thus, the average effect of a gene substitution, symbolized α, is equivalent to the regression of the genotypic value on the gene dosage, $b_{GN^{A1}}$. It may be shown that the covariance of the genotypic value and gene dosage $[\mathrm{Cov}(G)(N^{A1})]$ and the variance of the gene dosage ($V_{N^{A1}}$) are as follows: $\mathrm{Cov}(G)(N^{A1}) = 2pq[a + d(q - p)]$ and $V_{N^{A1}} = 2pq$. Thus,

$$\alpha = b_{GN^{A1}} = \frac{\mathrm{Cov}(G)(N^{A1})}{V_{N^{A1}}} = \frac{2pq[a + d(q - p)]}{2pq} = a + d(q - p).$$

That the average effect of a gene substitution is also equivalent to the difference between the average effects of A_1 and A_2 genes in the population may be seen from the following:

$$\alpha = \alpha_1 - \alpha_2 = q[a + d(q - p)] - \{-p[a + d(q - p)]\}$$
$$= (p + q)[a + d(q - p)]$$
$$= a + d(q - p).$$

We may now more conveniently express α_1 and α_2 in terms of α as follows:

$$\alpha_1 = q\alpha$$

$$\alpha_2 = -p\alpha.$$

Additive Genetic Value. Each gene in the genotype that influences a character has some average effect on that character. The sum of these average effects, where the summation is across both genes at each locus and across all loci for polygenic characters, is the *additive genetic value*, symbolized A. We shall consider here only two alleles at one autosomal locus.

Table 9.7

Additive genetic values (A) *and frequencies*

Genotype	Frequency	A
A_1A_1	p^2	$2\alpha_1 = 2q\alpha$
A_1A_2	$2pq$	$\alpha_1 + \alpha_2 = (q - p)\alpha$
A_2A_2	q^2	$2\alpha_2 = -2p\alpha$

SOURCE: After Falconer, *Introduction to Quantitative Genetics*. Ronald Press. Copyright 1960, p. 121.

Table 9.8

Offspring of A_1A_1 males when mating is at random

Type of mating	Frequency of type of mating	Frequency of offspring, by genotype and genotypic value	
		A_1A_1 a	A_1A_2 d
$A_1A_1 \male \times A_1A_1 \female$	p^2	1	0
$A_1A_1 \male \times A_1A_2 \female$	$2pq$	$\frac{1}{2}$	$\frac{1}{2}$
$A_1A_1 \male \times A_2A_2 \female$	q^2	0	1

The additive genetic values corresponding to the three possible genotypes are presented in Table 9.7. The A_1A_1 genotypes contain two A_1 alleles, with average effects α_1. Thus, the additive genetic value of A_1A_1 genotypes is $2\alpha_1 = 2q\alpha$. The additive genetic values of A_1A_2 and A_2A_2 may be obtained in a similar manner.

It is instructive to derive these additive genetic values using alternative approaches. For example, if A_1A_1 males mate with females chosen at random, the possible types of matings will occur with the frequencies indicated in Table 9.8. The offspring genotypes, genotypic values, and relative frequencies are also given. Thus, the mean genotypic value of offspring produced by A_1A_1 males (when mating is at random), expressed as a deviation from the population mean, may be calculated as follows:

$$p^2a + pqa + pqd + q^2d - [a(p - q) + 2dpq]$$
$$= pa + qd - [a(p - q) + 2dpq] = q[a + d(q - p)] = q\alpha.$$

In Table 9.7 it was shown that the additive genetic value of A_1A_1 individuals is $2q\alpha$. This is no coincidence; in general, the mean value of the progeny of an individual, expressed as a deviation from the population mean, provides an estimate of one-half the additive genetic value of that individual. Symbolically,

$$\tfrac{1}{2}A = (\overline{G}^0 - \overline{G})$$

or

$$A = 2(\overline{G}^0 - \overline{G}),$$

where $\overline{G}^0$ refers to the mean genotypic value of the progeny of an individual that had mated at random with a large number of members of the opposite sex. Using this approach we may empirically estimate the additive genetic value of individuals, even for polygenic characters.

The additive genetic value of an individual is also equivalent to the expected genotypic value based upon the gene dosage. The expected genotypic value $(\hat{G})$, expressed as a deviation from the mean, may be estimated from the following regression equation:

$$\hat{G} = b_{GN^{A1}}(N^{A1} - \overline{N}^{A1}),$$

where $b_{GN^{A1}} = \alpha$ and $\overline{N}^{A1} = 2p$.

For A_1A_1 genotypes,

$$\hat{G} = \alpha(2 - 2p) = 2\alpha(1 - p) = 2q\alpha,$$

in agreement with the indicated additive genetic value in Table 9.7. The additive genetic values of A_1A_2 and A_2A_2 may be obtained from the same expression upon substitution of appropriate values for N^{A1}, namely 1 and 0, respectively.

The mean additive genetic value may be obtained by summing the products of the additive genetic values and their respective frequencies listed in Table 9.7:

$$\overline{A} = (p^2)(2q\alpha) + (2pq)(q - p)\alpha + (q^2)(-2p\alpha)$$
$$= 2pq\alpha(p + q - p - q) = 0.$$

This is the expected result since these additive genetic values have been expressed as deviations from the mean.

Dominance Deviation.   As indicated in the previous section, the additive genetic value for some character measured on an individual may be defined as the sum of the average effects of that individual's genes. Thus, for a single locus, any departure of the additive genetic value from the genotypic value must be ascribed to dominance. Symbolically,

$$G = A + D,$$

Table 9.9

G, A, *and* D, *expressed as deviations from the mean*

Genotype	G	A	D	Frequency
A_1A_1	$2q(\alpha - qd)$	$2q\alpha$	$-2q^2d$	p^2
A_1A_2	$(q - p)\alpha + 2pqd$	$(q - p)\alpha$	$2pqd$	$2pq$
A_2A_2	$-2p(\alpha + pd)$	$-2p\alpha$	$-2p^2d$	q^2

SOURCE: After Falconer, *Introduction to Quantitative Genetics.* Ronald Press. Copyright 1960, p. 124.

where D is a *dominance deviation* due to the nonlinear interaction between alleles at a given locus. In order to obtain the dominance deviation corresponding to each genotype, we merely need to subtract the additive genetic value from the genotypic value. However, since additive genetic values are expressed as deviations from the mean, genotypic values must be similarly expressed.

The genotypic value of A_1A_1 individuals, expressed as a deviation from the mean, may be calculated as follows:

$$a - [a(p - q) + 2dpq] = a - ap + aq - 2dpq = a(1 - p + q) - 2dpq.$$

Recalling that $1 - p = q$, this expression reduces to:

$$2qa - 2dpq = 2q(a - dp).$$

In order to express G in terms of α, we may add and subtract dq within the parentheses as follows:

$$2q(a - dp + dq - dq) = 2q[a + d(q - p) - qd] = 2q(\alpha - qd).$$

Thus, both A and G have been expressed as deviations from the mean and in terms of α. The corresponding values of G, A, and D for each of the three genotypes are presented in Table 9.9. It may be noted from this table that each of the dominance deviations is expressed as a function of d; thus, if there is no dominance, i.e., if $d = 0$, each of the dominance deviations will be zero.

In Table 9.9, D is also expressed as a deviation from the mean; thus, $\overline{D} = 0$. Because both $\overline{A}$ and $\overline{D}$ equal zero, the covariance of A and D may be directly calculated as follows:

$$\text{Cov}(A)(D) = (2q\alpha)(-2q^2d)(p^2) + [(q - p)\alpha](2pqd)(2pq)$$
$$+ (-2p\alpha)(-2p^2d)(q^2)$$
$$= -4p^2q^3\alpha d + 4p^2q^2(q - p)\alpha d + 4p^3q^2\alpha d = 0.$$

Because $\text{Cov}(A)(D) = 0$, it follows that A and D are uncorrelated.

Table 9.10

*Genotypic values, additive genetic values, and
dominance deviations for open-field activity scores of
pigmented and albino mice (all values are expressed
as deviations from the mean)*

Genotype	G	A	D	Frequency
CC	0.5	1.0	−0.5	0.25
Cc	0.5	0.0	+0.5	0.50
cc	−1.5	−1.0	−0.5	0.25

As an example, let us again consider the case of albinism and open-field activity in mice, where $p = q = \frac{1}{2}$ and $a = d = 1$. As indicated in Figure 9.3, the mean genotypic value was found to be equal to 0.5. Thus, the genotypic value of CC mice, expressed as a deviation from the mean, is as follows:

$$a - \overline{G} = 1.0 - 0.5 = 0.5.$$

This result may also be obtained from substitution of the appropriate values into the expression in Table 9.9:

$$2q(\alpha - qd) = 2(\tfrac{1}{2})[1 - (\tfrac{1}{2})(1)] = 0.5,$$

where $\alpha = a + d(q - p) = 1 + 1(\tfrac{1}{2} - \tfrac{1}{2}) = 1$.

The additive genetic value of CC genotypes may also be obtained from the expression in Table 9.9:

$$2q\alpha = 2(\tfrac{1}{2})(1) = 1.$$

The dominance deviation may be determined either by subtraction $(0.5 - 1.0 = -0.5)$ or by substitution into the tabulated expression for **D** (Table 9.9):

$$-2q^2d = -2(\tfrac{1}{4})(1) = -0.5.$$

Corresponding values of **G**, **A**, and **D** for each of the three genotypes are presented in Table 9.10.

The values listed in Table 9.10 may also be presented graphically as in Figure 9.4. It may be seen that values of A fall upon a straight line. This is the regression line corresponding to $\hat{G} = b_{GN^{A1}}(N^{A1} - \overline{N}^{A1})$. Thus, the additive genetic values represent points on a line that has been derived to provide the best linear fit to the observed genotypic values, i.e., the sum of the squared deviations of G from A is at a minimum. Dominance deviations represent the deviations of the observed points about the regression line. As indicated in Chapter 3, we may use this relationship to partition the

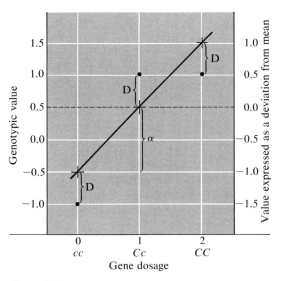

Figure 9.4

Genotypic values (black circles), additive genetic values (crosses), and dominance deviations (D) plotted as a function of the gene dosage. The average effect of a gene substitution (α) is also indicated.

genotypic variance into two parts: one due to variation of the expected genotypic values (additive genetic values) about the population mean and another due to variation of the observed values about the regression line (dominance deviations). This partitioning of the genotypic variance shall be considered in a later section in this chapter.

Epistatic Interaction Deviation. As noted in the previous section, $G = A + D$ when only one locus is considered. However, when genes at more than one locus influence a character, nonlinear interactions may occur among alleles at the different loci. This is the meaning of epistasis as defined in Chapter 2. For simplicity, consider only the two-locus case: At one locus, $G_1 = A_1 + D_1$ and at the other locus, $G_2 = A_2 + D_2$. When both loci are considered jointly, the over-all genotypic value may be partitioned as follows:

$$G = G_1 + G_2 + I,$$

where I symbolizes the deviation due to epistatic interactions. This may also be expressed in the following more general form:

$$G = A + D + I,$$

where G is the genotypic value due to all loci, A is the sum of the additive genetic values across all loci and D is the sum of the dominance deviations across all loci.

Again considering only two loci, the epistatic interaction deviations may be due to interactions between additive genetic values at the different loci, symbolized $I_{(A \times A)}$, to interactions between additive genetic values at one locus and dominance deviations at the other, $I_{(A \times D)}$, or to interactions between dominance deviations at the two loci, $I_{(D \times D)}$. It will be shown later in this chapter that the variances due to these different epistatic interaction deviations contribute differentially to the covariance of relatives.

Variance.   Each of the values and deviations previously defined has a corresponding variance, symbolized by V and an appropriate subscript. These variances and their symbols are indicated in Table 9.11.

In the absence of genotype-environment interactions, $P = G + E$; thus, $V_P = V_G + V_E + 2\mathrm{Cov}(G)(E)$. If G and E are uncorrelated, i.e., if environmental effects are distributed at random across genotypes, as is frequently true with laboratory animals whose environment is at least to some extent controlled, $V_P = V_G + V_E$. If genotype-environment interactions are present, a variance term corresponding to such interactions should also be included:

$$V_P = V_G + V_E + V_{(G \times E)}.$$

Because $G = A + D + I$, $V_G = V_A + V_D + V_I$. No covariance terms are necessary in this expression because A, D, and I have been derived in such a way as to be independent.

Expressions for V_A, V_D, and V_G due to segregation at a single autosomal locus may be easily calculated from the information presented in Table 9.9. Because A, D, and G are expressed as deviations from the mean, their variances may be obtained by merely summing the squared values or deviations, where each squared term is weighted by its frequency, i.e.,

$$V_A = p^2(2q\alpha)^2 + 2pq[(q - p)\alpha]^2 + q^2(-2p\alpha)^2$$
$$= 2pq\alpha^2(2pq + q^2 + p^2 - 2pq + 2pq)$$
$$= 2pq\alpha^2$$

and

$$V_D = p^2(-2q^2d)^2 + 2pq(2pqd)^2 + q^2(-2p^2d)^2$$
$$= 4p^2q^2d^2(q^2 + 2pq + p^2)$$
$$= (2pqd)^2.$$

Because

$$\mathrm{Cov}(A)(D) = 0, \quad V_G = V_A + V_D = 2pq\alpha^2 + (2pqd)^2.$$

Table 9.11

Summary of variances and their symbols

Variance	Symbol
Phenotypic	V_P
Genotypic	V_G
Additive genetic	V_A
Dominance	V_D
Epistatic	V_I
Environmental	V_E
Genotype-environment interaction	$V_{(G \times E)}$

It was shown previously that additive genetic values are equivalent to expected genotypic values based upon the gene dosage; thus, V_A may be thought of as the variance in G due to variation in gene dosage:

$$(r^2_{GN^{A1}})(V_G) = (b^2_{GN^{A1}})(V_{N^{A1}}/V_G)(V_G) = (b^2_{GN^{A1}})(V_{N^{A1}}) = \alpha^2 2pq = V_A.$$

The open-field activity values and deviations presented in Table 9.10 may be used to estimate V_G, V_A, and V_D due to segregation at the c locus:

$$V_G = \tfrac{1}{4}(0.5)^2 + \tfrac{1}{2}(0.5)^2 + \tfrac{1}{4}(-1.5)^2 = 0.75,$$

$$V_A = \tfrac{1}{4}(1)^2 + \tfrac{1}{2}(0)^2 + \tfrac{1}{4}(-1)^2 = 0.50,$$

$$V_D = \tfrac{1}{4}(-0.5)^2 + \tfrac{1}{2}(0.5)^2 + \tfrac{1}{4}(-0.5)^2 = 0.25.$$

Or, using our derived expressions:

$$V_G = 2pq\alpha^2 + (2pqd)^2 = \tfrac{1}{2}(1)^2 + [(\tfrac{1}{2})(1)]^2 = 0.75,$$

$$V_A = 2pq\alpha^2 = \tfrac{1}{2}(1)^2 = 0.50,$$

$$V_D = (2pqd)^2 = [(\tfrac{1}{2})(1)]^2 = 0.25.$$

It is interesting to note that although complete dominance was present in this example, $0.50/0.75 = \tfrac{2}{3}$ of the total genetic variance is additive. Thus, V_A may account for a substantial fraction of V_G, even in the absence of classical additive gene action $(d = 0)$.

The epistatic variance, symbolized V_I, may be partitioned according to the number of loci involved in the interactions discussed previously. Thus,

$$V_I = V_{I(A \times A)} + V_{I(A \times D)} + V_{I(D \times D)} + V_{I(A \times A \times A)} + \cdots .$$

The contributions of these individual components of V_I to the covariance of relatives will be shown in the next section.

Covariance of Relatives.   The genetic variances contribute differentially to the covariance of relatives. Knowledge of the extent to which this is so is necessary to understand how the observed resemblance of relatives may be used to obtain estimates of the genetic variance. Let us consider first the covariance of the phenotypic values of one offspring (P^o) and one parent (P), where data from many such points are available. The covariance of their phenotypic values may be expressed as follows:

$$\text{Cov}(P^o)(P) = \text{Cov}(G^o + E^o)(G + E).$$

In a previous section it was shown that the expected genotypic value of an offspring, expressed as a deviation from the mean, is equivalent to one-half the additive genetic value of its parent. Thus, we may express the genotypic value of the offspring in terms of the additive genetic value of the parent as follows:

$$\text{Cov}(P^o)(P) = \text{Cov}(\tfrac{1}{2}A + E^o)(A + D + E)$$
$$= \tfrac{1}{2}\text{Cov}(A)(A) + \tfrac{1}{2}\text{Cov}(A)(D) + \tfrac{1}{2}\text{Cov}(A)(E)$$
$$+ \text{Cov}(E^o)(A) + \text{Cov}(E^o)(D) + \text{Cov}(E^o)(E).$$

It was previously shown that $\text{Cov}(A)(D) = 0$. Thus, in the absence of a genotype-environment correlation, where $\text{Cov}(A)(E) = \text{Cov}(D)(E) = 0$,

$$\text{Cov}(P^o)(P) = \tfrac{1}{2}\text{Cov}(A)(A) + \text{Cov}(E^o)(E).$$

The covariance of a variable with itself is equal to the variance of that variable, i.e., $\text{Cov}(A)(A) = V_A$. Therefore, the above expression may be rewritten as follows:

$$\text{Cov}(P^o)(P) = \tfrac{1}{2}V_A + \text{Cov}(E^o)(E).$$

Although the epistatic interaction deviation has not been included in this derivation for simplicity of exposition, variance due to such deviations will be included in a summary table at the end of this section.

The environmental deviation, E, may be partitioned into the following parts:

$$E = E^c + E^w,$$

where E^c is the deviation due to environmental factors shared by relatives and E^w is due to environmental effects which are independent of relationship. Thus,

$$\text{Cov}(E^o)(E) = \text{Cov}(E^c + E^{w(o)})(E^c + E^w).$$

Since the E^W's of family members are defined as being uncorrelated, covariances involving these terms are zero. Therefore,

$$Cov(E^O)(E) = Cov(E^C)(E^C) = V_{E^C}.$$

Upon substitution into the above expression,

$$Cov(P^O)(P) = \tfrac{1}{2}V_A + V_{E^{C(OP)}},$$

where (OP) indicates that this is the variance due to environmental effects shared by parents and their offspring.

With laboratory animals, the environments can be randomized to some extent so that $V_{E^{C(OP)}}$ may be safely ignored. Although such is not the case with human data, this by no means indicates that the problem of estimating additive genetic variance is intractable. A similar problem exists with domestic farm animals where parents and offspring are reared on the same farm and environmental differences exist among farms. Animal breeders employ special statistical analyses that effectively remove the resulting environmental correlation (cf. Lush, 1940).

We may also obtain the covariance of the phenotypic value of an offspring with the average phenotypic value of its two parents (*midparental value*). We shall symbolize the phenotypic values of the male and female parents as P and P', respectively, and the midparental value by P^M. Thus,

$$Cov(P^O)(P^M) = Cov(P^O)\frac{(P + P')}{2} = \tfrac{1}{2}(CovP^OP + CovP^OP').$$

If the sexes have equal variance, if the character is autosomal, and if the environmental covariance of offspring and male parent is equal to that of offspring and female parent, then $CovP^OP$ should equal $CovP^OP'$. Under this condition,

$$Cov(P^O)(P^M) = Cov(P^O)(P) = \tfrac{1}{2}V_A + V_{E^{C(OP)}},$$

i.e., the covariance of offspring and midparent is equivalent to that of offspring and one parent. If any of these assumptions are not fulfilled, the covariances should be calculated separately for males and females. In general, the covariance of an individual with the mean of a number of relatives of the same kind is equivalent to the covariance of the individual with one of these relatives (Falconer, 1960), e.g., $Cov(\bar{P}^O)(P) = Cov(P^O)(P)$, where $\bar{P}^O$ is the mean phenotypic value of a number of offspring.

Half sibs are individuals that have one and only one parent in common. Thus, the mean genotypic value of a family of half sibs should equal one-half the additive genetic value of the common parent. With data from half-sib families (or full-sib families or sets of identical or fraternal twins), only one variable is available for analysis, i.e., data do not correspond to X and Y variables as they do for parents and offspring. In such a case, a univariate analysis is appropriate. An analysis of variance is performed so that components of variance attributable to differences among families and differences among individuals within the same family may be estimated. It may be shown that the component of variance among families is equivalent to the covariance of members of the same family and that this component of variance may be estimated from the variance of the family means when the means are known with exactness. Details concerning the estimation of components of variance and "intraclass" correlations are available in most intermediate level statistics books (see especially Haggard, 1958).

The covariance of members of half-sib families is thus estimated from the variance of the family means, which are expected to equal one-half the additive genetic value of the common parent, i.e.,

$$\mathrm{Cov(HS)} = \mathrm{V}(\tfrac{1}{2}A) = \tfrac{1}{4}V_A.$$

If the common parent is the father with the offspring being borne by different mothers and reared in different families, the variance due to common environment may perhaps be safely ignored.

Full sibs, of course, have both parents in common. Thus, the covariance of full sibs should be at least twice that of half sibs. In addition to containing twice the additive genetic variance, the covariance may be shown to contain one-fourth of the dominance variance. This should make some intuitive sense since one-fourth of the time full sibs would be expected to receive the same alleles from both parents and thus would have the same dominance deviation. Thus, the covariance of full sibs is as follows:

$$\mathrm{Cov(FS)} = \tfrac{1}{2}V_A + \tfrac{1}{4}V_D + V_{E\,C(FS)},$$

where $V_{EC(FS)}$ is that part of the environmental variance due to environmental effects that are common to members of the same full-sib family. Research with laboratory animals, especially with litter-bearing animals having full sibs as members of the same litter, indicates that $V_{EC(FS)}$ may be an important source of variance.

Genetically, fraternal or dizygotic (DZ) twins are no more alike than full sibs. Thus, the genetic covariance of DZ twins contains the same genetic variance as that of full sibs. However, the variance due to common environmental effects may be somewhat higher for DZ twins due to their sharing the same uterine environment and having a contemporaneous period of postnatal development. The components of the covariance of DZ twins are shown below:

$$\mathrm{Cov(DZ)} = \tfrac{1}{2}V_A + \tfrac{1}{4}V_D + V_{E\,C(DZ)}.$$

Table 9.12

*Coefficients indicating the contributions of additive, dominance,
and epistatic variances to the genetic covariance of various relatives*

Covariance	V_A	V_D	$V_{I(A \times A)}$	$V_{I(A \times D)}$	$V_{I(D \times D)}$	$V_{I(A \times A \times A)}$	$\cdots$
Cov(P^O)(P)	$\frac{1}{2}$	0	$\frac{1}{4}$	0	0	$\frac{1}{8}$	$\cdots$
Cov(P^O)(P^M)	$\frac{1}{2}$	0	$\frac{1}{4}$	0	0	$\frac{1}{8}$	$\cdots$
Cov(HS)	$\frac{1}{4}$	0	$\frac{1}{16}$	0	0	$\frac{1}{64}$	$\cdots$
Cov(FS)	$\frac{1}{2}$	$\frac{1}{4}$	$\frac{1}{4}$	$\frac{1}{8}$	$\frac{1}{16}$	$\frac{1}{8}$	$\cdots$
Cov(DZ)	$\frac{1}{2}$	$\frac{1}{4}$	$\frac{1}{4}$	$\frac{1}{8}$	$\frac{1}{16}$	$\frac{1}{8}$	$\cdots$
Cov(MZ)	1	1	1	1	1	1	$\cdots$
General	x	y	x^2	xy	y^2	x^3	$\cdots$

SOURCE: After Falconer, *Introduction to Quantitative Genetics*. Ronald Press. Copyright 1960, p. 157.

Identical or monozygotic (MZ) twins are genetically identical. Thus,

$$\text{Cov}(MZ) = \text{Cov}(G + E^C + E^{W(1)})(G + E^C + E^{W(2)})$$

$$= \text{Cov}(G)(G) + \text{Cov}(E^C)(E^C)$$

$$= V_G + V_{E^{C(MZ)}}$$

$$= V_A + V_D + V_{E^{C(MZ)}}.$$

As mentioned previously in this section, variance due to epistatic inter-action deviations also contributes to the covariance of relatives. These contributions to the genetic covariances previously discussed are sum-marized in Table 9.12. It will be noted that the coefficients of the various epistatic variances are simple functions of the coefficients for the additive and dominance variances. In addition, it may be noted that coefficients for the higher order epistatic interactions become quite small, except for MZ twins, and thus probably contribute relatively little to the observed resem-blance of relatives other than MZ twins.

Heritability.   Lush (1940) first defined *heritability* "as the fraction of the observed variance which was caused by differences in heredity." It was emphasized that heritability is a population parameter and thus is a function of both the character under study and the population in which it is measured. Lush (1949) later defined "heritability in the narrow sense" as the ratio of the additive genetic variance (variance due to the average effects of genes) to the phenotypic variance (variance due to individual differences in pheno-typic values) and "heritability in the broad sense" as the proportion of the phenotypic variance due to all sources of genetic variance, i.e., both additive and nonadditive. The former ratio is more conveniently referred to simply as heritability (Falconer, 1960) and that terminology shall be adopted here.

Heritability provides a useful alternative to the old nature-nurture dichotomy:

> Furthermore, it gradually came to be recognized that the question whether the nature or the nurture, the genotype or the environment, is more important in shaping man's physique and his personality is simply fallacious and misleading. The genotype and the environment are equally important, because both are indispensable. . . . The nature-nurture problem is nevertheless far from meaningless. Asking right questions is, in science, often a large step toward obtaining right answers. The question about the roles of the genotype and the environment in human development must be posed thus: To what extent are the *differences* observed among people conditioned by the differences of their genotypes and by the differences between the environments in which people were born, grew and were brought up? (Dobzhansky, 1964, p. 55)

Or, as Roberts has stated: "We need to know how much of the total variation (in a population) is due to various genetic causes, for it is axiomatic that the importance of a source of variation is proportional to the contribution it makes to the total variation" (Roberts, 1967a, p. 217). Thus, heritability in the broad sense (h_B^2) may be thought of as an index of the relative importance of gene differences as a cause of individual differences in a population, where

$$h_B^2 = \frac{V_G}{V_P},$$

and V_G and V_P are the genotypic and phenotypic variances.

In addition to having this descriptive property, heritability is also predictive. In the absence of a genotype-environment correlation, heritability in the broad sense is equivalent to the regression of the genotypic value of an individual on its phenotypic value:

$$b_{GP} = \frac{\text{Cov}(G)(P)}{V_P} = \frac{\text{Cov}(G)(G+E)}{V_P} = \frac{\text{Cov}(G)(G)}{V_P} = \frac{V_G}{V_P} = h_B^2.$$

If a positive correlation exists between G and E, h_B^2 will underestimate this regression:

$$b_{GP} = h_B^2 + \frac{\text{Cov}(G)(E)}{V_P} = h_B^2 + \frac{r_{GE}\sigma_G\sigma_E}{V_P} = h_B^2 + (h_B)(e)(r_{GE}),$$

where $e = (V_E/V_P)^{1/2}$. The maximum value that $(h_B)(e)$ may achieve is 0.5; thus, unless r_{GE} is relatively large, h_B^2 will not differ substantially from b_{GP}, even in the presence of a genotype-environment correlation.

Table 9.13

Phenotypic resemblance of relatives

Relatives	Covariance	Regression (b) or Correlation (t)
Offspring, 1 parent	$\frac{1}{2}V_A + V_{EC(OP)}$	$b_{P^0P} = (\frac{1}{2}V_A + V_{EC(OP)})/V_P \geq \frac{1}{2}h^2$
Offspring, midparent	$\frac{1}{2}V_A + V_{EC(OP)}$	$b_{P^0PM} = (\frac{1}{2}V_A + V_{EC(OP)})/\frac{1}{2}V_P \geq h^2$
Half sibs	$\frac{1}{4}V_A$	$t_{HS} = \frac{1}{4}V_A/V_P = \frac{1}{4}h^2$
Full sibs	$\frac{1}{2}V_A + \frac{1}{4}V_D + V_{EC(FS)}$	$t_{FS} = (\frac{1}{2}V_A + \frac{1}{4}V_D + V_{EC(FS)})/V_P \geq \frac{1}{2}h^2$
DZ twins	$\frac{1}{2}V_A + \frac{1}{4}V_D + V_{EC(DZ)}$	$t_{DZ} = (\frac{1}{2}V_A + \frac{1}{4}V_D + V_{EC(DZ)})/V_P \geq \frac{1}{2}h^2$
MZ twins	$V_A + V_D + V_{EC(MZ)}$	$t_{MZ} = (V_A + V_D + V_{EC(MZ)})/V_P \geq h_B^2$

SOURCE: After Falconer, *Introduction to Quantitative Genetics*. Ronald Press. Copyright 1960, p. 162.

In a similar manner, it may be shown that heritability (narrow sense), symbolized h^2, is equivalent to the regression of the additive genetic value of an individual (sum of the average effects of that individual's genes) on its phenotypic value, i.e., $h^2 = b_{AP}$, where A is the additive genetic value. Thus, except when r_{AE} is large, h^2 may be used to predict the additive genetic value of an individual (or the mean additive genetic value of a group), based upon the observed phenotypic value. In addition, since the mean genotypic value of the progeny of an individual is equal to one-half the additive genetic value of that individual, the phenotypic value of progeny and the response to selection may be predicted using this parameter. This predictive property of h^2 will be discussed in more detail in the next section of this chapter.

The square root of heritability is equivalent to the correlation between the additive genetic value of an individual and its phenotypic value:

$$r_{AP} = b_{AP}\frac{\sigma_P}{\sigma_A} = h^2(\frac{1}{h}) = h.$$

This is a reasonable result, since the square of this correlation should indicate the proportion of the variance in P due to variation in A, which by definition is h^2.

Although the contribution of the genetic variance to the covariance of relatives was derived in the previous section, resemblance of relatives is usually measured in terms of regression coefficients or correlations. For parents and offspring, the preferred statistic is the regression (Falconer, 1960). For sibs and twins, however, the intraclass correlation (t) is used. Because regressions and intraclass correlations are nothing more than covariances divided by appropriate variances, these parameters may be simply expressed in terms of the genetic variances derived in the previous section (see Table 9.13). It should be noted that variance due to epistatic interaction deviations has been omitted from the expressions in Table 9.13 because the contributions due to this source should be relatively small for most relatives except MZ (identical) twins. For reasons discussed earlier, variance due to common environmental effects (V_{Ec}) has also been omitted from the expression for half sibs.

From Table 9.13, it may be seen that h^2 may be estimated in a number of different ways. For example, h^2 may be estimated by doubling the regression of scores of offspring on the phenotypic values of individual parents. However, this method could lead to overestimates of h^2 if parents and offspring share important environmental effects. Although the covariance of offspring and one parent is equal to that of offspring and midparent (average score obtained by mother and father), the regression of offspring on midparent provides a direct estimate of h^2. This is due to the fact that the variance of the midparental value is one-half that of single parental scores when mating is at random. This method of estimating h^2 has several advantages, including its insensitivity to nonrandom mating (Falconer, 1960). That departures from random mating can change the genetic variance in a population will be discussed later in this chapter.

The correlation among half sibs must be multiplied by four to estimate h^2; thus, any errors of measurement will also be multiplied by four. As a consequence, estimates of h^2 based upon half-sib correlations tend to be quite variable except when sample sizes are relatively large. In animal breeding research, where records of hundreds of progeny "artificially" sired by hundreds of bulls are available, this method has been most useful.

Doubling the correlation of full sibs is likely to yield overestimates of h^2. The genetic covariance of full sibs includes one-fourth of the dominance variance (V_D) as well as half of the additive genetic variance. In addition, as mentioned previously, variation due to common environmental effects has been frequently found to be an important source of variance in full-sib families.

As examples of the use of these methods, we may again consider open-field behavior in mice. DeFries and Hegmann (1970) tested mice from 72 F_2 litters and then chose one male and two females at random from each litter. These mice were then mated at random with the restriction that each male be mated to two females; thus, the resulting 128 litters (841 mice) consisted of a number of half-sib and full-sib families. The heritability of open-field activity in mice may be estimated in several different ways from these data. As mating was at random and the variances were similar in males and females, the regression of offspring on midparent may be employed. The resulting estimate was 0.22 ± 0.09, where ± 0.09 is the *standard error*, a measure of confidence to be placed in the estimate (Falconer, 1963). Mice that have different mothers but the same father are half-sibs. Multiplying the half-sib correlation by four resulted in a heritability estimate of 0.14 ± 0.14. This estimate is somewhat lower than that obtained from the regression of offspring on midparent, but is also less reliable. Members of the same litter are full sibs. Doubling the full-sib correlation yields a value of 0.74, clearly in excess of the other estimates. This is likely due to the common environmental effects shared by members of the same litter, either prenatal, postnatal, or both.

Although regression estimates are preferred to correlations for parent-offspring comparisons, the correlation of single-offspring and single-parental scores also provides an estimate of $\frac{1}{2}h^2$. When midparental values or the means of scores of several offspring are involved, however, resulting correlations are not a simple function of h^2 (see DeFries, 1967).

In order to illustrate the estimation of h^2 using single-offspring and single-parent correlations, let us again consider the results of Skodak and Skeels (1949) on the resemblance of adopted children to their biological mothers presented in Chapter 8. The observed correlation (Figure 8.7) of about 0.35 suggests that the heritability of performance on IQ tests may be as high as $2(0.35) = 0.7$; however, due to the likelihood of positive assortative mating for intelligence (members of a mating couple being more similar to each other than would be the case if mating were entirely at random), this is probably an overestimate.

It may be seen from Table 9.13 that use of the MZ correlation or twice the DZ correlation will also result in overestimates of h^2. However, it might be argued that the MZ correlation may provide a valid estimate of heritability in its broad sense (h_B^2) if the twins are separated shortly after birth and reared in homes in which the environmental effects are uncorrelated. In such a case, the only possible source of $V_{E_{C(MZ)}}$ would be due to common prenatal maternal effects. In 1966, Burt reported that the correlation between members of 53 pairs of identical twins separated shortly after birth was 0.86 for an individually administered IQ test. Burt also obtained data concerning the occupational class of the breadwinner in the homes in which the separated twins were reared. In most cases, one of the twins was brought up by its natural parents, whereas the other twin was reared by foster parents. The correlation between the occupational classes of the parents and corresponding foster parents was essentially zero. Material and cultural conditions of the homes were also examined and, in general, neither measure was found to be significantly related to performance on the individual test. Although these results suggest that h_B^2 may be rather high for performance on individually administered IQ tests (at least in this population), non-specified environmental effects may still have contributed to the observed correlation.

As seen in Table 9.13, MZ and DZ correlations each contain a component of variance due to common environmental effects; however, the relative contributions of these components will differ if the environmental factors shared by MZ twins are more similar than those shared by DZ twins. Thus, $V_{E_{C(MZ)}}$ minus $V_{E_{C(DZ)}}$ is probably nonzero. Nevertheless, since t_{MZ} contains all of V_A, whereas t_{DZ} contains only $\frac{1}{2}V_A$ when mating is at random, it is possible to approximate h_B^2 by doubling the difference between these correlations:

$$h_B^2 \simeq 2(t_{MZ} - t_{DZ}) = 2\left[\frac{V_A + V_D + V_{E^{C(MZ)}}}{V_P} - \frac{(\frac{1}{2}V_A + \frac{1}{4}V_D + V_{E^{C(DZ)}})}{V_P}\right]$$

$$= \frac{V_A + (1.5)V_D + 2(V_{E^{C(MZ)}} - V_{E^{C(DZ)}})}{V_P}.$$

Although not indicated in this derivation, variance due to epistatic interaction deviations will also be over-represented in the numerator of this expression. Thus, when mating is at random, this method of estimating heritability will result in an overestimate if any of the following three conditions exist: (1) environmental factors are more similar for pairs of MZ than pairs of DZ twins; (2) dominance variance is nonzero; (3) epistatic variance is nonzero. However, under certain conditions, this method may yield underestimates of h_B^2. For example, when there is positive assortative mating (discussed later in this chapter), the covariance of DZ twins contains a fraction of V_A larger than one-half. Thus, doubling the difference between MZ and DZ correlations can result in either an underestimate or an overestimate of h_B^2.

Let us now consider some actual data. Vandenberg (1971) has recently summarized a number of twin studies in which intraclass correlations of MZ and DZ twins on the same IQ tests were reported. These correlations and their corresponding estimates of broad-sense heritability (h_B^2) are presented in Table 9.14. In addition, Holzinger's H index is also presented, where:

$$H = \frac{t_{MZ} - t_{DZ}}{1 - t_{DZ}}.$$

It may be shown that this ratio is equivalent to the following:

$$H = \frac{\frac{1}{2}V_A + \frac{3}{4}V_D}{\frac{1}{2}V_A + \frac{3}{4}V_D + V_{E^{W(DZ)}}},$$

where V_{E^W} is the within-family environmental variance and it is assumed that $V_{E^{C(MZ)}} = V_{E^{C(DZ)}}$. Thus, although Holzinger's ratio has frequently been used to provide an estimate of h^2, it is clearly inappropriate for this purpose. The mean H across the 13 studies in Table 9.14 is 0.64, whereas the mean h_B^2 is 0.52. If mating were at random for this character, we could justifiably conclude that h_B^2 must be 0.52 or less. However, as shown later in this chapter, positive assortative mating for this character does occur. Thus, we must regard the obtained value of 0.52 with some skepticism, keeping in mind that it could be either an underestimate or an overestimate of h_B^2.

Table 9.14

Summary of estimates for the heritability of performance on IQ tests using MZ and DZ twin correlations: h_B^2 *is broad-sense heritability estimated by doubling the difference between MZ and DZ twin correlations; H is Holzinger's index*

Country	Published report	t_{MZ}	t_{DZ}	h_B^2	H
U.S.A.	(1932) Day	0.92	0.61	0.62	0.80
England	(1933) Stocks & Karn	0.84	0.65	0.38	0.54
U.S.A.	(1937) Newman, Freeman, & Holzinger[a]	0.90	0.62	0.56	0.74
Sweden	(1952) Wictorin[a]	0.89	0.72	0.34	0.61
Sweden	(1953) Husén	0.90	0.70	0.40	0.67
England	(1954) Blewett	0.76	0.44	0.64	0.57
England	(1958) Burt	0.97	0.55	0.84	0.93
France	(1960) Zazzo	0.90	0.60	0.60	0.75
U.S.A.	(1962) Vandenberg[b]	0.74	0.56	0.36	0.41
U.S.A.	(1965) Nichols	0.87	0.63	0.48	0.65
England	(1966) Huntley	0.83	0.66	0.34	0.50
Finland	(1966) Partanen, Bruun, & Markkanen[c]	0.69	0.42	0.54	0.51
U.S.A.	(1968) Schoenfeldt[d]	0.80	0.48	0.64	0.62

SOURCE: After Vandenberg, 1971, p. 197.

[a]Average of 2 tests.

[b]Average of 6 tests, recalculated from twin differences.

[c]Average of 8 tests.

[d]Data for both sexes combined.

Vandenberg (1967) has also summarized the literature pertaining to the genetics of normal variation in personality characters. In one table, Vandenberg pools data from three separate studies that reported MZ and DZ correlations for performance on the Minnesota Multiphasic Personality Inventory (MMPI). Considered individually, each of the studies is relatively small. However, there is a total of 120 MZ twin pairs and 132 DZ twin pairs when all three studies are considered together. Due to the presence of sampling errors, sample sizes of at least this magnitude are required to provide reasonably reliable estimates of heritability. These pooled MZ and DZ twin correlations and the resulting estimates of h_B^2 and H are presented in Table 9.15. A wide range of values for h_B^2 is obtained, suggesting that some personality factors may be more heritable than others. However, the sample sizes are too small for much confidence to be placed in these differences and, furthermore, some of the observed variation may merely reflect difference in scale reliability. From Table 9.15 it may also be seen that H does not always result in estimates larger than h_B^2.

Table 9.15

Values of t_{MZ} *and* t_{DZ} *for combined MMPI samples from Gottesman (1963, 1965) and Reznikoff and Honeyman (1967) and estimated values of* h_B^2 *and* H

Item	t_{MZ} (N = 120 pairs)	t_{DZ} (N = 132 pairs)	h_B^2	H
Social introversion	0.45	0.12	0.66	0.37
Depression	0.44	0.14	0.60	0.35
Psychasthenia	0.41	0.11	0.60	0.34
Psychopathic deviate	0.48	0.27	0.42	0.28
Schizophrenia	0.44	0.24	0.40	0.27
Paranoia	0.27	0.08	0.38	0.21
Hysteria	0.37	0.23	0.28	0.19
Hypochondriasis	0.41	0.28	0.26	0.17
Hypomania	0.32	0.18	0.28	0.17
Masculinity-femininity	0.41	0.35	0.12	0.09
L scale	0.46	0.17	0.58	0.35
F scale	0.40	0.38	0.04	0.03
K scale	0.35	0.20	0.30	0.18
Es scale	0.41	0.41	0.00	0.00

SOURCE: After Vandenberg, 1967, p. 78.

The methods for estimating heritability outlined in Table 9.13 are by no means exhaustive. For example, data from inbred lines and their F_1, back-cross, and F_2 generations may be utilized (see Bruell, 1962). Since these are the generations used by Mendel, we may refer to this method as the *classical analysis*. When two highly inbred strains are crossed to produce F_1 hybrids, the variance observed within each of these three isogenic populations should be due to nongenetic causes. Thus, the phenotypic variance observed in each of these populations should provide an estimate of the environmental variance:

$$V_E = V_{P1} = V_{P2} = V_{F1},$$

where V_{P1} is the phenotypic variance observed in strain 1, etc. In addition, the phenotypic variance observed in the F_2 generation should reflect all sources of variation, i.e.,

$$V_{F2} = V_G + V_E.$$

As $V_{F2} - V_E = V_G$, h_B^2 may be estimated from the data of the isogenic and F_2 generations as follows:

$$h_B^2 = \frac{V_{F2} - V_E}{V_{F2}}.$$

Tests for the presence of significant epistatic interaction deviations in such data sets are available (cf. Rasmuson, 1961). If found to be an important source of variation in the raw data, a transformation may be available which so rescales the data that genotypic effects become additive across loci. When this assumption is fulfilled, it may be shown that the additive genetic variance may be estimated as follows:

$$V_A = 2V_{F2} - (V_{B1} + V_{B2}),$$

where V_{B1} and V_{B2} symbolize the phenotypic variances of the two backcross generations. Thus, h^2 may be estimated from the ratio of this estimate of V_A and the F_2 variance:

$$h^2 = \frac{V_A}{V_{F2}}.$$

By subtracting the estimates of V_E and V_A from the F_2 variance, an estimate of V_D may also be obtained.

Adequate data for a classical analysis were available in the study of open-field behavior in mice previously mentioned (DeFries and Hegmann, 1970). When the variances in the F_2 and backcross generations were utilized, the heritability of open-field activity was estimated to be approximately 0.43. This estimate is considerably higher than the values previously estimated from regression of offspring on midparental value and from four times the half-sib correlation, 0.22 and 0.14, respectively. One possible explanation for this higher estimate may be due to the use of parental strains that were markedly different in open-field behavior. Thus, although the reason for the difference between these estimates of h^2 is not known with certainty, segregation of blocks of genes with positive genotypic values and blocks of genes with negative genotypic values within the F_2 and backcross generations may result in overestimates of the genetic variance when the classical analysis is employed. More information concerning the partitioning of genetic variances from the data of inbred lines and their derived generations is presented in the monograph, *Biometrical Genetics* (Mather and Jinks, 1971).

The methods for estimating h^2 discussed thus far are all appropriate for characters that are continuously variable (quantitative). However, characters that do not manifest continuous variation may also have a quantitative genetic basis. For example, individual differences in susceptibility to many diseases that are not inherited in a simple Mendelian manner are nonetheless to some extent heritable. This may be inferred from the higher incidence of the disease among the relatives of affected individuals than in the general population. The heritability of such a character may be estimated by methods proposed by Falconer (1965), which assume that the all-or-none character has an underlying gradation of liability to the disease. Individuals having a liability above a certain value are assumed to exhibit the disease, whereas those whose liability is below this *threshold* value do not. Liability in this

Table 9.16

Heritability of liability to schizophrenia

Relationship to index case	Investigator and year	Incidence among relatives of index case		Falconer's h^2
		N	%	
Same-sex twins	Slater (1953b)			
	MZ co-twins	28/41	68	105% ± 8%
	DZ co-twins	11/61	18	106% ± 14%
	Gottesman and Shields (1966)			
	MZ co-twins	14/28	50	87% ± 9%
	DZ co-twins	4/34	12	86% ± 21%
	Kringlen (1966)			
	MZ co-twins	28/64	44	82% ± 6%
	DZ co-twins	12/100	12	86% ± 12%
Parents, sibs, and children	Ødegaard (1963)			
	Age corrected	84/832	10	79% ± 4%
	Erlenmeyer-Kimling et al. (1966)			
Sibs	Observed	131/2007	6.5	61% ± 3%
	Age corrected	131/1260.5	10	80% ± 2%
Aunts and uncles (second-degree relatives)	Ødegaard (1963)			
	Age corrected	81/1749	4.6	96% ± 8%

SOURCE: From Gottesman and Shields, 1967, p. 202.

view is thus an unobserved phenotype that is a function of both genotypic values and environmental deviations. Data pertaining to the incidence of the disease in the general population and among relatives of index cases are necessary to estimate the heritability of liability to the disease.

Gottesman and Shields (1967), proponents of a polygenic theory of schizophrenia, were the first to apply this approach to the study of psychopathology. Their results, assuming an incidence of schizophrenia in the general population of one percent, are summarized in Table 9.16. Although these results suggest that the heritability of liability to schizophrenia may be quite high, values greater than 100 percent indicate that some of the underlying assumptions (e.g., normality of liability distribution among relatives of the index case and equal variance to that of the general population) are unfulfilled when twin data are utilized. In addition, failure to meet the assumption of an absence of environmental effects common to index cases and their relatives and the assumption concerning absence of nonadditive gene effects may also have resulted in overestimates of h^2. More recent results (Gottesman and Shields, 1972), obtained by a somewhat improved method, indicate that the heritability of liability to schizophrenia is about 0.85. Alternative genetic models that have been proposed to explain the inheritance of schizophrenia will be considered in Chapter 11.

Selection. The heritability of a character is most useful for purposes of predicting the response to selection. When individuals selected to be parents for the next generation differ in gene frequency from that of the general population, a change in gene frequency will result. For polygenic characters, however, changes in the frequency of genes at individual loci are only rarely detected. Nevertheless, changes in the population mean are observable. We may predict the change in the mean that will accompany one generation of such selection as follows:

$$\bar{P}^O - \bar{P} = b_{P^OP^M}(\bar{P}^M - \bar{P}),$$

where $\bar{P}^M$ is the mean phenotypic value of individuals chosen to become parents of the next generation. The *response to selection,* symbolized R, is the difference between the mean of the offspring and that of the population mean in the previous (unselected) generation, i.e., $R = \bar{P}^O - \bar{P}$. The amount of selection exerted is measured by the difference between the mean of the selected individuals and the mean of the unselected population from which they were derived, i.e., $S = \bar{P}^M - \bar{P}$, where S is the *selection differential.* We have previously shown that the regression of offspring on midparent provides an estimate of h^2 (narrow sense). Thus, we may rewrite the expression for R in the following form:

$$R = h^2 S.$$

From this formulation, the response to selection or "gain" may be predicted from knowledge of the heritability of a character and the selection differential. Conversely, if a selection experiment has already been undertaken, the *realized heritability* may be estimated from the ratio of the observed response to the selection differential:

$$h^2 = R/S.$$

The relationship between these variables is illustrated in Figure 9.5.

Selection experiments were employed early in the history of behavioral genetics. In 1924, Tolman reported the results of two generations of selection for maze learning by rats. Tolman saw the genetic approach, and selective breeding particularly, as a tool for "dissecting" behavioral characteristics:

> The problem of this investigation might appear to be a matter of concern primarily for the geneticist. Nonetheless, it is also one of very great interest to the psychologist. For could we, as geneticists, discover the complete genetic

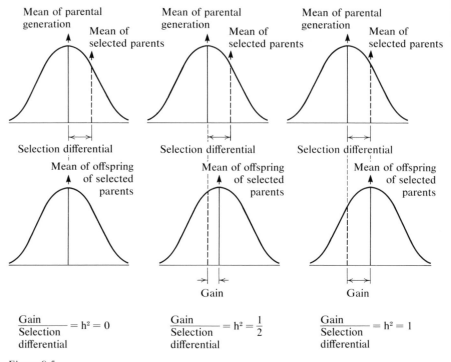

Figure 9.5

Relationship between selection differential, response to selection or gain, and realized heritability. (From Lerner, Heredity, Evolution, and Society, W. H. Freeman and Company. Copyright © 1968, p. 142.)

mechanism of a character such as maze-learning ability—i.e., how many genes it involves, how these segregate, what their linkages are, etc.—we would necessarily, at the same time, be discovering what psychologically, or behavioristically, maze-learning ability may be said to be made up of, what component abilities it contains, whether these vary independently of one another, what their relations are to other measurable abilities, as, say, sensory discrimination, nervousness, etc. The answers to the genetic problem require the answers to the psychological, while at the same time, the answers to the former point the way to those of the latter. (1924, p. 1)

As his own contribution toward this end, Tolman began with a diverse group of eighty-two rats, which were assessed for learning ability in an enclosed maze. Using as a criterion for selection "a rough pooling of the results as to errors, time, and number of perfect runs," nine male and nine female "bright" rats were selected and mated with each other. Similarly, nine male and nine female "dull" rats were selected to begin the "dull" line. The offspring of these groups constituted the first selected generation. These animals were then tested in the maze and selection was made of the brightest of the bright and the dullest of the dull. These two groups of selected animals were mated among themselves (brother × sister) to provide the second selected generation of "brights" and "dulls."

The results were quite clear in the first generation, with the bright parents having bright progeny, and the dull parents dull progeny. Due to the completeness of the data presented by Tolman (1924), it is possible to estimate the realized heritability for each of the three characters selected: errors, 0.93; time, 0.57; and number of perfect runs, 0.61. However, due to small sample size, these estimates are subject to large standard errors. The difference between "brights" and "dulls" decreased, however, in the next generation, primarily because of a drop in efficiency of performance of the bright strain. These second generation results were, of course, disappointing, and Tolman examined various possible explanations. In the first place, the maze used turned out to be a not particularly reliable measuring instrument. Secondly, it was suggested that the mating of brother with sister might have led to what was known as inbreeding degeneration—a phenomenon quite commonly encountered in genetic work.

To facilitate further investigation, an automatic, self-recording maze was developed by Tolman in collaboration with Jeffress and Tryon (1929). With the new maze, which provided superior control of environmental variables and which proved to be highly reliable, Tryon began the selection procedure again, starting with a large and highly heterogeneous "foundation stock" of rats. The energies of Tolman himself were taken up in the development of his theory of learning, and he did no further actual experimentation on behavioral genetics. Nevertheless, he made a continuing contribution to the field by insisting on the importance of heredity in his well-known H.A.T.E. (Heredity, Age, Training, Endocrine, drug, vitamin conditions) list of individual-difference variables.

In Tryon's experiment, selection was based upon the total number of entrances into blinds from days 2 to 19, following a preliminary run of 8 days to acquaint subjects with the maze. Intentional inbreeding was again practiced. The results are shown in Figure 9.6. A fairly consistent divergence between the bright and dull strains may be noted through generation 7, at which time there was practically no overlap between the distributions of the two groups, i.e., the dullest bright rats were about equal in performance to the brightest dull rats. Little or no additional response to selection was observed in the later generations. From an unpublished paper in the series "Studies in Individual Differences in Maze Ability," Tryon provided sufficient information for Tyler (1969) to calculate the realized heritability, which was found to be 0.21.

Although Tryon's selection experiment is a classic in experimental behavioral genetics, the design suffers from several inadequacies that have been perpetuated in more recent selection research. As indicated, deliberate

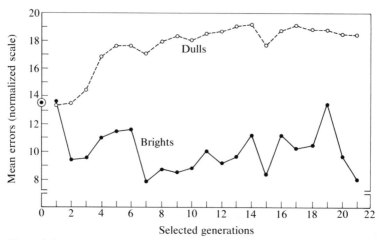

Figure 9.6

The results of Tryon's selective breeding for maze brightness and maze dullness. (From McClearn, 1963, p. 213.)

inbreeding was practiced by both Tolman and Tryon. Although one of the objectives of these studies was to produce highly inbred lines with uniform behavioral differences, inbreeding may impede the response to selection. Inbreeding results in a decrease in genetic variance within lines and, thus, a decrease in the potential selection response. In addition, inbreeding is almost always accompanied by a reduction in fertility, resulting in a decrease in the selection differential.

Another inadequacy of this experimental design is the lack of an unselected control group. When such a group is included, effects of intergeneration environmental influences may be measured. In addition, the response to selection in the high and low lines may each be measured by their deviation from that of the control group. In this manner, the degree of asymmetry of response to selection may be ascertained. Finally, selected and control lines should each be replicated. Since considerable intergeneration variability is encountered in selection experiments, the reliability of the result may be indicated by the inclusion of replicate selected lines. More importantly, fortuitous correlations between the character under selection and other characters of interest may often occur when only one high line and one low line are being maintained. As indicated in Chapter 6, correlations across two strain means are of relatively little value in mechanism oriented research. However, if similar associations are noted in each of two or more replicates, the correlation is much more likely to be indicative of a causal relationship.

These refinements have been incorporated into a selection experiment for open-field activity in mice that is currently in progress in our laboratory.

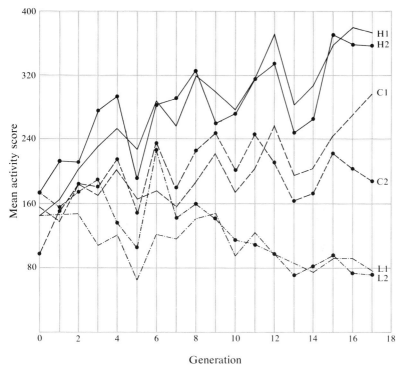

Figure 9.7
Response to selection for open-field activity in mice.

The foundation population for selection was the F_3 generation derived from an initial cross of two highly inbred strains, BALB/cJ and C57BL/6J. Six closed lines have been formed, two selected for high open-field activity, two selected for low activity, and two randomly mated to serve as controls. Within-litter selection is practiced, i.e., a male and a female are saved for breeding from each litter, in order to minimize inbreeding. Selected animals are then mated at random within each line and an effective population size of ten mating pairs per line is maintained.

The direct response to selection is shown in Figure 9.7. Data from a total of 8,029 mice are summarized in this figure. In general, a clear, relatively symmetrical response to selection has been realized. However, considerable intergeneration variability exists that tends to affect all lines in the same direction. Since the response is relatively symmetrical, effects of these temporary environmental effects may be removed by plotting the difference in activity scores between the high and low lines. From this divergence of response, presented in Figure 9.8, a more systematic trend is evident.

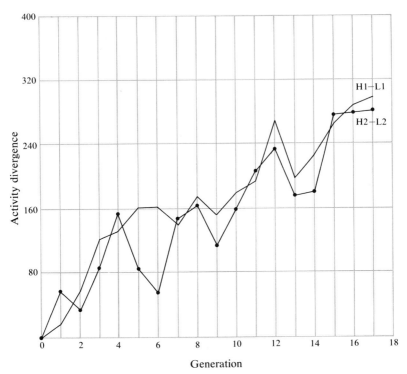

Figure 9.8
Divergence of response to selection for open-field activity in mice.

Laboratory conditions were essentially the same for the first five genera-tions of the selection study and the heritability studies previously discussed. Thus, a check on the validity of these earlier estimates may be provided by comparing the response during the first five generations of selection to the cumulative selection differential during this period. The response should be compared to the cumulative selection differential, rather than the selection differential for a given generation, as the response is a measure of the cumu-lative gain due to selection across all generations. The resulting realized heritability estimate is 0.26 ± 0.03, in close agreement with that estimated from the regression of scores of F_3 subjects on their F_2 midparental values (0.22). Thus, if the validity of an estimate of heritability is indicated by its agreement with the realized heritability (the proof of the pudding in this case), the results of this study indicate that the regression method may be more appropriate than the classical analysis.

In addition to the experiments discussed above, selection has been shown by various workers to be effective in producing strain differences in the fol-lowing: spontaneous activity, open-field behavior, and saccharine preference in rats; susceptibility to audiogenic seizures, wildness and tameness, aggressiveness and alcohol preference in mice; aggressiveness and mating behavior in chickens; and geotaxis, phototaxis, mating speed, and activity in *Drosophila*, to name but a few. This great success of selective breeding is indicative of the considerable genetic variation that exists for behavioral characters in segregating populations. More importantly, these results indicate that selective breeding can generate strains ideal for mechanism related research in a wide range of organisms.

Fundamental Theorem of Natural Selection. When estimating realized heritability, it is essential that the correct selection differential be used. If we only consider the mean value of the selected parents and not the number of offspring that they contribute to the next generation, systematic biases may occur in the data due to nonrandom contribution of the parents. For example, individuals with more extreme phenotypes may be less fit and hence may contribute fewer offspring to the next generation than individuals with more intermediate phenotypes. In order to adjust for the possible nonrandom differential fertility of selected parents, a weighted selection differential may be calculated as follows:

$$S = \left(\frac{1}{n}\right)\sum\left[\frac{f_i}{\bar{f}}(X_i - \overline{X})\right],$$

where f_i is the number of progeny produced by the i-th parental pair, $\bar{f}$ is the average contribution of all parental pairs, X_i is the midparental value of the i-th parental pair, and n is the number of parental pairs. This procedure was employed in analysis of the open-field-activity selection experiment described above.

This formula for the weighted selection differential also facilitates the derivation of an important result obtained by Fisher (1930) many years ago. Although Fisher's derivation was complicated, the result was elegantly simple: Fisher showed that the change per generation in relative fitness is equal to its additive genetic variance. We may now easily derive this result using the methods of quantitative genetics (Falconer, 1966). Let $F_i = f_i/\bar{f}$ symbolize relative fitness, i.e., the relative contribution of the i-th parental pair to the population constituting the next generation. The weighted selection differential for relative fitness is as follows:

$$S_F = \left(\frac{1}{n}\right)\sum\left[\frac{f_i}{\bar{f}}(F_i - \overline{F})\right]$$
$$= \left(\frac{1}{n}\right)\sum[F_i(F_i - \overline{F})]$$
$$= \frac{\sum F_i^2 - [(\sum F_i)^2]/n}{n}$$
$$= V_{P_F}.$$

The response to selection, i.e., the change in relative fitness per generation (R_F), may thus be estimated as follows:

$$R_F = h_F^2 S_F$$
$$= (V_{A_F}/V_{P_F})(V_{P_F})$$
$$= V_{A_F}.$$

For a species well adapted to its particular ecological niche, fitness must be relatively constant from generation to generation; thus, additive genetic variance for characters that are major components of fitness must be near zero for such populations. Empirical evidence agrees with this expectation, i.e., the heritabilities of such characters of obvious fitness value as fertility are generally low (Falconer, 1960). It must be recognized, however, that a low heritability may also be indicative of an unreliable test instrument.

As a corollary to this theorem, most of the genetic variance associated with fitness characters in stable populations must be nonadditive (Roberts, 1967b). Since nonadditive gene action is indicated by the deviation of the F_1 from the midparental value, fitness characters are expected to manifest considerable *heterosis*, or hybrid vigor (Bruell, 1964b; 1967). Heterosis may be thought of as the mirror image of inbreeding depression since characters that are subject to inbreeding depression also display considerable heterosis when different inbred lines are crossed. Thus, we may characterize fitness characters in stable populations as being those that have low heritabilities and that manifest considerable heterosis and inbreeding depression. Application of these criteria to behavioral characters will be undertaken in Chapter 10.

Correlated Characters.   Thus far in this chapter we have utilized information on only one character of each individual. However, several characters may be measured on each individual and subjected to quantitative genetic analysis. If two characters (X and Y) are measured on each individual in a population and a correlation is observed, this phenotypic correlation may be due to either genetic or environmental causes. Among the genetic causes, pleiotropy is the most interesting since it results in permanent correlations between characters. Due to recent admixture of populations, temporary associations between genes at different loci may sometimes occur (genetic disequilibrium), also resulting in genetic correlations. It is easy to visualize how environmental effects may give rise to correlations between characters. A favorable diet, for example, may result in higher values for both characters X and Y, whereas an unfavorable diet may be accompanied by depressed values for both characters.

It may be shown that the observed phenotypic correlation between two characters may be partitioned into its genetic and environmental parts as follows:

$$r_{P(XY)} = h_X h_Y r_{A(XY)} + e_X e_Y r_{E(XY)},$$

where $r_{A(XY)}$ is the genetic correlation (correlation between the additive genetic values for two characters on the same individuals in a population) and $r_{E(XY)}$ is the corresponding environmental (plus nonadditive genetic) correlation.

The expression just given indicates that r_A and r_E do not merely add up to r_P. Instead, each is weighted by corresponding values of $h_X h_Y$ or $e_X e_Y$. As a consequence, inferences about the relative magnitude of r_A and r_E are difficult to make from $r_{P(XY)}$ alone, even when such correlations are observed in both segregating and isogenic populations. This problem has recently been discussed by McClearn, Wilson, and Meredith (1970).

Genetic correlations may be estimated by methods that are perfectly analogous to those used to estimate heritability. If a set of data permits heritability estimation, genetic correlations may be estimated from the same data set if two or more characters are measured on each individual. For example, consider data from parents and their offspring:

$$r_{A(XY)} = \frac{\text{Cov}(P_X)(P_{Y_0})}{\sqrt{[\text{Cov}(P_X)(P_{X_0})][\text{Cov}(P_Y)(P_{Y_0})]}}.$$

This is the so-called "cross-covariance" method (Falconer, 1960), where character X of the parent is compared to character Y of the offspring in the numerator of the above equation. This method was employed in the genetic analysis of open-field behavior in mice discussed previously. From the cross-covariance of F_3 offspring and their F_2 midparental values, a genetic correlation between open-field activity and defecation scores of -0.76 ± 0.14 was obtained. This very high genetic correlation indicates that activity and defecation in the open-field-test situation are probably influenced by many of the same genes.

In addition to serving as an index of the extent to which two characters are influenced by the same sets of genes, r_A may be used to predict the change in a correlated character (Y) that accompanies direct selection for character X. This correlated response in Y, symbolized CR_Y, may be predicted as follows:

$$CR_Y = r_{A(XY)} h_X h_Y \frac{\sigma_{P_Y}}{\sigma_{P_X}} S_X.$$

As with direct selection, the correlated response observed during selection may be used to estimate the realized genetic correlation. During the course of selection for open-field activity in mice, a correlated response in

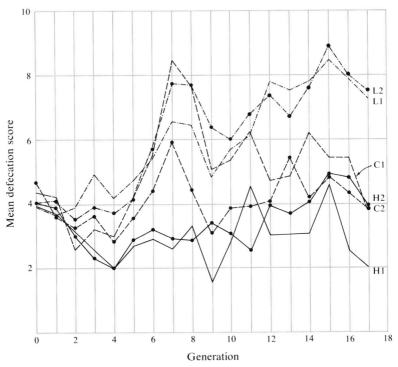

Figure 9.9
Correlated response in open-field defecation during selection for activity.

open-field defecation scores was noted. This correlated response is shown in Figure 9.9. From the data of the first five generations of selection (same laboratory conditions as when data were obtained during the F_2 and F_3 generations), the realized genetic correlation was found to be -0.86 ± 0.14. This is in relatively close agreement to the estimate obtained from the cross-covariance of F_3 offspring and their F_2 midparental values (-0.76), again indicating that these behaviors are influenced by many of the same genes.

Broadhurst (1967) has selected for open-field behavior in rats and in a similar fashion has noted a negative genetic correlation between open-field activity and defecation scores. However, in his experiment, selection was based upon the defecation response and activity (ambulation) was the corre-lated character. The direct and correlated responses to selection are shown in Figures 9.10 and 9.11, respectively. These Maudsley "reactive" and "nonreactive" strains have been assayed for a large number of behavioral and physiological correlates. In a summary by Eysenck and Broadhurst in 1964, these strains were reported to have differed significantly on 24 of 32 different behavioral tests administered during the course of selection and on

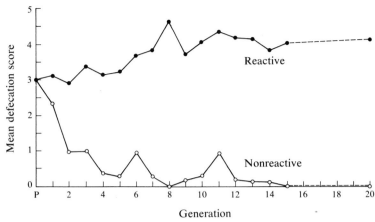

Figure 9.10

Progress of selection for high and low defecation in the rat—the Maudsley emotionally reactive and nonreactive strains. After the fifteenth generation, selection was suspended for five generations. Measurement at the twentieth generation showed little loss of divergence. (From Broadhurst, 1967, p. 126.)

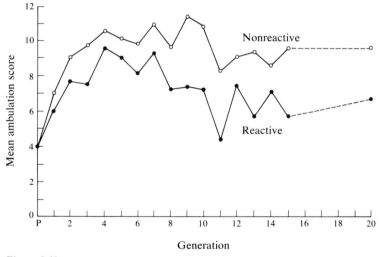

Figure 9.11

Graph of the ambulation scores of the selectively bred rats illustrating the effect upon ambulatory activity in the open-field test of the bidirectional selection practiced for high and low defecation—reactive and nonreactive strains respectively. (From Broadhurst, 1967, p. 130.)

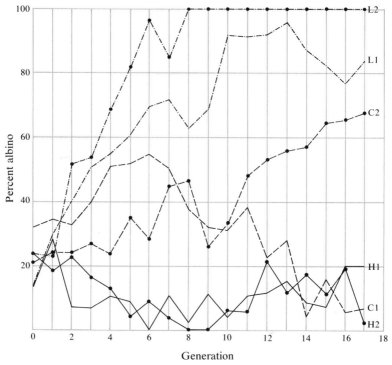

Figure 9.12

Change in the genotypic frequency of albinism as a function of selection for open-field activity in mice.

19 of 24 physiological measures. Unfortunately, for the reasons outlined earlier, without additional research it is not possible to ascribe a causal relationship to observed correlations in an experiment in which only one high line and one low line are being selected.

Major Genes in Polygenic Systems. Open-field behavior in mice is clearly influenced by genes at many loci. Although minimum estimates of the number of loci that influence a character are crude at best (see Chapter 5), estimates of 3.2 and 7.4 for activity and defecation, respectively, indicate that both characters are probably polygenic (DeFries and Hegmann, 1970). Nevertheless, as discussed in Chapter 5 and earlier in this chapter, a major gene effect on open-field behavior has been found. In the brightly illuminated open-field test situation, albino mice are less active and have higher defecation scores than pigmented animals. If albinism has a major effect on open-field activity, the frequency of albinism would be expected to change as a function of selection. As shown in Figure 9.12, changes in the frequency of albinism are in the expected directions. Fixation has been achieved for albinism in one low-active line (L_2) and its frequency is high in the other. Due to the difficulty of eliminating a rare recessive gene, the lines selected for high open-field activity will probably continue to segregate for albinism for many generations.

These results indicate that the techniques of single-gene analysis and quantitative genetic analysis may be fruitfully combined. As a result of the quantitative genetic analysis of open-field behavior, estimates of the additive genetic variance of open-field activity and defecation due to all gene effects are available. As indicated earlier in this chapter, identification of a major gene effect permits calculation of the additive genetic variance in the character due to segregation at this single locus. From the ratio of these two variances, we find that segregation at the c locus accounts for 12 percent of the additive genetic variance in open-field activity and 26 percent of the additive genetic variance in defecation.

It may be hoped that further research in this area will lead to the identification of additional single-gene effects within polygenic systems. It is entirely conceivable that a large fraction of the genetic variance in quantitative characters may be due to segregation at only a few loci. As may be seen from the derivations in this section, it has not been necessary to assume that all gene effects within polygenic systems are small, equal in magnitude, or even additive in effect. Thus, a combination of the methods of single-gene analysis and quantitative genetic analysis may provide a powerful means of dissecting behavior into more basic units.

Systems of Mating

In the preceding section of this chapter, it was assumed that random mating was occurring with regard to the character under study. In this section, some of the consequences of departures from random mating will be examined.

Inbreeding.   The mating of individuals more closely related than average — that is, *inbreeding* — is measured by Wright's (1922) coefficient of inbreeding:

$$F_x = \sum [(\tfrac{1}{2})^{n_1+n_2+1}(1 + F_a)],$$

where F_x is the coefficient of inbreeding of some individual x, n_1 is the number of generations from one parent of x back to an ancestor that is common to both parents, n_2 is the number of generations from the other parent of x back to the common ancestor, F_a is the coefficient of inbreeding of the common ancestor, and Σ refers to the summation of the contributions obtained after considering each ancestral contribution individually.

The coefficient of inbreeding has several different interpretations. It may be thought of as the percentage decrease in heterozygosity in an individual, relative to that existing in members of some base or founding population. Thus, inbreeding by itself will influence genotypic frequencies in populations, but not gene frequencies. F_x may also be thought of as the probability that a pair of alleles carried by an individual are *identical by descent*, i.e., are replicates of a particular gene carried by a common ancestor. The coefficient of inbreeding was originally defined by Wright (1921) as the correlation between uniting gametes.

Wright (1922) also proposed a measure of relationship:

$$R_{yz} = \frac{\Sigma[(\frac{1}{2})^{n_1+n_2}(1+F_a)]}{\sqrt{(1+F_y)(1+F_z)}},$$

where R_{yz} is the coefficient of relationship of two individuals, y and z, n_1 is the number of generations from y back to a, n_2 is the number of generations from z to a, and F_y and F_z are the coefficients of inbreeding of y and z. The coefficient of relationship is formally equivalent to the correlation between the additive genetic values for a given character measured on each of two individuals; however, it may be shown that the coefficient of relationship between two individuals is the same for all characters, i.e., R_{yz}, like F_x, is not character specific. The coefficient of relationship is also equal to the probability that two individuals will have replicates of the same genes carried by a common ancestor.

Let us now consider the pedigrees in Figure 9.13. The path pedigree illustrated in Figure 9.13 (a) represents the mating of two unrelated individuals, y and z. Thus, by definition, F_x must equal zero. Recalling our expression for F_x, what are the appropriate values for n_1 and n_2 in this case? Since no common ancestor is indicated in this pedigree, n_1 and n_2 must be assumed to be very large, *not* zero. When n_1 and n_2 are very large, F_x approaches zero.

In Figure 9.13 (b), y and z are half sibs, i.e., y and z have one (and only one) parent in common. In this case, j is the common ancestor and there is only one chain of paths that connect y and z through this common ancestor; thus, $n_1 = n_2 = 1$, and

$$F_x = (\tfrac{1}{2})^{1+1+1}(1+F_j) = (\tfrac{1}{2})^3 = \tfrac{1}{8},$$

since j is not indicated in the pedigree as being inbred. When y and z are half sibs,

$$R_{yz} = \frac{(\tfrac{1}{2})^{1+1}(1+F_j)}{\sqrt{(1+F_y)(1+F_z)}} = \tfrac{1}{4},$$

since $F_y = F_z = F_j = 0$ in this example.

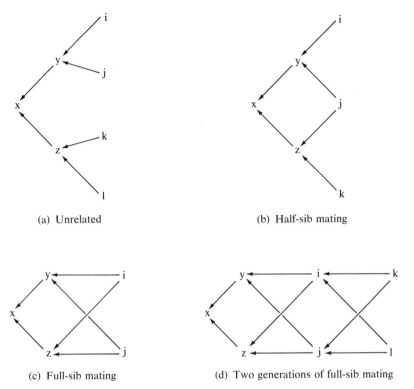

(a) Unrelated

(b) Half-sib mating

(c) Full-sib mating

(d) Two generations of full-sib mating

Figure 9.13
Hypothetical path pedigrees illustrating various levels of inbreeding.

When y and z are full sibs, as in Figure 9.13 (c), two common ancestors must be considered:

$$F_x = (\tfrac{1}{2})^{1+1+1}(1 + F_i) + (\tfrac{1}{2})^{1+1+1}(1 + F_j) = 2(\tfrac{1}{2})^3 = \tfrac{1}{4},$$

when the common ancestors are not inbred. As expected, the coefficient of relationship of full sibs is one-half. That is,

$$R_{yz} = \frac{(\tfrac{1}{2})^{1+1}(1 + F_i) + (\tfrac{1}{2})^{1+1}(1 + F_j)}{\sqrt{(1 + F_y)(1 + F_z)}} = \tfrac{1}{2},$$

where $F_i = F_j = F_y = F_z = 0$.

Highly inbred strains of animals are usually produced by successive generations of full-sib mating. Two generations of full-sib mating are illustrated in Figure 9.13 (d). In this case, y and z have four common ancestors. In addition, y and z may be connected through k and l in each of two ways.

For example, y and z may be connected through k by the chain of paths: y-i-k-j-z and y-j-k-i-z. All of these contributions must be considered when evaluating F_x:

$$F_x = (\tfrac{1}{2})^{1+1+1}(1+F_i) + (\tfrac{1}{2})^{1+1+1}(1+F_j) + 2(\tfrac{1}{2})^{2+2+1}(1+F_k) + 2(\tfrac{1}{2})^{2+2+1}(1+F_l)$$

$$= 0.375,$$

where i, j, k, and l are all noninbred. In this case, however, y and z are inbred, since the parents of y and z are full sibs. Thus,

$$R_{yz} = \frac{2(\tfrac{1}{2})^2 + 4(\tfrac{1}{2})^4}{\sqrt{(1+\tfrac{1}{4})(1+\tfrac{1}{4})}} = 0.6.$$

With regular systems of mating, recurrence equations may be developed that express the coefficients of inbreeding in a given generation as a function of inbreeding in previous generations. For successive generations of full-sib mating,

$$F_t = \tfrac{1}{4}(1 + 2F_{t-1} + F_{t-2}),$$

where F_t is the coefficient of inbreeding after t generations of full-sib mating, F_{t-1} is the coefficient of inbreeding in the previous generation, etc. Thus, with only one generation of full-sib mating,

$$F_1 = \tfrac{1}{4}(1 + 0 + 0) = \tfrac{1}{4},$$

as shown above. With two generations of full-sib mating,

$$F_2 = \tfrac{1}{4}[1 + 2(\tfrac{1}{4}) + 0] = 0.375.$$

After three generations,

$$F_3 = \tfrac{1}{4}[1 + 2(\tfrac{3}{8}) + \tfrac{1}{4}] = 0.50, \text{ etc.}$$

The coefficients of inbreeding resulting from 20 generations of full-sib mating are presented in Table 9.17. In addition, the probability of fixation is also indicated. The coefficient of inbreeding refers to the percentage decrease in heterozygosity in individuals. For inbred lines, however, we are more interested in the probability that all individuals in a line are homozygous for the same allele. The probability of fixation shown in Table 9.17 may be regarded as being a minimum value, since it was assumed that the maximum number of different alleles (four) at a locus were present in the initial sib mating. If fewer different alleles were present, the probability of fixation would be higher. On the other hand, natural selection for heterozygosity may impede the approach to fixation. From Table 9.17 it may be seen that both the coefficient of inbreeding and the probability of fixation are asymptotically

Table 9.17

Coefficients of inbreeding and probabilities of fixation resulting from successive generations of full-sib mating

Generation (t)	F_t	Probability of fixation
1	0.250	0
2	0.375	0.063
3	0.500	0.172
4	0.594	0.293
5	0.672	0.409
6	0.734	0.512
7	0.785	0.601
8	0.826	0.675
9	0.859	0.736
10	0.886	0.785
11	0.908	0.826
12	0.926	0.859
13	0.940	0.886
14	0.951	0.908
15	0.961	0.925
16	0.968	0.940
17	0.974	0.951
18	0.979	0.960
19	0.983	0.968
20	0.986	0.975

SOURCE: After Falconer, *Introduction to Quantitative Genetics*. Ronald Press. Copyright 1960, p. 91.

approaching one after 20 generations of full-sib mating. For mice, the term "inbred strain" is reserved for strains that are products of at least 20 generations of full-sib mating. Thus, members of such strains are likely to be homozygous for the same alleles at a very high proportion of their loci. In fact, most commercially available pedigreed inbred mice have been brother-sister mated for 50–100 or more generations.

When populations have few members, inbreeding may result even when mating is at random. It may be shown that the change in inbreeding per generation, symbolized ΔF, may be estimated as follows:

$$\Delta F \cong \frac{1}{2N_e},$$

Table 9.18

Genotypic frequencies under inbreeding

Genotype	Frequency	Genotypic value
A_1A_1	$p^2 + pqF$	$+a$
A_1A_2	$2pq(1 - F)$	d
A_2A_2	$q^2 + pqF$	$-a$

where N_e is the *effective population size* and is a function of the numbers of breeding males (N_m) and females (N_f) in the population:

$$N_e = \frac{4N_mN_f}{N_m + N_f}.$$

Thus, upon substitution into the above equation and rearrangement, it may be seen that

$$\Delta F \cong \frac{1}{8N_m} + \frac{1}{8N_f}.$$

That a behavior (social dominance) may result in a small effective population size and, hence, in an increased level of inbreeding, will be discussed in Chapter 10.

If there is nonadditive gene action at a locus, inbreeding will bring about a change in the mean genotypic value. Only effects at a single locus will be considered here. The genotypes, their values, and frequencies under inbreeding are indicated in Table 9.18. Since F refers to the percentage decrease in heterozygosity, it is reasonable that the frequency of heterozygotes should be reduced by a factor of F and equally distributed in the homozygous classes. The products of the frequencies and the genotypic values, summed across the three genotypes, yield the following mean genotypic value:

$$\overline{G} = [a(p - q) + 2dpq] - 2pqdF,$$

where the term enclosed in brackets is the mean genotypic value under random mating. Thus the change in the mean genotypic value for a single locus is a function of gene frequency, level of dominance, and inbreeding. Since pq is at a maximum when $p = q = \frac{1}{2}$, the greatest change in the mean will occur when the gene frequencies are intermediate. In addition, when complete dominance or overdominance exists at a locus, the change will be greater than when dominance is incomplete. Finally, as the change in the mean is a linear function of F, the change will be greater for higher levels of

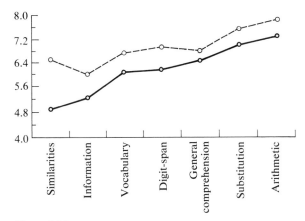

Figure 9.14

Scores on seven subtests of the WAIS achieved by thirty-eight children of first cousins plotted on the solid line; those of forty-seven matched controls, on the broken line. (From Vandenberg, 1971, p. 200. After Cohen et al., 1963.)

inbreeding. Note that the sign of the last term is negative; this indicates that if "increasing" alleles are dominant (i.e., if d is positive), the mean should decrease as a function of inbreeding. Since the mean phenotypic value for fitness characters does decrease with inbreeding (the well-documented phenomenon of inbreeding depression), it follows that genes that convey higher fitness values tend to be dominant in expression.

Relatively little is currently known about the effects of inbreeding on behavioral characters. However, information from several independent sources suggests that inbreeding tends to result in lower IQ scores (Vandenberg, 1971). First, the risk of mental retardation is more than 3.5 times as high among children of cousin marriages as among unrelated controls (Böök, 1957). In addition, children of cousin marriages perform less well on the average on subtests of the Wechsler intelligence test than children of unrelated spouses (Cohen et al., 1963). These results are summarized in Figure 9.14. Finally, results from a large-scale investigation in Japan by Schull and Neel (1965) also reveal deleterious effects of inbreeding on IQ. A Japanese version of the Wechsler intelligence test (WISC) was administered to 486 children of first-cousin marriages, 191 children of first cousins once removed, 188 children of second-cousin marriages, and 989 children whose parents were unrelated. The average effect corresponding to a 10 percent increase in inbreeding is shown in Table 9.19, after effects of differences in age and socioeconomic class were removed. Although there is some variability among subtests in terms of the magnitude of the effect, all means are depressed by inbreeding. As there is about a 7 percent reduction in mean

Table 9.19

Effect of consanguinity on WISC IQ scores

Subtest	Depression as percent of outbred mean	
	Boys	Girls
Information	8.1	8.5
Comprehension	6.0	6.1
Arithmetic	5.0	5.1
Similarities	9.7	10.2
Vocabulary	11.2	11.7
Picture completion	5.6	6.2
Picture arrangement	9.3	9.5
Block design	5.3	5.4
Object assembly	5.8	6.3
Coding	4.3	4.6
Mazes	5.3	5.4
Verbal score	8.0	8.0
Performance score	5.1	5.1
Total IQ	7.0	7.1

SOURCE: From Vandenberg, 1971, p. 200; data from Schull and Neel, 1965.

IQ corresponding to a 10 percent increase in inbreeding, children of full-sib matings (where $F_x = 25\%$) would be expected to score about $(2.5)(7\%) = 17.5\%$ less than children of unrelated parents.

Inbreeding affects the variance of a character as well as the mean. In general, when a population is subdivided into a number of inbred lines, the variance decreases within lines but increases among lines. The total additive and dominance variances in such a population are indicated by the complicated expressions shown below:

$$V_A = 2pq(1+F)\left[a + d(q-p)\left(\frac{1-F}{1+F}\right)\right]^2$$

$$V_D = 4pqd^2\left(\frac{1-F}{1+F}\right)[pq(1-F)^2 + F].$$

It may be seen that these expressions reduce to the more familiar forms of V_A and V_D when $F = 0$, i.e.,

$$V_A = 2pq[a + d(q-p)]^2$$

and

$$V_D = (2pqd)^2,$$

Table 9.20

Additive and dominance variance
in a population corresponding
to various levels of inbreeding,
where dominance is complete
$(a = d = 1)$ and $p = q = \frac{1}{2}$

F	V_A	V_D
0.00	0.50	0.25
0.1	0.55	0.25
0.2	0.60	0.24
0.3	0.65	0.23
0.4	0.70	0.21
0.5	0.75	0.19
0.6	0.80	0.16
0.7	0.85	0.13
0.8	0.90	0.09
0.9	0.95	0.05
1.00	1.00	0.00

when $F = 0$. Because of the complexity of the more general expressions, values of V_A and V_D are presented in Table 9.20 corresponding to various levels of inbreeding. Only the special case of $p = q = \frac{1}{2}$ and complete dominance $(a = d = 1)$ is considered. Under these conditions, it may be seen that the additive genetic variance doubles when F goes from zero to one, whereas the dominance variance goes to zero. More importantly, this table indicates that the variances change very little when F is 0.1 or lower. In present-day human populations, F is almost always less than 0.04, even for very small breeding isolates (Kuse, unpublished); thus, although changes in the mean have been observed to accompany inbreeding, the change in the population variance due to this departure from random mating should be negligible.

Assortative Mating. In contrast to the effects of inbreeding in man, assortative mating may greatly increase the variance of polygenic characters (Crow and Felsenstein, 1968). *Assortative mating* is the mating of individuals who are phenotypically more similar (or dissimilar) than would be expected if mating were entirely at random. Thus, assortative mating (like random mating) is character specific, whereas inbreeding and relationship are not. For this reason, assortative mating may result in a substantial increase in the variance for a character in a population, but is not likely to result in a substantial over-all reduction in heterozygosity.

Since inbreeding and assortative mating have different consequences, it is important to distinguish between them in empirical studies. An observed correlation between mates could be due to inbreeding, assortative mating,

Table 9.21

Phenotypic correlations between mates for various intelligence test scores

Item	Source	N pairs	r
Stanford-Binet	Burks, 1928	174	0.47 ± 0.04
Otis	Freeman et al., 1928	150	0.49 ± 0.04
Army Alpha	Jones, 1928	105	0.60 ± 0.04
Progressive Matrices	Halperin, 1946	324	0.76
Various tests	Smith, 1941	433	0.19 ± 0.03
Vocabulary	Carter, 1932	108	0.21 ± 0.06
Arithmetic	Carter, 1932	108	0.03 ± 0.06
Mental Grade	Penrose, 1933	100	0.44

SOURCE: After Spuhler, 1967, p. 262.

or some combination of both (Lewontin, Kirk, and Crow, 1968). However, since the level of inbreeding in human populations is relatively low, this probably presents no serious bias in studies of assortative mating in man.

Spuhler (1967, 1968) and Vandenberg (1972) have both reviewed the literature concerning assortative mating for physical and behavioral characters in man. The highest correlations between husband and wife are those for age and are in the range 0.51–0.99. Approximately one-third of the 290 correlations for physical characters summarized by Spuhler (1968) were in the range 0.1–0.2. The second and third most common ranges of correlations were 0.0–0.1 and 0.2–0.3. Only 35 correlations were negative and only two of these were significant: number of illnesses, −0.35; and pulse after exercise, −0.20. Thus, it appears that in general there is some positive assortative mating for physical characters in man; however, the observed level is relatively low.

Among behavioral characters, most personality-rating correlations between mates were found to be in the 0.1–0.2 range, comparable to values observed with the physical characters. Correlations for cognitive measures, however, were found to be considerably higher. The correlations between mates for performance on various intelligence tests, from samples each including at least 100 pairs, are summarized in Table 9.21. Although these relatively high values reflect to some extent level of school attainment, they may nonetheless be of sufficient magnitude to result in an increased genetic variance. This may in fact serve as a partial explanation for the finding of a relatively high heritability for performance on IQ tests.

In this chapter, the theory of population and quantitative genetics has been developed and applied to a very limited sample of behavioral characters. In Chapter 10, characters of more obvious evolutionary significance shall be considered.

Evolution and behavior

As discussed in Chapter 1, Darwin's formulation of his evolutionary theory caused great furor and vigorous contention. Although the ensuing years have seen his principal themes well substantiated, organized opposition to the theory continues to this day (Wade, 1972). Some of this opposition is motivated by religious conviction and some by genuine misunderstanding:

> One sees mention of the "Darwinian dogma" (a phrase without meaning) or of the "theory" of evolution, as though it were a hypothesis, a good guess. There is a difference between the "theory of evolution" and "evolutionary theory," for the latter means the whole body of knowledge and explanation tying the known facts together into one system. . . . the facts bearing on evolution known today, compared with those known a century ago, are as a whale to a mouse, and all these facts fall neatly into place in general evolutionary theory. (Howells, 1967, p. 16)

That behavior has been shaped by evolutionary processes is consistent with this body of knowledge. It is equally clear that behavioral processes influence evolution (Parsons, 1967). The complexities of the mutual interaction are substantial and subtle, and it is only for convenience of exposition that we shall consider the evolution of behavior separately from the effects of behavior upon the evolutionary process.

Evolution of Behavior

Natural Selection. Darwin, as we have seen, had no doubt but that evolution molded behavior as much as it did bodily form. The manner in which it did so, of course, had to be left undescribed for decades—until the basic workings of the genetic mechanism became known and the principles of population genetics came to be applied to the evolutionary process. In the previous chapter, the power of artificial selection to alter phenotypes was discussed. Basically, the phenomenon of natural selection is the major systematic process that shapes evolution. Phenomena related to small sample sizes and chance occurrences, the so-called stochastic processes, may also result in changes in gene frequency. As discussed later in this chapter, these processes may influence the rate of evolution. However, by definition these processes are random and thus nonsystematic. Natural selection, on the other hand, can generate organisms of increasingly fine-tuned adaptiveness to their environments. Under artificial selection imposed by man, a high selection differential is usually applied in order to accelerate the selection response. Under natural selection, however, extremely subtle and minute selective advantage or disadvantage can result in evolutionary change. Only a very slight alteration in the probability of leaving offspring is required, because natural selection has such a staggering amount of time within which to work.

In order to deal with this enormous expanse of time, geologists, archeologists, and paleontologists have provided a time scale of geologic eras and periods whose relationships are displayed in Figure 10.1. This figure also indicates the approximate times of origin of various classes of life. This evolutionary time scale is so far beyond human experience that analogies are almost indispensable to comprehension. Lerner has provided one such device in the form of a calendar analogy (Lerner, 1968). If we were to set the origin of earth, approximately 5 billion years ago, as January 1, and set the present as midnight, December 31, we should find that the origin of life would be on about May 26, marine invertebrates would first appear on November 24, and land plants on December 5. Dinosaurs would be on the earth from December 16 until December 25. Man's appearance would have occurred on this scale at 10:15 p.m. on December 31. About three-quarters of a second before midnight, we would note the emergence of the first civilization. Washburn (1968) has used a length analogy: The duration of life on earth can be represented by six football fields lined end to end (600 yards). Mammals would emerge on the last fifteen yards, man on the last goal line. Man is, in brief, a very recent evolutionary experiment.

That natural selection has a staggering amount of time in which to work by no means implies that rapid changes cannot occur if the necessity arises. The best studied case of rapid evolution is that of industrial melanism of certain species of moths in England. Although both light and dark (melanic) forms have been known for some time, the light variant was most common

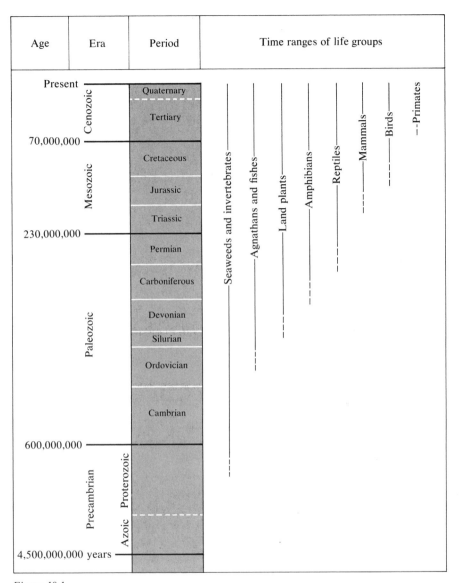

Figure 10.1
Time scale of geologic eras and periods. (After Oakley and Muir-Wood, 1967, p. 63.
Courtesy of the Trustees of the British Museum [Natural History].)

until about the middle of the nineteenth century. Since then the dark form
has begun to replace the light form, especially in heavily polluted areas near
factories. Kettlewell (1965) has studied this phenomenon in some detail and
has discovered that the two forms are differentially visible in polluted and

in unpolluted areas to birds that prey on them. Light moths on a tree trunk in an unpolluted area are almost invisible, while dark moths are clearly visible (see Figure 10.2). In contrast, in areas where tree trunks are darkened by airborne pollutants, dark forms are less visible than light forms. Release and recapture studies of the light and dark forms in both polluted and unpolluted environments have provided convincing evidence that natural selection has been responsible for the spread of the dark form.

The moth example illustrates that natural selection is imposed by the environment. Thus, relative fitness is a function of the ability to avoid predators, to find mates, to obtain food, to resist disease, and so on, under a given set of environmental circumstances. However, as Manning (1965) has stated, an organism may modify its environment by its behavior:

> By its behaviour an animal ceases to be a merely passive subject of natural selection. Even the most humble creature can influence its own evolution by its behaviour. Man uses his brain to create for himself an amenable environment. He thereby makes one basic set of physiological and structural adaptations serve, with minor modifications, from the Poles to the Equator. To a lesser extent and by much more stereotyped methods, the nest-building social insects also create their own environments, as do beavers when they dam up a stream. These are conspicuous examples of behaviour controlling environment, but in fact all animals possess in some measure the ability to choose the habitat in which to live and so adjust their environment to suit their physiology. For instance, most invertebrates have simple behavioural mechanisms which serve to move them out of dry places into moist, to avoid extremes of heat and cold, and so on. By its choice of habitat, an animal determines what food is available to it, how far it competes with its own or other species, and perhaps who it mates with. Changes to these and many other factors will, in turn, modify the selective forces acting on the animal. (p. 125)

Regardless of the environment, the ultimate criterion of fitness is the number of progeny left by an individual. In human terms, the fittest are not necessarily the brawniest nor the prettiest nor the brightest nor the socially most valuable; what counts in Darwinian fitness is *only* the number of viable progeny left behind. It is extremely important to keep this distinction between Darwinian fitness and social value in mind.

Although we usually think of natural selection as operating by means of the differential reproduction of individuals within a population, differences in fitness between one population and another or one group of organisms and another may also occur. As an example of a character that may have been subjected to *intergroup selection*, consider social dominance. As Wynne-Edwards (1963) has argued, order is maintained in a group with a stable dominance hierarchy with little risk of injury to its members, once the hierarchy has been established. In addition, social dominance may serve as an effective means of population control. Thus, such groups may have a

Figure 10.2

Left: *Peppered moths on a tree trunk in an unpolluted area. The almost invisible light form is below and to the right of the melanic form.* **Right:** *The visible and invisible forms are reversed in a polluted area. (From the experiments of H. B. Kettlewell, University of Oxford.)*

higher probability of surviving and transmitting their genes to succeeding generations than those in which a stable dominance hierarchy is not maintained.

The importance of group selection as a force in evolution is still a subject of some disagreement. Nevertheless, it provides a reasonable explanation for the evolution of characters of seemingly little fitness value to the individual. For a number of years, evolutionary biologists have been concerned with explaining the evolution of altruistic behaviors, i.e., those actions in which individuals expose themselves to risk of injury or death in order to benefit other members of their species (Dobzhansky, 1970). Many examples are well known, including parental defense of the young, issuing of alarm calls, and so on. Although the fitness value of the altruist may be lessened by such actions, groups that contain altruists may have a considerable advantage over those that do not. Natural selection against antisocial behavior may also occur via group selection. Murder, robbery, and rape (*especially* rape) might be considered to be of positive fitness value to the individual who successfully commits such acts. However, a high incidence of such behavior within a group might seriously diminish the fitness of the group as a whole and, hence, would be selected against.

One other particularly salient point about evolution is that it operates without foresight. Much of the early speculation and theorizing about evolutionary processes was impeded by teleological conjectures that sought ways in which the ultimate goal of an evolutionary trend could operate upon the trend itself. It has become clear that natural selection can only work with the raw material at hand, which in this context means the gene pool of the species. A given evolutionary problem, such as, say, care of young, may be "solved" in a variety of different ways depending upon the starting point of each developing species when the problem arises. As Pittendrigh (1968) has put it, it is

> from such an understanding of selection that we eventually perceive adaptive organization in its true light: not as . . . perfection demanding an intelligent designer, but rather as a patchwork of makeshifts pieced together, as it were, from what was available when opportunity knocked, and accepted in the hindsight, not the foresight, of natural selection. (p. 400)

Speciation.   The cumulative effect of very small advantages, when selection has been sustained and persistent, can gradually modify a character in a population. Many traits may be changing concurrently, and if some of the trends are incompatible with others, compromises must be reached. Eventually, a species may change to such a point that, in comparison with earlier versions of the same lineage, we choose to call it a separate species. Thus, although it is elementary, it may be rather startling to contemplate that if a person traces his ancestry back far enough, he would encounter creatures that he would be unwilling to acknowledge as human. Indeed, if he went back far enough, his ancestors would not even be recognizable mammals. Yet in no particular generation would there be a discontinuity. For each generation, the parents would be no more dissimilar from their children than ours are from us.

The environment is not constant, of course, over space or over time. Regional differences in the demands of environment result in geographically separated members of a species being exposed to different selection pressures, which in turn can lead to development of varieties, races, or subspecies. If this trend proceeds far enough, speciation occurs. That is, groups that were once just regional varieties can gradually become more and more divergent until they can no longer mate successfully with each other. It is important to note that a restriction in gene flow among such groups is a requisite of speciation. This may result from geographical isolation or from reproductive isolation. A number of different reproductive isolating mechanisms have been identified (Dobzhansky, 1970), including many behavioral ones.

Speciation may occur as a result of a change in the environment or as a result of a shift of an existing species into new or previously unoccupied niches. Consider again the species of finches that Darwin studied on the Galápagos Islands (Figure 10.3). Although these various species are all believed to be of common origin, considerable differentiation has occurred among them. They differ most obviously in body size and beak structure, necessary adaptations to the different niches in which they are found.

Speciation such as that observed in Darwin's finches may be initiated by a behavioral change. As Mayr (1965) has stated:

A shift into a new niche or adaptive zone is, almost without exception, initiated by a change in behavior. The other adaptations to the new niche, particularly the structural ones, are acquired secondarily (Mayr 1958, 1960). With habitat and food selection — behavioral phenomena — playing a major role in the shift into new adaptive zones, the importance of behavior in initiating new evolutionary events is self-evident. Sibling species, in spite of their morphological similarity, often show remarkable behavioral differences. Most recent shifts into new ecological niches are, at first, unaccompanied by structural modifications (Robson and Richards 1936). Where a new habit develops, structural reinforcements follow sooner or later. . . . If the new behavior adds to fitness, it will be favored by selection and so will be all genes that contribute to its efficiency. That new habits occur all the time in natural populations is abundantly documented in the natural history literature. A particularly striking example is the recently developed habit of British titmice, mostly *Parus major* (Fisher and Hinde 1948), of opening milk bottles and drinking the cream. If the milk bottles had been a natural unoccupied niche, it

Figure 10.3
The 14 species of Galápagos and Cocos Island finches. The species (a) *is a woodpecker-like finch that uses a twig or cactus spine instead of his tongue to dislodge insects from tree bark crevices;* (c), (d), *and* (e) *are insect-eaters;* (f) *and* (g) *are vegetarians;* (h) *is the Cocos Island finch. The birds on the ground eat mostly seeds. Note the powerful beak of* (i), *which lives on hard seeds. The birds are shown about one-third size. (From Lack, "Darwin's finches." Copyright © 1953 by Scientific American, Inc. All rights reserved.)*

is evident that a selection pressure would have been set up on one hand for the titmice to develop a more efficient milk-bottle opener, and for the milk bottles to become less easily opened, assuming the milk bottles to be organic material that could be modified with the help of selection. (pp. 604–605)*

Temporal changes in the selection pressures imposed by the environment may cause evolution to proceed for awhile in one direction, and for awhile in another, and so on. If the gene pool of a species does not have sufficient alleles that are compatible with continued life and reproduction in an altered environment, the species becomes extinct. Strictly speaking, in actuarial terms, extinction is the expected fate of a species. There are many more extinct species than extant ones. Some species, of course, became adapted to particular environmental milieus that have remained quite constant for substantial portions of the evolutionary calendar. Alligators and coelocanths, for example, have survived for millions of years without appreciable morphological change. Most species, however, eventually disappear from view either by evolving into new species or by extinction.

Man's Position in the Evolutionary Scheme of Things.   Naturally enough, the focus of our interest is on man and the manner in which his evolutionary history can be related to his present behavioral properties. To approach this problem it is necessary to examine the vertebrate context of man's evolution. Figure 10.4 shows man to be a member of the group of mammals, formally known as the class Mammalia, which, at the present time, as indicated by the width of the grey band representing it, is one of the more prevalent groups. Some two hundred million years ago, mammals originated from the then-prevalent reptiles shortly before birds did. The reptiles themselves originated from the amphibians, who in turn were derived from bony fish. The bony fish were and remain a most successful and prevalent form of life derived from the cartilaginous fishes more than four hundred million years ago.

Of special interest is the order Primates—that is, man and his closest mammalian relatives, the monkeys and the apes. Napier (1970) has provided a compilation of structural and behavioral properties that collectively define the primates:

1. Preservation of a primitive mammalian structure of limbs, e.g., retention of the five-digit pattern of hands and feet, and of the clavicle, the radius and the fibula—bones that are reduced or absent in some groups of mammals.

2. A progressive freedom in mobility of the digits, especially the thumb and the big toe.

3. The replacement of sharp claws by flattened nails associated with the development of sensitive touch pads on the tips of the digits.

---

*Reproduced with permission from *Animal Species and Evolution*. Copyright 1965. Harvard University Press.

4. A progressive shortening of the snout.

5. An increase[d] frontality of the eyes associated with the development of binocular vision.

6. Reduction in the apparatus and function of smell.

7. The reduction in number of teeth and the preservation of a simple molar cusp pattern.

8. Expansion and elaboration of the brain, particularly those regions concerned with vision, muscular coordination, tactile appreciation, memory and learning.

9. Progressive development of truncal uprightness.

10. Progressive elaboration of the placenta, with particular respect to the intimacy of the blood circulation between the mother and the fetus.

11. A greater dependency in locomotion on the forelimbs at the expense of the hindlimbs.

12. Prolongation of prenatal and postnatal life periods.

13. Increase in body size.

14. The development of a complex social system involving, progressively, a greater number of individuals whose hierarchal status must be taken into account. (p. 40–41)

Even with these common features, of course, the species of primates differ widely one from the other. There are two distinct suborders, the Prosimii and the Anthropoidea, with the former being the less "advanced" and including treeshrews, lemurs, lorises, and tarsiers. The Anthropoidea are generally more advanced than the Prosimii, but may themselves be subdivided into less and more advanced. The former are the new world, or Platyrrhini, monkeys, found in Central and South America and the latter are the Catarrhini, including all of the old world monkeys, the hominoid apes, and man. The phylogeny of the primates according to Washburn (1960) is given in Figure 10.5. Not only are the relationships among living species displayed in this figure, but it is also apparent that being a primate is no guarantee of evolutionary success. As can readily be seen, a large number of primate lines have ended in extinction.

Recent developments in molecular biology have facilitated quantitative analyses of the closeness of relationship of man and his living primate relatives. Several techniques are now available, one of which compares the amino acid sequence in the same proteins of different species. Figure 10.6 shows the numbers of amino acid differences found among the hemoglobins of man, the chimpanzee, the gorilla, and the rhesus monkey. The much closer relationship of man to African apes than to old world monkeys has also been shown from amino acid sequence comparisons of other proteins.

Another technique involves the hybridization of purified DNA from different species. In one study (Martin and Hoyer, 1967), DNA fragments

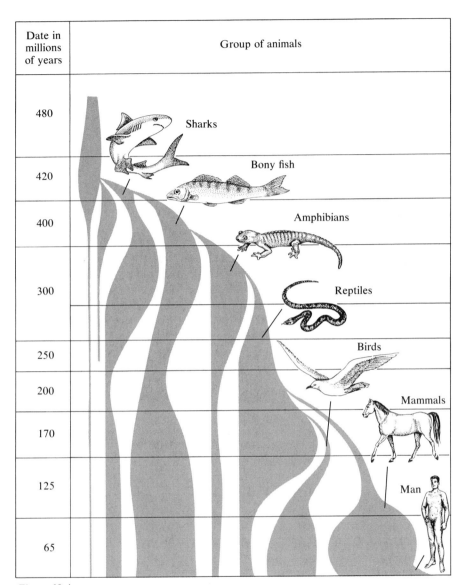

Figure 10.4
Relative importance of various vertebrates since the Cambrian period. (After Hardin,
Biology: Its Human Implications, W. H. Freeman and Company. Copyright © 1949.)

from chimpanzees and rhesus monkeys were tested for their ability to com-
pete with human DNA fragments for binding to unfragmented human DNA.
The competitive abilities of chimpanzee and human DNA were found to
differ only slightly (9 percent), in contrast to a rather marked difference

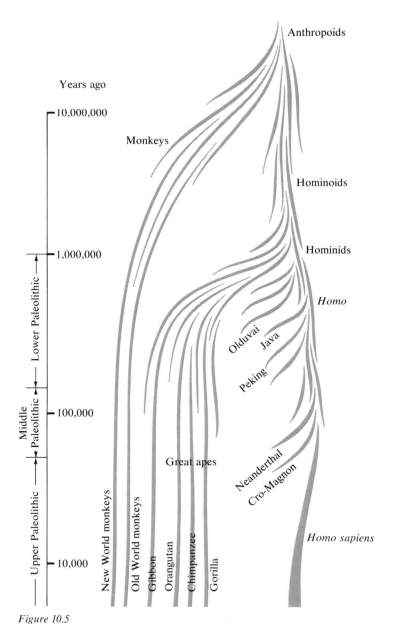

Figure 10.5

Pedigree of Homo sapiens *and his relatives on an exponential time scale. (From Washburn, "Tools and human evolution." Copyright © 1960 by Scientific American, Inc. All rights reserved.)*

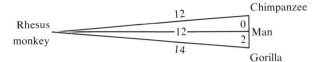

Figure 10.6
Comparison of hemoglobins of primates. The numbers of differences between species in amino acid sequence are given. (After Wilson and Sarich, 1969, p. 1090.)

between rhesus monkey and human DNA (34 percent). These results indicate that the base sequence of chimpanzee and human DNA and, hence, the genetic information encoded therein, is much more similar than that of rhesus monkey and human DNA.

The representation of primate phylogeny in Figure 10.5 is a considerable oversimplification. A much higher level of magnification is required if we wish to examine hominid evolution in any detail. The basic evidence for constructing hominid phylogeny is, of course, fossil material. In Darwin's time, very little of this evidence was available, although enough was on hand to suggest the prior existence of another type of man than ourselves. The evidence since that time has accumulated remarkably, and an extensive number of finds are now available for classification: Peking man, Java man, Heidelberg man, Swanscombe man, Rhodesian man, Solo man, Cromagnon man; the Pithecanthropines, Australopithecines, and so on. Unfortunately, there has been a profusion of names as a result of the tendency to assign each new find to a different species or genus. This trend, which led to great confusion in attempting to establish the phylogenetic relationships among these finds, has recently been reversed and the consensus appears to be emerging that the main groups of concern in hominid phylogeny are *Australopithecus, Homo erectus, Homo neanderthalensis,* and *Homo sapiens.* Representations of the skulls of these groups are shown in Figure 10.7. By no means is there universal agreement about the evolutionary relationships of these men and premen and new information is becoming available almost daily. One of the currently accepted interpretations of hominid phylogeny is shown in Figure 10.8 The principal point is that it is unlikely that a single line could accurately represent the relationships among all of the fossil forms. As radiating branches probably gave rise to some of the species, many of the fossils we discover are possibly not in the main line of human evolution, but represent instead evolutionary experiments that didn't work out.

Approaches to the Study of Behavioral Evolution. One approach to behavioral evolution is comparative. Examination of the behavioral repertoires of living representatives of related species that originated at different

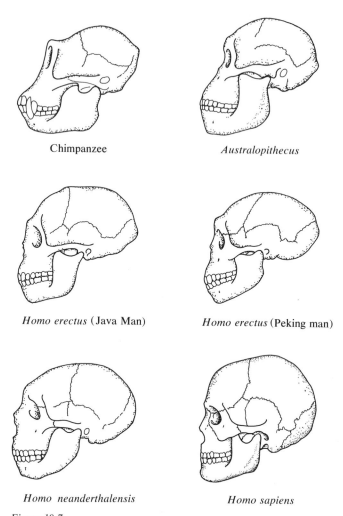

Chimpanzee   *Australopithecus*

Homo erectus (Java Man)   *Homo erectus* (Peking man)

Homo neanderthalensis   *Homo sapiens*

Figure 10.7

*Five hominid skulls shown with that of an anthropoid ape for
comparison. (From Merrell, Evolution and Genetics: The Modern
Theory of Evolution. Copyright © 1962 by Holt, Rinehart, and
Winston, Inc. Reprinted by permission of Holt, Rinehart and
Winston, Inc.)*

times might provide insights into behavioral phylogeny. However, numerous
difficulties are associated with this method. In the first place, the living representatives of a species have been subject to recent evolutionary forces and
may have changed substantially from the ancestral forms from which the
species evolved. Behavior of contemporary birds, for example, may not be

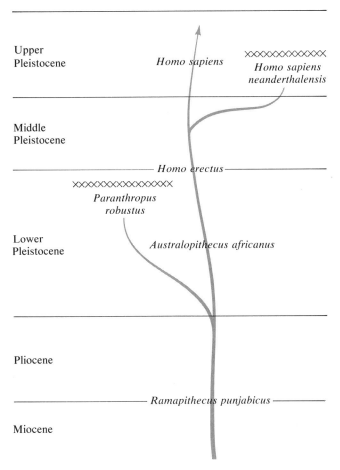

Figure 10.8
One of the current interpretations of hominid phylogeny. (From
Brace, Nelson and Korn, Atlas of Fossil Man. Copyright © 1971
by Holt, Rinehart and Winston, Inc. Reprinted by permission of
Holt, Rinehart and Winston, Inc.)

at all representative of archaeopteryx, the first bird to evolve from reptiles. In the second place, behavior is more difficult to measure and assign to a scale that permits comparisons than are morphological characteristics, and homologies are much less easily ascertained.

As an example of this comparative approach, let us consider the study of the evolution of hearing by Masterton, Heffner, and Ravizza (1969). Audiograms were obtained from four species of primitive mammals (opossum, hedgehog, tree shrew and bush baby) and compared with those of monkeys, apes, and men. These four species were chosen for comparison because of

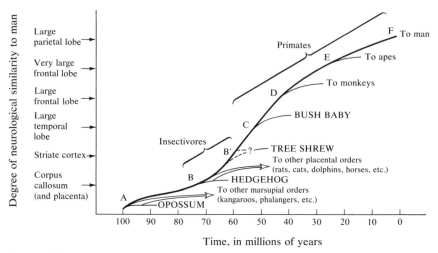

Figure 10.9

Phylogenetic relationship between some living mammals and mammals in man's ancestral lineage. A, B, B', C, and so on represent ancestors whose characteristics can be inferred through comparison of animals in phyletic sequence. (From Masterton et al., 1969, p. 967.)

their varying neurological similarity to man and because of the presumed phylogenetic relationships (Figure 10.9). A number of auditory characteristics were considered by the researchers, but only one will be discussed here: high-frequency sensitivity. In Figure 10.10, the frequencies above which sound cannot be perceived by opossum, hedgehog, tree shrew, bush baby, macaque, chimpanzee, and man are compared. In general, it may be seen that the upper limit of hearing is high and relatively constant among the primitive mammalian species. However, a marked drop in sensitivity to high frequencies may be noted in the more recent primates. These results suggest that high-frequency hearing was fairly commonplace among ancient mammals, but became diminished during later stages of phylogenetic development. The reason for this loss is unknown, but the researchers speculated that it may be related to the ability to discern where a sound comes from. In primitive mammals having close-set ears, high-frequency hearing may be essential for accurate sound localization. Primates with wider spaced ears, however, are able to locate sounds accurately at lower frequencies. Alternatively, the development of binocular vision in more recent primates may reduce for them the importance of the auditory system for localization.

The other general approach to the study of behavioral evolution is to examine the fossil record. In this case, the obvious problem is that, to use a cliché, behavior does not fossilize. In some rare instances, evidence of such behavior as burrowing may be found. More usually, however, it is necessary

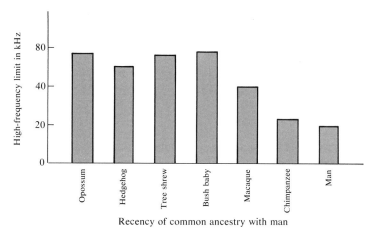

Figure 10.10

High-frequency hearing limits in a phyletic sequence of animals. (From Masterton et al., 1969, p. 974.)

to deduce or infer behavioral functions from the structural evidence and there is inevitably considerable scope for misinterpretation.

Nonetheless, it is possible to paint in broad strokes some general features of behavioral evolution. Romer (1958), for example, has identified a number of critical junctures in animal evolution that must have had important behavioral consequences. These include the development of locomotion, which permitted associated changes in food and which required increasingly elaborate sensory, motor, and neural integrative machinery; the development of jaws, which led further to active pursuit of prey and opened up numerous new ecological niches to the possessors of the jaws; the development of lungs and the subsequent movement of life onto land; and the development of parental behavior associated with the shift from egg laying to bearing live young. Romer notes that nursing behavior had important potentialities in that it favored the establishment of family groups and, through the prolonged association of parent and young, the beginnings of education.

Harlow (1958) has also taken a broad overview in examining the evolution of learning, and concludes that the points of greatest evolutionary change may not have been where a naive first view would suggest:

It is interesting to look at the evolution of learning from the anatomical point of view. We make the assumption that learning is primarily a function of the nervous system, or at least that complexity of learning is intimately related to the developing complexity of the nervous system. If we were to examine learning, using the same kind of evidence that we use for assessing locomotion in the evolution of the horse—the anatomical record—we would be struck by a number of facts. Between the Protozoa and the Coelenterata there must be a vast evolutionary gulf, for the members of the one phylum possess no nerve cells whereas the members of the other do. Between the Coelenterata and the Platyhelminthes there must be another separation, but one of lesser magnitude. In both kinds of organisms the mechanisms associated with coordination and adjustment is neural, but in the flatworm we find a new kind of organized structure, the cephalic ganglion, and this particular structure and its elaborations are going to characterize all higher nervous systems from here through man. Between the flatworm and the dogfish there is also a gulf, but a lesser one than either of the two previous separations; in this instance we have more neurons and some relatively small changes in their physical elaborations in the forebrain, suggesting an increasing differentiation of their single, basic function. From dogfish to man the separation is very slight: the number of neurons has increased, and the process of structural differentiation has continued. From monkey to man there is essentially no difference other than a very slight tendency to continue the evolutionary trends previously noted. (p. 271–272)

Indeed, it is becoming increasingly doubtful that man has any behavioral attributes that are truly unique. At one time it was suggested that the use of tools was reliably diagnostic of man, but the observations of van Lawick-Goodall (1971) have shown that chimpanzees not only use but actually make tools. Similarly, symbolic language was once thought to be the exclusive property of man. Studies on the communication of bees and dolphins, and the remarkable mastery of sign language by a chimpanzee (Premack and Premack, 1972) make this assertion less and less tenable.

The comparative approach to the study of the evolution of behavior, as represented by comparative psychology and ethology, is a rapidly growing field, and an extraordinary amount of information has been accumulated on behavioral characteristics of a wide array of species. For present purposes, some of the most interesting of these researches have been the field studies on various species of primates. This type of research has accelerated rapidly during the past fifteen years, and has revealed a richness and diversity of primate behavior that had been hitherto unsuspected. DeVore (see Washburn, Jay, and Lancaster, 1965), for example, described baboons hunting and killing small mammals. The widely held theory that sexual attraction constitutes the principal basis for primate social groupings has been strongly challenged by a number of observations showing seasonal variation in sexual behavior. Intricate dominance relationships have been identified, with alliances among males or between male and female, and with social rank of the mother influencing the position and behavior of her young in the society. The play of the young has been given a new perspective by these studies as well; indeed, Washburn, Jay, and Lancaster (1965) feel that play is so important that "a species that wastes the emotions and energies of its young by divorcing play from education has forfeited its evolutionary heritage – the biological motivation of learning."

Table 10.1

Cranial capacity of man and his relatives

Species	Extant		Average brain size, in cubic centimeters
	From (yrs ago)	To (yrs ago)	
Australopithecus	2,000,000	600,000	300–600
Homo erectus	1,000,000	500,000	700–1,200
Homo sapiens neanderthalensis	150,000	40,000	1,200–1,500
Homo sapiens sapiens	50,000	Present	1,350

SOURCE: After Lerner, *Heredity, Evolution, and Society.* W. H. Freeman and Company. Copyright © 1968, p. 64.

By the systematic study of tools and of animal-bone fragments associated with fossilized bones of early hominids, it is possible to arrive at some fairly detailed statements about some aspects of their lives. As shown in Table 10.1, *Australopithecus*, for example, who weighed about 60–90 pounds, had a brain comparable in size to that of a modern ape (350–750 cubic centimeters). The foramen magnum, where the spinal cord exits from the brain, was set quite far forward, implying that the animal had an upright posture. Other evidence from hipbone and pelvis remains makes clear that the Australopithecines were thoroughly erect animals walking with a gait similar to our own except, possibly, for a slight waddle. The primate characteristic of upright posture, releasing the forelimbs for manipulation, had already progressed far by the time *Australopithecus* appeared upon the evolutionary scene. Furthermore, his teeth were similar to our own although the molars were rather massive. The canine teeth were reduced in size from those of his forebears. As large canines function chiefly as weapons, we might infer that *Australopithecus* was able to compensate for reduced canines by using his hands to manipulate weapon tools. The tools first used in this way were probably naturally formed stones or broken bones. Evidently, at least one species of *Australopithecus* (*A. africanus*) was a meat eater. Another animal form, called *Paranthropus*, now thought by many to be another species of the genus *Australopithecus* and described as *A. robustus*, was most likely vegetarian. Members of this species were heavier than *A. africanus*, weighing perhaps 120 pounds. Coexisting with *A. africanus* for a while, *A. robustus*, as may be seen by reference to Figure 10.8, became extinct during the early (lower) Pleistocene.

Homo erectus, as exemplified by Peking man and Java man, had a larger cranial capacity than *Australopithecus*. There is much evidence that *H. erectus* made and used tools. Many of the known *H. erectus* skulls have had large holes bashed in, suggesting cannibalism either for food or, possibly, as part of some religious ritual. Something of a forehead with brow ridges was developing in *H. erectus*, a smaller face, and the beginnings of a chin. The first sign of the use of fire is associated with fossils of these men.

Neanderthal man is classified as either *Homo neanderthalensis* or *H. sapiens neanderthalensis*, the latter implying a mere subspecific difference from modern man. He had a large skull accommodating a brain as large or larger than modern man's. Neanderthal men were approximately five feet tall on the average and very heavily built, with barrel chests and short, heavy, and somewhat bowed limbs. One of their distinguishing characteristics was an extremely heavy brow ridge above the eyes. These men were cave dwellers, and many of the fossil remains were discovered in caves (or in strata deposited where formerly there were caves). Among Neanderthal remains has been found evidence of formal burials, which implies the development of religious beliefs. Very finely worked tools and implements of a variety of shapes for specialized purposes are characteristic relics of Neanderthal man. It is clear from the remains of animal bones associated with their own remains that they were superb hunters.

Modern man, *Homo sapiens*, has a brain no larger than Neanderthal man's, but the brow ridges have receded. He possesses a high vertical forehead and a strongly developed chin. Art, in the form of cave drawings, was developed by early representatives of *H. sapiens*. It is not clear whether *H. sapiens* competed directly with Neanderthal man, perhaps exterminating him, whether they interbred, or whether *H. sapiens* were simply very much more efficient at occupying the ecological niche. In any case, *H. sapiens neanderthalensis* became extinct during upper Pleistocene.

Importance of Hunting.   The invention, development, and refinement of hunting has played a central and pivotal role in man's behavioral development. Washburn and Lancaster (1968) have provided a detailed examination of the many ramifications of the hunting mode of life. They observe that:

> Human hunting is made possible by tools, but it is far more than a technique, or even a variety of techniques. It is a way of life, and the success of this adaptation (in its total social, technical, and psychological dimensions) has dominated the course of human evolution for hundreds of thousands of years. In a very real sense our intellect, interests, emotions, and basic social life—all these are evolutionary products of the success of the hunting adaptation. When anthropologists speak of the unity of mankind, they are stating that the selection pressures of the hunting and gathering way of life were so similar and the results so successful that populations of *Homo sapiens* are still fundamentally the same everywhere. (p. 213)

The hunting of large animals requires not only the efficient use of tools, but also a high degree of coordination and cooperation. Men capable of cooperating to kill large animals were capable as well of waging effective war on other men. Indeed, Bigelow (1969) has suggested that the strongest selection pressure on brain size in man was the presence of other men. The selection pressure, then, was for increasing ability to cooperate for conflict. The reader is referred to Lee and DeVore (1968) for a more comprehensive treatment of the behavioral consequences that followed from the hunting-gathering way of life.

Although cooperation, parental behavior, tool use, and intelligence are all of obvious evolutionary significance, a more objective method of assessing the fitness value of different behavioral characters is desirable. As discussed in Chapter 9 (see also Roberts, 1967b), a genetic analysis of a character may yield insights into its evolutionary importance.

Genetics of Fitness Characters. In stable populations, fitness characters are expected to have low heritabilities and most of the observed genetic variance should be nonadditive; thus, such characters should display considerable heterosis and inbreeding depression (Bruell, 1964a, 1964b). Since more is currently known about the genetics of animal behavior than human behavior, we shall chiefly refer to infrahuman species in this section. As an example, consider the extensive analysis of nest building in mice by Lynch (1971) and Lynch and Hegmann (1972). In order to assess nest building quantitatively, the investigators provided individual mice with a preweighed amount of cotton in their food hoppers and, twenty-four hours later, the cotton utilized by each subject to build its nest was removed from the home cage and the cotton remaining in the food hopper was weighed. This procedure was repeated for each of five successive days, yielding nesting scores corresponding to the total amount of cotton utilized throughout the five-day test period.

The heritability of this character was estimated in several different ways (inbred strain comparison, classical analysis, parent-offspring regression). Although some of the methods employed are likely to yield overestimates, an average of 0.12 was obtained, a relatively low heritability. Further research indicated that most of the observed genetic variance is nonadditive. The nesting scores of inbred strains and their F_1 crosses are shown for each of three separate experiments in Figure 10.11. Not only does each of the F_1 means exceed the midparental values (the criterion for presence of heterosis), but the F_1 mean also exceeds the higher parental mean in each case. Thus, there is considerable nonadditive genetic variance for this character. This conclusion is substantiated by the results of an inbreeding study still in progress (Lynch and Hegmann, unpublished). During five generations of full-sib mating, the nesting scores of wild mice were found to drop an average of 12 percent for each 10 percent increase in the coefficient of inbreeding. These findings of a low heritability, considerable heterosis, and inbreeding depression are clearly compatible with the hypothesis that nest building is a major component of fitness in mice.

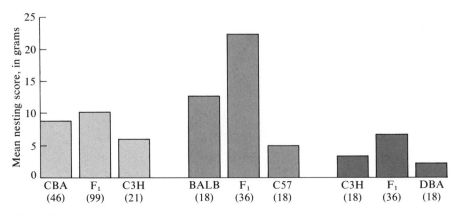

Figure 10.11

Nesting scores for members of inbred strains and their F_1 *hybrids in each of three experiments. Values of* N *are given in parentheses. (After Lynch, 1971, pp. 44 and 46.)*

Aggression is another character that is seemingly of fitness value. As discussed later in this chapter, social dominance is almost certainly a major component of relative fitness in mouse populations. Aggression, however, has often been measured in highly artificial situations. For example, "bouts" are conducted in which pairs of males are placed in neutral chambers and subjectively scored for intensity of fighting. In such situations, large strain differences are usually observed (McClearn and DeFries, In Press).

The most comprehensive genetic analysis of mouse aggression has been provided by Lagerspetz, who reported the results of seven generations of selective breeding for this character in 1964. Her results are summarized in Figure 10.12. From the data presented by Lagerspetz (range of selected males, etc.), it is possible to obtain an approximate estimate of the realized heritability of this character. The resulting estimate of 0.36 is higher than that expected for fitness characters. Perhaps aggression as measured in mouse "bouts" is not a major factor in determining social dominance in natural populations.

It was shown in Chapter 9 that the broad sense heritability of performance on IQ tests is probably in the range of 0.4 to 0.6. Although inbreeding depression for this character indicates that some of the genetic variance is nonadditive, the sizeable correlations observed between parents and offspring (Erlenmeyer-Kimling and Jarvik, 1963) suggest that most of the genetic variance is additive. It seems almost certain that intelligence was of enormous fitness value in man's evolutionary past. The evidence for high heritability of IQ, therefore, suggests strongly that IQ tests are not measuring that type of intelligence for which natural selection has been acting ever since the hominid line emerged. To state the matter another way, intelligence as we define and measure it today has perhaps been of only trivial signifi-

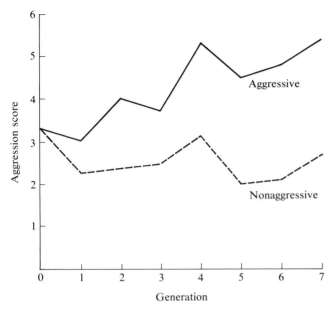

Figure 10.12

Aggression scores of male mice selectively bred for high and low aggressiveness. (After Lagerspetz, 1964, p. 53.)

cance in man's evolution. Numerical ability and verbal ability may have had little relevance to reproductive success prior to the advent of agricultural societies, only some 10,000 years ago. Perhaps performance on an IQ test constructed to correlate highly with success in a hunting and gathering society would be found to have a low heritability in modern man. Whether IQ has a fitness value today, positive or negative, has been a matter of considerable debate. The issue is discussed further in the next chapter.

Behavior as a Force in Evolution

As discussed earlier in this chapter, behavior is not only a product of evolution, it is also an important force in the evolutionary process. Two examples will be considered in this section—the rare-male advantage in *Drosophila* and social dominance in male mice, both of which are examples of *sexual selection* (competition among members of one sex for those of the opposite sex), a concept discussed in detail by Darwin (1859, 1871).

Rare-male Advantage. The greater relative reproductive success of rare males was independently discovered by Claudine Petit and Lee Ehrman. Petit (1951) first discovered this phenomenon in a study of the mating success of two strains of *Drosophila*. One strain was wild type, whereas the

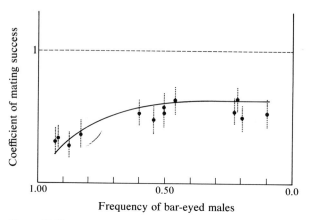

Figure 10.13

Sexual selection in Drosophila melanogaster. *Reproductive success of bar-eyed males. See text for explanation. (From Petit and Ehrman, 1969, p. 205.)*

other was bar eyed, a sex-linked condition that changes eye morphology. The two strains were permitted to breed freely for a number of generations. During the course of the experiment, randomly chosen females were occasionally separated from the population and allowed to lay their eggs in individual vials. By examining the offspring of these females, it was possible to determine whether the male with which the female had mated was wild type or bar eyed.

The coefficient of mating success, K, may be used to describe the results of Petit's experiment:

$$K = \frac{(A/a)}{(B/b)},$$

where a is the number of mutant males in the population, A is the number of females mated by mutant males, b is the number of wild-type males, and B is the number of females mated by wild-type males; thus, K is the ratio of the number of females mated per mutant male to that of females mated per wild-type male. If the reproductive success of the two types of males is equal, K will equal one. If K is less than one, the mutant males are at a disadvantage. Conversely, if K is greater than one, the mutant males mate more females than would be expected simply on the basis of their frequency.

The coefficient of mating success observed by Petit during the course of her experiment is graphed in Figure 10.13. In this study, the frequency of bar-eyed males fell from 93 percent of the total male population early in the experiment to 6 percent, due to their relatively low mating success. Although

Table 10.2

Coefficient of mating success, K, *of* Drosophila pseudoobscura *males of California strain as a function of their frequency in the mating population*

Pairs per population	Number of experimental runs	K
23 California: 2 Texas	5	0.28
20 California: 5 Texas	6	0.45
12 California: 12 Texas	7	1.10
10 California: 15 Texas	11	0.65
5 California: 20 Texas	7	2.40
2 California: 23 Texas	10	49.34

SOURCE: After Ehrman, 1966, p. 334.

K is less than one throughout the experiment, the reproductive success of bar-eyed males is frequency-dependent, i.e., the mating success of the mutant males increased as they became rarer. In contrast, the reproductive success of mutant and wild-type females was relatively constant throughout the experiment and was independent of their frequency in the population.

Rather than relying on progeny testing as an index of mating success, Ehrman has employed direct behavioral observation in her studies of the rare-male advantage in *Drosophila*. Females and males are placed into a mating chamber in which they are observed for several hours. Males and females from each of two strains are placed together, resulting in four possible mating combinations. Ehrman has found that the rare male is at a reproductive advantage in a number of different test situations: when the two strains possess different chromosome arrangements; are of different geographic origins; are raised in different temperatures; are mutant versus wild-type; are positively versus negatively geotactic, etc. Only rarely has the effect not been observed.

A sample of data from Ehrman's experiments is presented in Table 10.2. In this experiment, pairs of *Drosophila pseudoobscura* from different geographic origins were compared. In calculating the K values, the strain from California was assumed to be the mutant strain; thus, a low K value indicates that the California strain is at a reproductive disadvantage, relative to the Texas strain. From the tabulated values it may be seen that the Texas strain is at a reproductive advantage when its numbers in the population are low. However, this advantage disappears when the strains are more-or-less equal in frequency. When the California strain becomes rare, its relative mating

Table 10.3

Number of matings by male Drosophila pseudoobscura *of two strains present in equal frequencies in upper chamber as a function of type of male in lower chamber*

Type in lower chamber	*Matings by males in upper chamber*		χ^2	p <
	AR	CH		
CH	63	39	5.65	0.05
AR	37	64	7.22	0.01

SOURCE: From Ehrman, "Simulation of the mating advantage of rare Drosophila males," *Science,* 167, 905–906, 1970. Copyright © 1970 by the American Association for the Advancement of Science.

success increases. Thus, males from both strains are at a reproductive advantage when they are rare. In contrast, the ratio of California : Texas females mated in this experiment conformed very closely to the ratio of females introduced, indicating that the mating advantage associated with rarity does not obtain for females.

Female *Drosophila* appear to be passive during courtship, yet they clearly exercise some discrimination. Males, on the other hand, are indiscriminately active and attempt to mate with anything resembling another *Drosophila,* including other males, females of other *Drosophila* species, dead or etherized flies, and even inanimate objects. Thus, for there to be sexual selection for rare males, females must receive and process information about the relative frequency of various kinds of males available to them during courtship and prior to mating. The possible cues that may be involved in this discrimination have recently been investigated by Ehrman (1970a, 1972), utilizing a double-compartment mating chamber. Flies of two kinds are placed in the top chamber and those of one kind are placed below, the chambers being separated by a taut layer of coarse cheesecloth. Ehrman has found that the rare-male advantage among flies in the upper chamber may be eliminated by placing in the lower chamber many males of the strain that is rare in the upper. Conversely, a reproductive advantage may be created for males of one of two strains present in equal numbers in the upper chamber by placing males of the other strain in the lower. Data from one experiment in which a rare-male advantage was thus artificially created are summarized in Table 10.3. In this experiment, twelve pairs of each of two strains of *Drosophila pseudoobscura* were introduced into the upper chamber and large numbers of one strain, and then of the other in a second run, were placed in the lower chamber. The strains (CH and AR) differ in chromosomal arrangement.

Ehrman has not been able to create this artificial rare-male advantage when two layers of cheesecloth about one centimeter apart are used to separate the chambers. With two layers, physical contact between the flies of the two chambers is eliminated and transmission of airborne olfactory and acoustical cues is probably reduced. Subsequent research (Ehrman, 1972) suggests that olfactory cues may be important in the discrimination. When flies were killed, ground up, and various chemical compounds were extracted from them and placed in the lower chamber, it was found that one extract, probably containing a lipid, was effective in manipulating the reproductive success of males in the upper chamber.

The finding of the rare-male advantage is an important contribution to our knowledge of evolutionary biology since this form of frequency-dependent selection provides another mechanism for maintaining a balanced polymorphism, discussed in Chapter 9. If the selective advantage of a gene increases as its frequency decreases, the allele will be maintained in the population, even if it is selected against when at intermediate frequencies. This provides a reservoir of genetic variation upon which selection may act in the face of changing environmental circumstances and thus may be of considerable evolutionary significance.

Ehrman (1970b) has demonstrated how the rare-male advantage may result in a balanced polymorphism. One hundred pairs of *Drosophila pseudoobscura* (80 pairs of orange-eyed flies and 20 pairs of purple-eyed flies in experiment 1 and 20 orange:80 purple in experiment 2) were placed in bottles for 24 hours and allowed to mate. Males were then discarded and each female was placed in an individual vial where she laid her eggs. Eye color of resulting offspring permitted identification of the type of male with which each female had mated. To eliminate all other selective factors, each generation utilized 100 mating pairs and the proportion of orange- to purple-eyed flies was dictated by the matings of the previous generation. Thus, if 80 females were found to have mated with 60 orange-eyed males and 20 purple-eyed males in generation n, generation n + 1 would begin with $(60/80)(100) = 75$ orange-eyed pairs and 25 purple-eyed pairs. The results of 10 generations of such simulated selection are summarized in Table 10.4. From these results it may be seen that the rare-male advantage results in an initial increase in the frequency of the rare allele in both experimental populations that continues until a stable equilibrium is achieved at intermediate frequencies.

Social Dominance. As in *Drosophila,* sexual selection in mammals may also have important effects on population structure. Recent research concerning the "fine structure" of populations of house mice (*Mus musculus*) indicates that stable populations of this species consist of a mosaic pattern of *demes,* i.e., small, local breeding communities (Bruell, 1970), and that little effective gene exchange occurs among demes (Selander, 1970). Deme formation appears to be largely due to male territoriality, with each deme

Table 10.4

Approach to equilibrium in a simulated selection experiment due to the rare-male advantage in Drosophila pseudoobscura

	Eye color of parental pairs			
Generation	Experiment 1		Experiment 2	
	Orange	Purple	Orange	Purple
1	80	20	20	80
2	60	40	29	71
3	68	32	38	62
4	56	44	35	65
5	31	69	41	59
6	63	37	50	50
7	62	38	52	48
8	60	40	50	50
9	50	50	44	56
10	46	54	47	53

SOURCE: After Ehrman, 1970b, p. 346.

consisting of a dominant male, several subordinate males, and several females (Reimer and Petras, 1967).

When a population is subdivided into demes, random genetic drift may become a powerful force in determining differences in gene frequency from one deme to another. In addition, since favorable combinations will occasionally occur by chance within a few demes, evolutionary trial and error is facilitated. The extent to which this population pattern is important in the evolutionary biology of a species is a function of the effective population size (N_e), defined in Chapter 9. Even if all adults of breeding age contribute equally to the gene pool of a mouse deme, N_e will be relatively small. However, if the contribution is unequal, that is, if some adults produce more offspring than others, N_e will be smaller still.

Recent research (DeFries and McClearn, 1970, 1972) clearly indicates that the reproductive success of dominant and subordinate males is by no means equal. Several experiments concerning the genetics of social behavior in laboratory mice have been conducted utilizing the "triad" paradigm. Rather than employing 20-minute "bouts," standardized social living units were constructed that facilitated behavioral observations over extended periods of time. Each unit is constructed of three standard mouse cages connected by a Y-shaped plastic manifold that permits free access among cages. Adequate food, water, and bedding are placed in each cage. A photograph of an assembled triad is shown in Figure 10.14.

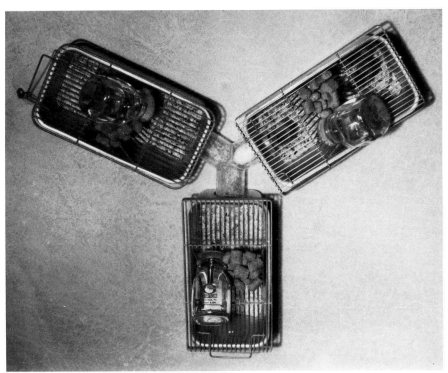

Figure 10.14

Triad unit used in studies of the social behavior of laboratory mice. (From DeFries and McClearn, in Evolutionary Biology, Vol. 5 (edited by Dobzhansky, Hecht, and Steere), Appleton-Century-Crofts, p. 285. Copyright © 1972 by the Meredith Corporation.)

In our first experiment, in each of 22 triads three males from different inbred strains were housed for two weeks with three inbred females. At the end of the two-week period, females were placed in individual cages until resulting litters, if any, were old enough to be classified by coat color. Combination of parental males and females was such that paternity could be ascertained by coat color of offspring.

Males usually began to fight within a few minutes of being placed in a triad and the typical result was that within 24–48 hours one male emerged as the dominant one. During the period of active fighting, the tails and hindquarters of most of the subordinate males were wounded. Evaluations of social dominance based upon observation of behavioral interactions indicated that receiving the fewest tail wounds (skin punctures) was an excellent index of social dominance.

Strain differences in social dominance of males were observed during this experiment. Males of strains A and BALB/c tended to be dominant,

Table 10.5

Reproductive success of dominant male mice in the triad paradigm in each of three experiments

Experiment	Number of litters	Number of litters sired by dominant male	% of all litters sired by dominant male
1	61	56	92
2	42	40	95
3	76	70	92

C57BL males were more-or-less intermediate, and DBA/2 males were most frequently subordinate. The differences in reproductive success of dominant and subordinate males was even more striking. In 18 of 22 triads, the dominant male sired all of the resulting litters and in all triads it sired at least one litter. As indicated in Table 10.5, of the 61 litters produced in this first experiment, 56 (92 percent) were sired by the dominant male. Because in about one-half of the triads at least one of the subordinate males died before the end of the two weeks, it might be argued that the reproductive advantage of the dominant male was a direct result of his killing the subordinates. By considering gestation length and time of birth, however, it was found that only three of the litters were conceived during periods when only the dominant male was alive in the triad. Of the remaining 58 litters, 53 (91 percent) were sired by dominant males. Thus, a dominant male appears to sire more than 90 percent of the offspring in his triad and this reproductive advantage does not depend upon his killing the subordinate males.

This experiment was repeated using outbred (HS) females derived from an initial cross of eight inbred strains. Because of the greater difficulty of ascertaining paternity for offspring borne by these females, only two inbred males were placed with two or three females in each triad in this second experiment. In spite of this difference, the reproductive success of dominant males was similar to that observed in the first experiment. Table 10.5 shows that, of the 42 litters conceived in the 20 triads of the second experiment, 40 (95 percent) were sired by dominant males. When 10 litters were excluded on the basis of the subordinate male's being dead at the time of conception, 30 of the remaining 32 litters (94 percent) were still found to be sired by the dominant male. These results again indicate that a dominant male sires more than 90 percent of the litters in his triad and, as outbred females were used, further demonstrate that this reproductive advantage is independent of type of female.

In a third experiment, the generality of the relationship between social dominance and reproductive success was tested using outbred males. In 36 triads, F_1 males were paired with HS males, and in 5 triads, F_1 males were paired with inbred males. In each triad, two or three females (inbred or outbred) were housed with the males. Neither of the males in one triad was found to have a tail wound and this triad was thus excluded from further consideration. In the remaining 40 triads, a total of 76 litters was conceived, 70 (92 percent) of which were sired by dominant males, 3 (4 percent) were sired by subordinate males, and 3 were of mixed paternity. The three litters of mixed paternity were produced by females from two triads. In each of these triads, few tail wounds were found on either male, indicating that the dominance hierarchy was not well established.

If the association between social dominance and reproductive success observed in the triad paradigm also exists in natural populations of mice, the effective population size within demes is very small, perhaps less than four. This effective population size would be sufficiently small to permit random genetic drift to exert a major force in determining differences in gene frequency at the local level (Levin, Petras, and Rasmussen, 1969) and would facilitate the differential success of established demes in starting new breeding units by migration. Although almost nothing is known about the opportunity for gene exchange among demes during periods of out migration, the subdivision of the population into demes, due in large part to behavior, could play an important role in the evolutionary biology of the species.

The relationship between social dominance and reproductive success in mice may not be all that different from what it once was in man. For example, Neel (1970) and his colleagues have studied some of the most primitive Indians of South America and have observed that these polygynous Indians live in small villages and that there is a marked genetic heterogeneity among villages. In four villages studied in detail, a highly disproportionate number of grandchildren were born to a few headmen, the four males with the most grandchildren in these villages being two father-son combinations. Thus, as in mouse demes, the chance of inbreeding is greatly increased in such villages. There exists, furthermore, a possible selective mechanism for intelligence. Although no data are currently available, Neel speculates that dominant males (headmen) in a village may be more intelligent than subordinates.

In this chapter we have briefly discussed the evolution of man and have shown how behavior may serve as a force in evolution. That man is now capable of directing his own further evolution will be discussed in the next and final chapter.

Chapter 11

Behavioral genetics and society

In Chapter 1, we noted Theognis' indignation over his observation that, for considerations of money, the good would marry evil and the evil marry good. We also noted Plato's and Aristotle's concerns over the circumstances that would favor the maintenance or increase of quality of the citizenry. However, the beginning of a systematic and integrated approach to the general problem of the hereditary quality of human populations is attributable to Galton (1869). Recognizing the powerful implications of the arguments that he had adduced with respect to the genetics of behavioral characteristics, he announced in the introduction to his work, *Hereditary Genius*:

> I propose to show in this book that a man's natural abilities are derived by inheritance, under exactly the same limitations as are the form and physical features of the whole organic world. Consequently, as it is easy, notwithstanding those limitations, to obtain by careful selection a permanent breed of dogs or horses gifted with peculiar powers of running, or of doing anything else, so it would be quite practicable to produce a highly-gifted race of men by judicious marriages during several consecutive generations. (1869, p. 1)

Elsewhere, in rather less restrained language, Galton put the point as follows:

If a twentieth part of the cost and pains were spent in measures for the improvement of the human race that are spent in the improvement of the breed of horses and cattle, what a galaxy of genius might we not create! We might introduce prophets and high priests of civilisation into the world, as surely as we can propagate idiots by mating *cretins*. Men and women of the present day are, to those we might hope to bring into existence, what the pariah dogs of the streets of an Eastern town are to our own highly-bred varieties. (Pearson, 1924, p. 78)

The Rise of the Eugenics Movement

In 1883, Galton coined the name *eugenics* to apply to the improvement of mankind by appropriate matings, and the furtherance of eugenic goals became the underlying theme around which most of his subsequent work was oriented. Eugenics was defined variously in Galton's subsequent publications. In a lecture given to the Sociological Society in 1904, Galton stated that, "Eugenics is the science which deals with all influences that improve the inborn qualities of a race; also with those that develop them to the utmost advantage." As we recall that the word "genetics" had not yet been introduced, and that "race" was frequently used to refer to species, we may paraphrase this definition to characterize eugenics as the study of forces that could improve the genetic properties of the species and those environmental circumstances that would optimize their phenotypic expression. It might be noted that this optimizing of phenotypic expression through environmental manipulation is similar to Lederberg's (1963) concept of *euphenics*, i.e., improving the phenotype by utilizing the advances of modern biology in the practice of medicine, especially in the treatment of genetic ills. Galton was much more concerned with the genetic than the environmental aspects of this discipline, however, and focussed most of his energies on the hereditary features.

To further his eugenic goals, Galton was instrumental in the founding of the Eugenics Society, in 1907, and the Eugenics Laboratory, in 1911. The laboratory was to be concerned with basic research on the influence of heredity on human characteristics, and the society was intended to be a propagandistic and activist organization, bringing the scientific findings to the attention of the public, and influencing legislation and social attitudes appropriately.

The establishment of these organizations in Great Britain was followed quickly by the establishment, in 1912, of a Eugenics Record Office in the United States. The Eugenics Research Association was established in 1913, and the Eugenics Committee of the United States, subsequently changing its name to the American Eugenics Society, in 1921. This latter was an activist group with aims very similar to those of the Eugenics Society of Great

Britain. The movement spread far and wide. Similar societies were established in France, Germany, Italy, Norway, Russia, China, and Japan.

The research that was supported by agencies concerned with eugenics varied greatly in quality. Some would be judged by present-day standards to be worthless, while other studies constitute part of the early foundation of human genetics. In respect to application, many early eugenicists suffered from excessive enthusiasm. Although Galton himself had warned against premature application of the basic science findings, the reforming zeal of many of the early workers led them to ambitious plans and exaggerated claims. It is difficult from the vantage point of the present to appreciate the enormous interest in, and the extraordinary effort expended on, eugenics topics. To illustrate the point, it may be noted that Holmes, in 1924, published a 514-page *bibliography* of matters of interest to eugenics.

What were the concerns that prompted this incredible outpouring of research findings, speculation, and exhortation? A principal objective was the reduction of inherited defect. The types of defect that were regarded to be of particular importance were feeble-mindedness, epilepsy, insanity, tuberculosis, and deafness. Another principal concern was with the problems of crime, delinquency, prostitution, pauperism, and vagrancy. Alcohol and venereal disease were both under suspicion as "poisoners of the germ plasm" (today we would say "mutagenic agents"). Birth control, infant mortality, war, and urban life were examined as potential selective factors influencing the relative proportions of individuals of different types and classes in the population. The possibility was examined that medical advances, which permitted individuals who otherwise might die to reproduce, might have a deleterious effect on the genetic constitution of the population. Differences among races in attributes of interest and the possible consequences of racial mixture were the subject of extensive debate. An observed differential in reproduction, with individuals of lower intelligence having larger families than those of higher intelligence, was viewed with great alarm for the implication was that there would be a steady decline in the average intelligence of the population.

The action programs designed to respond to these perceived problems can be categorized as either positive or negative. Galton himself had emphasized the positive aspects of eugenics — that is, mechanisms to promote "good" marriages. He suggested, for example, that fellowships be established to enable and encourage the "well bred" to marry young and thus to be able to produce more children during their reproductive lifetimes. In general, however, more emphasis was given by early eugenicists to negative eugenics, to programs and procedures that would prevent certain kinds of individuals from reproducing. These negative measures were motivated largely from a fear of intellectual deterioration of the population either through the unrestrained reproduction of "feeble-minded" or from the immigration of "inferior" races. The studies that popularized the former concern were those of the notorious families of the Jukes and the Kallikaks, discussed in Chapter 1.

The anxiety that the population's quality might be reduced by immigrants was sparked in large measure by the results of large-scale psychometric testing applied to draftees during World War I. In general, the results of this early application of standardized intelligence tests revealed higher average mental ages of the recruits whose heritage could be traced to northern and western Europe, with lower average mental ages of those whose lineage was from southern and eastern Europe. The susceptibility of these tests to educational and other cultural factors was largely ignored, and the observed differences were taken to indicate genetic "inferiority" of the latter groups in respect to intelligence.

One of the negative measures championed particularly for the feeble-minded was isolation. The principle of isolation was simply that by sequestering feeble-minded in institutions, their opportunity for reproduction would be reduced. The efforts of the early eugenicists in this regard were instrumental in achieving adequate and humane institutional care for the retarded. Another mechanism of negative eugenics was sterilization. Programs of sterilization were not unanimously supported by those in the movement (see Haller, 1963), but proponents of sterilization were sufficiently persuasive that a number of sterilization laws were enacted. The first was introduced in Indiana in 1907. This law made mandatory the sterilization of confirmed criminals, idiots, imbeciles, and rapists in state institutions when recommended by a board of experts. Prior to World War I, fourteen other states, California, Connecticut, Iowa, Kansas, Michigan, Nebraska, New Hampshire, New Jersey, New York, North Dakota, Oregon, South Dakota, Washington, and Wisconsin, enacted similar laws. Comparable programs became established in various other countries: Denmark, Finland, Germany, Sweden, Norway, Iceland, Switzerland, and Mexico. These laws were not just formalities; many individuals were sterilized under their provisions. For example, in California between 1909 and 1929, there were 6,255 sterilizations of insane and feeble-minded individuals. The program had enthusiastic supporters, including many social workers who felt the results of the program to be salubrious.

The laws did not go without challenge, however, and a number of test cases were taken to court. Perhaps the best known case was that of *Buck vs. Bell* (274 U.S. 200), which was decided by the United States Supreme Court in 1927. Carrie Buck, a feeble-minded girl who had already borne one illegitimate child, was chosen in 1924 for the first sterilization to be performed under a Virginia law. In a well known statement concerning the case, Justice Oliver Wendell Holmes said:

> We have seen more than once that the public welfare may call upon the best citizens for their lives. It would be strange if it could not call upon those who

already sap the strength of the State for these lesser sacrifices, often not felt to be such by those concerned, in order to prevent our being swamped with incompetence. It is better for all the world, if instead of waiting to execute degenerate offspring for crime, or to let them starve for their imbecility, society can prevent those who are manifestly unfit from continuing their kind. The principle that sustains compulsory vaccination is broad enough to cover cutting the Fallopian tubes. . . . Three generations of imbeciles are enough. (Buck vs. Bell, p. 207)

This approval of the Supreme Court injected new vigor into the sterilization program throughout the country.

Another negative eugenic mechanism was restriction on marriage. In some cases these measures were directed against insane, feeble-minded, or epileptics, and by the middle 1940's, some 41 states of the United States had laws prohibiting marriage of individuals in these categories. Other laws were directed against miscegenation. Particularly in the southern United States many laws were enacted prohibiting the marriage of whites with blacks and, in some cases, with orientals. In Germany, in 1935, the Nuremberg law prohibited the marriage of Jews and non-Jews.

Another method that seemed to have potential for protection of quality of a population was to prevent the immigration of undesirable persons. To this end, in the United States in 1924, the Immigration Restriction Act was passed, which restricted greatly the number of immigrants that would be received in the country from southern Europe and from Asia, with strong preference given to those immigrants from northern and western Europe.

Disenchantment with Eugenics

As we have seen, the eugenics movement consisted both of basic researchers —geneticists—and of those anxious to bring about social change in ways expected to improve the gene pool of the species. At no time were all the geneticists in support of the eugenics movement, but for an appreciable period of time a large number of eminent human geneticists did support the cause. Gradually, disenchantment set in. In large measure, this was due to an increasing understanding of the principles of genetics that revealed the fallacies of many of the more ambitious eugenics programs.

The Question of Efficacy. An increasing understanding of the basic principles of population genetics revealed that the gene pool of a species is a very conservative and stable system, and that gene frequencies were unlikely to be rapidly changed through differential reproduction. Thus, things were not in the critical condition that some of the earlier eugenics alarmists had believed; likewise, many of the proposed remedies came to be seen as being much less effective than they had been imagined to be. A negative eugenics program that would select against individuals with a trait determined by a deleterious dominant gene might be reasonably effective. That is, all of the individuals carrying the dominant gene would be identifiable in the population and could be persuaded from passing their genes on to the next generation. Therefore, the only source of that condition in the next generation would be new mutation. Selection against rare recessive homozygotes posed an entirely different problem, however. Because most of the homozygotes are the progeny of matings of heterozygotes, it would have an almost trivial effect upon the gene frequency. Thus, if the frequency of the recessive allele for phenylketonuria is approximately 0.01, it would require approximately 100 generations to reduce the frequency by one-half (i.e., 0.005) if no homozygotes reproduced. Clearly, unless there are ways of preventing heterozygotes from breeding, there is little point in a negative eugenics program with respect to phenylketonuria. A parallel argument could be made for all conditions of recessive origin. For these and other reasons, particularly in the United States, the academic geneticists largely abandoned the eugenics programs. As Lerner (1968) described the situation, "By the 1920's American eugenics degenerated into a mixture of pseudo-science, Bible belt religion, extreme reactionary politics, and racism, so that the very term became repulsive to geneticists" (p. 269).

The Nazi Carnage.   The realization of the relative ineffectiveness of many of the proposed eugenics measures resulted in diminished scholarly support; an even more powerful repellant, however, was provided by the gross distortion of knowledge of human genetics in the service of political programs in Germany during the Nazi regime. Racial myths constituted a core of Nazi political philosophy. These did not appear *de novo* in Hitler's time, however. People have discriminated against each other on the basis of group membership, probably for all of man's history. In many languages of "primitive" groups, for example, the name that the people have for themselves is synonymous with man; all others are regarded as inferior. In classical times, things were more explicit. Aristotle, naturally enough, thought very highly of the Greeks:

Having spoken of the number of the citizens, we will proceed to speak of what should be their character. This is a subject which can be easily understood by any one who casts his eye on the more celebrated states of Hellas, and generally on the distribution of races in the habitable world. Those who live in a cold climate and in Europe are full of spirit, but wanting in intelligence and skill; and therefore they retain comparative freedom, but have no political organization, and are incapable of ruling over others. Whereas the natives of Asia are intelligent and inventive, but they are wanting in spirit, and therefore they are always in a state of subjection and slavery. But the Hellenic race, which is situated between them, is likewise intermediate in character, being high-spirited and also intelligent. Hence it continues free, and is the best-governed of any nation, and, if it could be formed into one state, would be able to rule the world. (1952, pp. 531–532)

In the mid-nineteenth century, a work entitled *Essay on the Inequality of Human Races* was published by Count De Gobineau (1853–1855). According to Gobineau, the Teutons, also called Aryans, were the superior people of the world, responsible for all of civilization's advances. These Aryans were to be found among the aristocracy of many European countries, but lamentably they were declining in number and influence. An Englishman named Houston S. Chamberlain expanded Gobineau's racist concepts. Chamberlain, so enamored of German *Kultur* that he moved to Germany and became a citizen of that country, believed that the critical element in the history of civilization was race. The teutonic peoples he regarded to be the very highest in innate endowment, and to be the sole hope for the future of mankind. This philosophy he published in *The Foundations of the Nineteenth Century* at the very end of that century (1899; English edition, 1910) and Adolph Hitler made this thesis a key point of his own world view. Identifying the Aryan race as superior and as being concentrated particularly in Germany, he provided a fanciful history of the Aryans in which they are described as being nearly the sole producers of all that is important in human culture. In their doing so, strong measures had been necessary:

> Thus, the road which the Aryan had to take was clearly marked out. As a conqueror he subjected the lower beings and regulated their practical activity under his command, according to his will and for his aims. But in directing them to a useful, though arduous activity, he not only spared the life of those he subjected; perhaps he gave them a fate that was better than their previous so-called "freedom." As long as he ruthlessly upheld the master attitude, not only did he really remain master, but also the preserver and increaser of culture. For culture was based exclusively on his abilities and hence on his actual survival. As soon as the subjected people began to raise themselves up and probably approached the conqueror in language, the sharp dividing wall between master and servant fell. The Aryan gave up the purity of his blood and, therefore, lost his sojourn in the paradise which he had made for himself. He became submerged in the racial mixture, and gradually, more and more, lost his cultural capacity, until at last, not only mentally but also physically, he began to resemble the subjected aborigines more than his own ancestors. For a time he could live on the existing cultural benefits, but then petrifaction set in and he fell a prey to oblivion. (Hitler, 1943, pp. 295–296)*

However, all was not lost. Enough of the good Aryan heredity was left in some peoples, notably the Germans, that by careful policies they could restore things to rights and the Aryan, German people could attain what

---

*Reproduced from *Mein Kampf* (translated by R. Manheim) courtesy of Houghton Mifflin Company.

Hitler regarded as their rightful role as the master people of humanity. This Aryan mythology flew in the face of anthropological knowledge about man's evolutionary origins. The term Aryan, for example, properly refers to a linguistic grouping, not to any recognized racial grouping. Unhindered by any necessity of basing their racial policy on fact, the Nazis were able to make some astonishing reversals in attitude. The Italians were generally regarded by the Aryan theorists as being of inferior Mediterranean stock until the alliance between Mussolini and Hitler made it necessary to embrace the Italians as fellow supermen. In Chamberlain's conception, the Teutons, the Celts, and the Slavs were all descended from a single pure stock. The willingness of the Nazis to accept the Slavs in this fashion was directly correlated with their political relations with the Soviet Union. The Norwegians, originally hailed as having a high proportion of Aryan "blood," were demoted after their valiant though brief resistance to the Nazi invasion. Finally, with the advent of the alliance between Berlin and Tokyo, it became necessary for the Nazi race propagandists to claim a certain kind of genetic brotherhood between Japanese and Germans.

Particularly despised were the Jews, and anti-Semitism became a pervasive feature of German life. Beginning with the "Nuremberg law," which prohibited marriage of Jews and non-Jews and deprived Jews of the right of citizenship, there was a systematic escalation of harrassment, including removal of Jewish children from elementary schools, expropriation of property without compensation, mass arrests, and pogroms culminating in the establishment of the concentration camps and the systematic extermination of the Jews as Germany's "final solution" to the "Jewish problem."

Actually, the origin of a "euthanasia" program in Germany was not based exclusively on considerations of genocide. It began as a brief letter of authorization from Hitler (Remak, 1969, pp. 133–134):

Berlin, September 1, 1939

Reichsleiter Bouhler and Dr. Brandt

are authorized to extend the responsibilities of physicians still to be named in such a manner that patients whose illness, according to the most critical application of human judgment, is incurable, can be granted release by euthanasia.

s. Adolf Hitler

Under a program code named Aktion T4, hospitals throughout Germany were required to report patients suffering from schizophrenia, epilepsy, encephalitis, numerous other diseases, criminal insanity, those not German by nationality, those not of German "blood," and those who had been in an institution for longer than five years. After review of the cases, those selected for extermination were moved to special camps where, beginning in the early part of 1940, they were killed (Remak, 1969).

Thus, the overwhelming horror of Hitler's genocidal policy grew out of a program of negative eugenics. It is little wonder that the revulsion at the extermination policy generalized to eugenics, which had been twisted to support it. Eugenics became, and to a considerable degree remains, a bad word throughout much of the world.

Perennial Problems

Although eugenics fell into disrepute, many of the concerns of the early eugenicists persisted in one or another form. Indeed, in modern guise, they define some of the most active areas of research in human behavioral genetics today. In this section, results of recent research pertaining to some of these perennial problems will be briefly reviewed.

Psychoses. As indicated earlier in this chapter, early eugenicists regarded the problem of insanity to be of particular importance. Clearly, this problem is still with us. Over half of the hospital beds in the United States are occupied by patients affected with some form of mental illness and about one-half of these are diagnosed as schizophrenic; thus, schizophrenia is a most serious public health problem. The incidence of this disease in the population at large is about one percent. Although this percentage may not seem large, multiplication by 200 million indicates that about 2 million persons currently living in the United States either have the condition or will fall victim to it sometime during their lifetimes.

Schizophrenia is clearly a familial disease. Relatives of schizophrenics have a much higher incidence of the disease than members of the general population; the closer the relationship to the index case, the higher the index of schizophrenia.

In a now classic study, Kallmann (1946) reported a concordance of 69 percent for MZ co-twins of schizophrenic probands and a concordance of 10 percent for DZ co-twins of schizophrenics. A number of subsequent studies applying the twin method to schizophrenia in England, Japan, Denmark, Norway, and Finland have appeared. The results of these studies are presented in Table 11.1. It may be seen that the results obtained from the United States, the United Kingdom, and Japan are in good accord. Results from Denmark, Norway, and Finland generally give evidence of a smaller hereditary contribution to schizophrenia. Attempts to account for these discrepant results have led to an extensive analysis of differences in diagnostic criteria in different countries, the biases introduced by different types of sampling procedure (resident hospital population, consecutive admission, or twin registry), the possibility that schizophrenia is really a heterogeneous

Table 11.1

Uncorrected concordances for schizophrenia in twin studies reported since 1946

Investigator	Year	Country	MZ co-twins of schizophrenics		DZ co-twins of schizophrenics	
			Affected/Total	*%*	*Affected/Total*	*%*
Kallmann	1946	USA	120/174	69	53/517	10
Slater	1953	UK	24/37	65	10/112	9
Inouye	1961	Japan	33/55	60	2/17	12
Tienari	1963	Finland	0/16	0	2/21	10
Kringlen	1966	Norway	19/50	38	13/94	14
Gottesman & Shields	1966	UK	10/24	42	3/33	9
Fischer	1966	Denmark	3/10	30	0/8	0

SOURCE: After Shields et al., 1967, p. 393.

Table 11.2

Relationship between severity of and concordance for schizophrenia in MZ twin pairs (table entries indicate concordance rates in co-twins according to degree of impairment in probands)

Degree of impairment	Different criteria of severity in proband		
	(1) ≥ 1 year in hospital	*(2)* ≥ 2 years in hospital	*(3)* Not working, <6 months out of hospital
Mild	20% (2/10)	29% (4/14)	17% (2/12)
Severe	67% (12/18)	71% (10/14)	75% (12/16)

SOURCE: From Gottesman, 1968, p. 44.

complex of psychotic conditions, and other problems (see Rosenthal, 1970). The extent to which the data might be influenced by these factors can be illustrated by results of analyses of two of the most recent studies. Gottesman (1968) analyzed separately the concordances of MZ co-twins of mild and severe proband cases from the subject pool originally described by Gottesman and Shields (1966). Table 11.2 presents the results for three increasingly strict criteria of proband severity. For each of these criteria, concordance is substantially higher for severe schizophrenia than it is for mild schizophrenia. Clearly then, the severity of the cases chosen for study can affect the results obtained. It has been suggested, for example, that Kallmann's case material consisted mostly of severe or chronic cases

Table 11.3

Concordance for schizophrenia as a function of "strictness" of diagnostic criteria

Author	MZ		DZ	
	"Strict" schizophrenia	Including "borderline" cases	"Strict" schizophrenia	Including "borderline" cases
Kringlen (1966)	14/50 (28%)	19/50 (38%)	6/94 (6%)	13/94 (14%)
Gottesman & Shields (1966)	10/24 (42%)	13/24 (54%)	3/33 (9%)	6/33 (18%)
Fischer et al. (1968)	5/21 (24%)	10/21 (48%)	4/41 (10%)	8/41 (19%)

SOURCE: After Fischer et al., 1969, p. 984.

(Shields, Gottesman, and Slater, 1967), and that this might account for the relatively higher MZ concordance obtained by Kallmann than by subsequent investigators.

Another diagnostic matter of considerable importance concerns the strictness of the definition of schizophrenia. Fischer and colleagues (1969) reviewed several studies in which it was possible to analyze the cases using both a "strict" definition of schizophrenia and a broader definition that included "borderline" cases. The results of three of the most recent studies are shown in Table 11.3, where it can be seen that concordances for both MZ and DZ pairs are higher when the broader definition of schizophrenia is employed.

Whether these considerations of diagnostic criteria fully account for the generally lower concordance for schizophrenia of MZ twins in the Scandinavian studies remains for further research to clarify. It is possible, for example, that populations differ in frequencies of genes that may modify the expression of the schizophrenic phenotype (Shields, 1968).

Most recently, quite compelling evidence on the hereditary basis for schizophrenia has been provided by the application of adopted-child methods. Heston (1966) identified a group of subjects born to schizophrenic mothers who had been permanently separated from their biological mothers during the first month of life and reared in foster or adoptive homes. As control subjects, another group of children were chosen who had also been separated from their biological mothers before they were one month old and reared in foster or adoptive homes, but whose biological parents showed no record of psychiatric disturbance. When they were adults, the subjects were assessed by various means, including psychiatric interviews and reviews of school, police, medical and Veterans Administration records, and then evaluated by a team of clinicians. The results of this study are summarized in Table 11.4.

The greater number of cases of schizophrenia among experimental subjects ($p = 0.024$) than among controls provides definitive evidence for the genetic basis of this disorder. However, in addition to an excess of schizophrenia per se among experimental subjects, there was also an excess of various other psychiatric disorders. Heston uses the term "schizoid" to refer to the schizophrenic-like disorders found in relatives of schizophrenics and to the persons manifesting such disorders. Other investigators have also found a higher percentage of such disorders among the relatives of schizophrenics than in the general population, and the term "schizophrenic spectrum" (Rosenthal et al., 1968) has been coined to encompass these various disorders including schizophrenia.

Several clinical features of schizoid individuals have been listed by Heston (1970): Antisocial behavior is common among males, including impulsive crime, unreasoning assault, and poorly planned theft. Social isolation, heavy consumption of alcohol, and sexual deviance are commonly reported. Both male and female schizoids have been described as eccentric and suspicion-ridden recluses. Among females, incapacitating attacks of panic or unreasoning fear when faced with ordinary social challenges have been reported. Both the schizoid and the schizophrenic are characterized by rigidity of thinking, blunting of affect, anhedonia, exquisite sensitivity, suspiciousness, and a relative poverty of ideas, although these characteristics are more prominent among schizophrenics. Schizoids do not show the well-marked thought disorders, delusions, or hallucinations of schizophrenics, but micropsychotic episodes may occur.

Heston suggests that the fuzziness in definitions and diagnoses of schizophrenia may be due to the biological unreality of the distinction between schizoidia and schizophrenia. Twin data are utilized to support the hypothesis that schizoidia and schizophrenia are manifestations of the same underlying genetic disease. The incidence of schizophrenia and other significant psychiatric disorders among identical co-twins of schizophrenic index cases is summarized in Table 11.5. These data indicate that monozygotic co-twins of schizophrenics are almost as likely to be schizoid as schizophrenic. Since monozygotic twins are genetically identical, the schizoid and schizophrenic disorders may be different expressions of the same genotype. The range of expression from schizoid to schizophrenic within identical twin pairs would then be due to environmental factors. This interpretation is highly consistent with recent biochemical evidence (Wyatt, Murphy, Belmaker, Cohen, Donnelly, and Pollin, 1973): Monoamine oxidase activity in blood platelets of both schizophrenic and nonschizophrenic co-twins was found to be lower than that of normal controls. Activity of this enzyme was somewhat higher on the average, however, in the nonaffected twins. Enzyme activity was

Table 11.4

Results of a study of persons born to schizophrenic mothers and reared in adoptive or foster homes, and of controls born to normal parents and similarly reared

Item	Control	Experi-mental	Exact probability (Fisher's test)
Number of subjects	50	47	
Number of males	33	30	
Age, mean (years)	36.3	35.8	
Number adopted	19	22	
MHSRS, means[a]	80.1	65.2	0.0006
Number with schizophrenia	0	5	.024
Number with mental deficiency (I.Q. <70)[b]	0	4	.052
Number with antisocial personalities	2	9	.017
Number with neurotic personality disorder[c]	7	13	.052
Number spending more than 1 year in penal or psychiatric institution	2	11	.006
Total years incarcerated	15	112	
Number of felons	2	7	.054
Number serving in armed forces	17	21	
Number discharged from armed forces on psychiatric or behavioral grounds	1	8	0.021
Social group, first home, mean[d]	4.2	4.5	
Social group, present, mean[d]	4.7	5.4	
IQ, mean	103.7	94.0	
Years in school, mean	12.4	11.6	
Number of children, total	84	71	
Number of divorces, total	7	6	
Number never married, >30 years of age	4	9	

SOURCE: After Heston, "The genetics of schizophrenic and schizoid disease," *Science,* 167, 249–256, 1970. Copyright © 1970 by the American Association for the Advancement of Science.

[a]The MHSRS is a global rating of psychopathology moving from 0 to 100 with decreasing psychopathology. Total group mean, 72.8; S.D., 18.4.

[b]One mental defective was also schizophrenic; another had antisocial personality.

[c]Considerable duplication occurs in the entries under "neurotic personality disorder"; this designation includes subjects diagnosed as having various types of personality disorder and neurosis whose psychiatric disability was judged to be a significant handicap.

[d]Group 1, highest social class; group 7, lowest.

Table 11.5

Data on monozygotic co-twins of schizophrenics

Investigator and year	Pairs (No.)	Schizophrenia (No. pairs)	Other significant abnormality[a] (No. pairs)	Normal, or mild abnormality (No. pairs)
Essen-Möller (1941)	9	0	8	1
Slater (1953b)	37	18	11	8
Tienari (1968)	16	1	12	3
Kringlen (1967)	45	14	17	14
Inouye (1961)	53	20	29	4
Gottesman & Shields (1966)	24	10	8	6
Kallmann[b]	174	103	62	9
Total	358	166 (46.4%)	147 (41.1%)	45 (12.6%)

SOURCE: After Heston, "The genetics of schizophrenic and schizoid disease," *Science,* 167, 249–256, 1970. Copyright © 1970 by the American Association for the Advancement of Science.

[a]Investigators' diagnoses: ? schizophrenia, schizophreniform, transient schizophrenia, reactive psychosis, borderline state, schizoid, suicide, psychopathic, neurosis, and variations of these diagnoses.

[b]From Shields, Gottesman, and Slater (1967).

found to be highly correlated between members of the same twin pairs and to be inversely correlated with a measure of the degree of the severity of the schizophrenic disorder.

Kety, Rosenthal, and colleagues (Kety et al., 1971; Rosenthal et al., 1971) have applied variations of the basic adopted-child method in investigations of schizophrenia in Denmark. Of the various comparisons made in these studies, two may be singled out for further discussion. The adoptees' family method, illustrated in Figure 11.1, compares the incidence of schizophrenia in biological relatives and in adoptive relatives of schizophrenic offspring with the incidence among relatives of nonschizophrenic offspring. (For convenience, only parents are shown in the figure, but meaningful data can be obtained as well from siblings and half-siblings.) Environmental theory, emphasizing the importance of learning from parents or parent figures whose behavior is itself disorganized, might predict that adoptive relatives of schizophrenic persons would show a higher incidence of schizophrenia than would adoptive relatives of nonschizophrenic persons. On the other hand, a genetic theory would not predict any such differences in incidence. The data showed 2 out of 74 adoptive parents of schizophrenics and

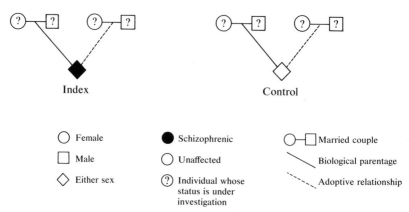

Index   Control

○ Female   ● Schizophrenic   ○—□ Married couple

□ Male   ○ Unaffected   ╲ Biological parentage

◇ Either sex   ⑦ Individual whose   ⁝⁝⁝ Adoptive relationship
                     status is under
                     investigation

Figure 11.1
Research design of adoptee's family method.

3 out of 83 adoptive parents of nonschizophrenics to be themselves schizophrenic—a nonsignificant difference, as would be expected on the basis of
the genetic model. The genetic model is further substantiated by observations on the biological relatives. Most environmental theories would not
predict any difference in incidence here, whereas a genetic theory would
predict a higher incidence of schizophrenia among biological relatives of
schizophrenic adopted children than among biological relatives of normal
adopted children. The actual results were that 13 out of 150 of the biological
relatives of schizophrenic children and 3 out of 156 of the biological relatives
of normal children were schizophrenic. This difference is statistically significant and constitutes further support for a genetic interpretation.

Figure 11.2 shows a complementary method of analysis, the adoptees
study method, in which the relevant information is the prevalence of schizophrenia in adopted children of nonschizophrenic and of schizophrenic
biological parents. This, it may be seen, was Heston's basic design. Rosenthal's (1971) study differs from Heston's (1970) in that Heston's mothers

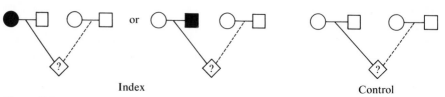

Index   Control

Figure 11.2
Research design of adoptees study method. See Figure 11.1 for key to symbols.

Table 11.6

Percentages of first-degree relatives found to be schizophrenic or schizoid

Relationship	Number of individuals	Schizo- phrenia[a] (%)	Schizoid (%)	Total: schizoid plus schizophrenic (%)
Children[b]	1000	16.4	32.6	49.0
Siblings[c]	1191	14.3	31.5	45.8
Parents[c]	2741	9.2	34.8	44.0
Children of two schizophrenics[d]	171	33.9	32.2	66.1

SOURCE: From Heston, "The genetics of schizophrenic and schizoid disease," *Science,* 167, 249–256, 1970. Copyright © 1970 by the American Association for the Advancement of Science.

[a] Age-corrected rates.

[b] From Kallmann (1938).

[c] From Kallmann (1946).

[d] From Kallmann (1938), Kahn (1923), Schulz (1940), and Elsässer (1952).

were schizophrenic while pregnant, thus permitting speculation that some toxin incident to the mother's condition (or some postnatal sequelae) might have acted as an environmental agent in the etiology of the child's subsequent psychosis, whereas most of the mothers in Rosenthal's group were not schizophrenic while pregnant with the probands. A genetic hypothesis would predict the incidence of schizophrenia to be lower in adopted children whose biological parents were not schizophrenic than in adopted children whose biological parents were schizophrenic. Because both types of offspring were reared by nonschizophrenic adoptive parents, many theories of psychological etiology would predict that there would be no differences in incidence of schizophrenia in the two groups of children. The results, showing that none of the 47 control offspring and 3 out of 39 offspring of schizophrenic biological parents became schizophrenic, provide still further support for the genetic model.

Data from several investigations concerning the incidence of schizophrenia and schizoidia among first-degree relatives of schizophrenics are summarized in Table 11.6. As indicated in this table, approximately one-half of the first-degree relatives (children, siblings, and parents) of schizophrenics manifest either schizophrenic or schizoid disease. From these results, Heston concludes that schizoidia and schizophrenia are due to one basic genetic disease and that the mode of transmission conforms to that expected of a single-locus autosomal dominant gene. For any genetic disease, the expected concordance among identical twins is 100 percent. As shown in

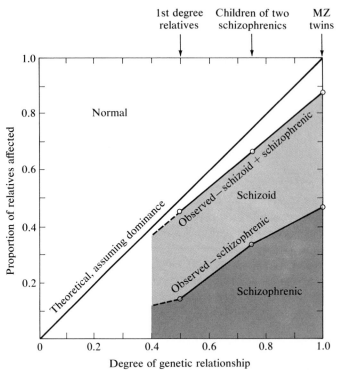

Figure 11.3

Observed and expected proportion of schizoids and schizophrenics among relatives. (From Heston, "The genetics of schizophrenic and schizoid disease," Science, 167, 249–256, 1970. Copyright © 1970 by the American Association for the Advancement of Science.)

Table 11.5, the observed incidence among monozygotic co-twins of schizophrenic index cases is 87.5 percent. For characters due to a simple autosomal dominant gene, 50 percent of the first-degree relatives of index cases should manifest the disease. The observed incidence of schizophrenic and schizoid disease among first-degree relatives of schizophrenics is about 46 percent (Table 11.6). When both parents are heterozygous for the dominant allele, 75 percent of the resulting children should be similarly affected. The observed incidence of schizophrenia and schizoid disease among children of two schizophrenics is 66 percent (Table 11.6). Finally, among grandchildren of heterozygotes, 25 percent would be expected to have the disorder. The observed incidence (Kallmann, 1938) is 27 percent. The fit of the observed data to that expected on the basis of an autosomal-dominant-gene model is summarized in graphic form in Figure 11.3.

Several biases may be reflected in these data. For example, most of the observed incidences are lower than the corresponding expected values. This may be due to a failure of the model or it may be that some of the affected are not diagnosed as such, but are, rather, counted as normals. In addition, schizophrenics produce fewer offspring on the average than do normals. It may follow from this that the schizophrenics who do reproduce tend to be those having a relatively mild expression of the disease.

It is important to note that Heston does not discount the possibility of modifying factors in his hypothesis. For example, the higher incidence of the disease among identical co-twins of severely afflicted index cases than among identical co-twins of those less strongly affected suggests that there may be an influence of modifying genes, environmental effects, or possibly both. In addition, although Heston's model accounts for many of the observed data, there are some features that may be better accommodated by a polygenic threshold model such as that proposed by Gottesman and Shields (see Chapter 9). A recent review and evaluation of the various models proposed to account for the inheritance of schizophrenia has been provided by these authors (Gottesman and Shields, 1972).

Although schizophrenia has been the most thoroughly studied of the psychoses from the genetic point of view, a small but growing literature is accumulating concerning manic-depressive illness as well. Winokur, Clayton, and Reich (1969) have recently reviewed some of the early family and twin studies and reported further data from their own recent family study. The data strongly implicate a genetic factor in the etiology of manic-depressive psychosis, and Winokur proposes that the pattern of transmission of the condition can be explained by the hypothesis of a single X-linked dominant gene with incomplete penetrance. Some support for this hypothesis is provided by further studies suggesting a linkage between the hypothesized manic-depressive locus and that for colorblindness, but other research results (Perris, 1968) are in disagreement with this suggestion. Kringlen (1967) has provided a summary of the relevant twin studies. As may be seen in Table 11.7, the data consistently show higher concordance for MZ than for DZ twins, and thus support an interpretation of a heritable component of manic-depressive psychosis. Much further research is needed on this topic, particularly in view of the evidence that the unipolar syndrome (manifesting depressive state only) and the bipolar syndromes (having both manic and depressive states) may have different genetic bases (Slater and Cowie, 1971).

Alcoholism. Alcoholism, regarded by the early eugenicists as part of a syndrome of hereditary degeneration including also criminality, insanity, epilepsy, and feeble-mindedness, has remained a social issue of great importance. Indeed, alcoholism is one of the principal public health problems in the United States today, with estimates of the number of those dependent

Table 11.7

Concordance in twins of manic-depressive patients

Investigator	Year	Country	MZ		DZ	
			N	% Con- cordance	N	% Con- cordance
Luxenburger	1928	Germany	4	75	13	0
Rosanoff et al.	1935	United States	23	70	67	16
Kallmann	1950	United States	23	96	52	26
Slater	1953b	England	8	50	30	23
da Fonseca	1959	England	21	75	39	39
Harvald & Hauge	1963	Denmark	15	60	40	5
Kringlen	1967	Norway	6	33	9	0

SOURCE: After Kringlen, 1967, pp. 73 and 93.

upon alcohol in the country ranging as high as 10 million (Brown, 1970). In view of the magnitude of the problem, only a rather disappointingly small body of knowledge has accumulated concerning the possibility that genetic factors are of importance in the etiology of alcoholism. The tendency for alcoholism to "run in families" has been generally noted in folklore, and is substantiated by a number of systematic investigations into the distribution of alcoholism within families. A particularly extensive investigation was made by Åmark (1951), who studied several large samples of alcoholics and their relatives in the Swedish population. The morbidity risks for first-degree relatives of alcoholic probands and for adult males and females randomly drawn from the population are given in Table 11.8. Two features of these data are particularly clear. The incidence of the condition is higher in males than in females, and is also higher among relatives of probands than in the population at large. The latter observation is not, of course, definitive evidence for a genetic component to the condition, since intrafamily environmental factors could easily generate a similar distribution.

Roe (1954) approached the problem with an adopted-child study, comparing persons reared as adopted children who had an alcoholic biological parent with other adopted children among whose biological parents there was no evidence of alcohol abuse. No significant difference was found in the incidence of alcoholism between these two classes of adopted persons, with 7 percent of those of alcoholic biological parentage and 9 percent of the control group displaying alcoholism. Certain features of this study have been criticized, however. Goodwin and co-workers (1973), for example, have pointed out that Roe's subjects had not yet lived through the age of risk. Furthermore, females, for whom alcoholism is a low risk, constituted a large part of the sample. Only 21 of the adoptees with alcoholic biological parent-

Table 11.8

Morbidity risks for various relatives of alcoholic probands and for controls in a Swedish population

Relationship	Morbidity risk
Brother	21%
Father	26%
Sister	0.9%
Mother	2%
None — male controls	3.4%
None — female controls	0.1%

SOURCE: After Åmark, 1951, p. 90.

age, and 15 of the controls, were male. Thus, the sample size of subjects at risk for alcoholism was small. The two types of adoptees also experienced differential types of placement. Control adoptees were adopted at a younger age and more were placed in urban environments than the index adoptees. In a study that avoided these difficulties, Goodwin et al. (1973) used the adoptees study method with Danish subjects. Fifty-five male index cases were obtained. All of them had a biological parent diagnosed as alcoholic, all were adopted by a nonrelative prior to 6 weeks of age, and none had known subsequent contact with biological parents. A matched control sample of 78 males similarly adopted but without alcoholism in their biological parentage was also obtained. The adoptees' ages ranged from 23 through 45 with a mean of 30 for both groups. No significant differences were found between the homes of the two groups in terms of economic status or psychopathology of the adoptive parents. Information on personality, psychopathology, and, specifically, alcoholism was obtained from the adoptees by psychiatric interview and examination of law enforcement records. A strict criterion of alcoholism was employed that required, in addition to heavy drinking, that the subject experience social or marital problems, difficulties with employment, encounters with the law, and various physiological symptoms due to alcohol. It was found that, of the index cases born to alcoholic biological parents but not reared by them, 18 percent became alcoholic at some time during their lives; only 5 percent of the control adoptees became alcoholic. It is interesting to note that the two groups did not differ in the incidence of heavy drinking alone.

Kaij (1957) utilized the twin method in investigating the drinking habits of co-twins of alcoholic probands in Sweden. Drinking habits were scored on a scale ranging from 0, for complete abstainers, to 4, for chronic alcoholics. In brief, the drinking habits of monozygotic co-twins were much more similar than were those of DZ co-twins. Another large-scale twin study has been reported by Partanen, Bruun, and Markkanen (1966). This investigation utilized a general sample of 172 MZ and 557 DZ male twin pairs from the Finnish population. Questionnaires concerning the use of alcoholic beverages were completed by the subjects, and the results were subjected to a multivariate analysis that yielded three dimensions: *density*, which is an index of the frequency of alcohol use; *amount*, which is an index of the amount typically consumed on single drinking occasions; and *lack of control*, an index of "dependency." Computing H as the ratio of the difference between the within-pair variances of DZ and MZ twins to the within-pair variance of DZ twins yielded evidence of a heritable component for density and amount, but not for lack of control.

Schuckit and his colleagues (1972a; 1972b) have used a modification of a half-sibling approach in studying the etiology of alcoholism. The half-sibling relationship makes several interesting comparisons possible: Children of alcoholic biological parentage may be raised in homes that have or do not have alcoholic parent figures, and children not having alcoholic biological parentage may be raised by an alcoholic foster parent. Theories of environmental etiology of alcoholism often suggest the behavior of an alcoholic parent figure to be of critical importance. In Schuckit's (1972a) material, 42 half-siblings of alcoholic probands themselves had an alcoholic biological parent. Twenty-three lived with that parent and 19 did not. Ten of the former (44 percent) themselves became alcoholic, and 8 of the 19 who did not live with their alcoholic biological parent (42 percent) also became alcoholic. This comparison suggests that the influence of living with an alcoholic biological parent is relatively unimportant. The question may be broadened by inquiring about the influence of alcoholic parent *figures*, whether biological or adoptive. Of the 150 half-sibs in the study, 28 were alcoholic and 122 were not. Ten of the former (35 percent) lived with an alcoholic parent figure, and 30 of the latter (25 percent) lived with an alcoholic parent figure. These percentages do not differ significantly, in accord with previous results.

However, the important environmental etiological factors may not be from parental example, but may be sociocultural ones that might be expected to influence children reared in the same family, living in the same neighborhood, attending the same schools, and so on. Seventy-four of the half-sibs in the study shared their childhood with the proband alcoholic subjects. In spite of this sharing of childhood experiences, only 4 of the 66 subjects not having an alcoholic biological parent (6 percent) became alcoholic, whereas 2 of the 8 subjects having alcoholic biological parents (25 percent) became alcoholic. This comparison suggests the relatively greater importance of biological parentage than of shared childhood with a half-sibling who became alcoholic. A final comparison was made of those 28 half-siblings who also became alcoholic and the 122 who did not. Of the former group, 18 (65 percent) had an alcoholic biological parent, and of the latter group only

Table 11.9

Correlations on intelligence measures for MZ and DZ twins reared together and MZ twins reared separately

Group	Newman et al. (1937)		Shields (1962)		Burt (1966)	
	Correlation	N	Correlation	N	Correlation	N
MZ together	0.88	50	0.76	44	0.92	95
MZ apart	0.77	19	0.77	44	0.87	53
DZ together	0.63	50	0.51	28	0.55	127

24 (20 percent) had an alcoholic biological parent. Thus, all of the comparisons suggest a preponderant role of heredity in the etiology of alcoholism.

We have already reviewed (see Chapter 6) some research dealing with alcohol-related behavior in animals. Together with the human data just presented, the results suggest the importance of further research into the biological basis of this very considerable social and individual problem.

Intelligence. Intelligence was a key concern of the early eugenicists and it remains today a central focus of behavioral genetic research. Some of the available data on the genetics of intelligence were previously presented in Chapter 9. Reference again to Table 9.14 will show the unanimous finding of higher resemblance of MZ than of DZ twins in IQ performance in a variety of populations. It may further be recalled that the average estimate of broad-sense heritability from those reported studies is about 0.5.

Considerable effort has been expended in the variation of the twin method that compares the similarities of MZ twins reared together and MZ twins reared apart. The classical study of this type is that of Newman, Freeman, and Holzinger (1937), who were able to report on 19 cases in which MZ twins had been separated and reared apart from each other. The resemblance of these twins, that of MZ twins reared together, and that of DZ twins reared together is expressed in terms of correlations in the first column of Table 11.9. The resemblance of MZ's reared apart is intermediate to the other groups. These results have been a prominent feature in discussions of the "nature-nurture controversy" ever since their publication, with some authors emphasizing that rearing apart reduced the resemblance of MZ twins, thus showing an effect of environment, and others pointing out that, even when reared apart, MZ twins were more alike than DZ twins reared together, thus showing the importance of hereditary factors. That both genetic and environmental sources of variation may be operative in intelligence is, of course, a commonplace expectation in contemporary models of quantitative inheritance. The evidence on the relative importance of these two sources of variation from the Newman et al. study has been challenged,

Table 11.10

Correlations between IQ of children and parental mental age for adopted and control children

Mental age	Adopted children	Control children
Father	0.07	0.45
Mother	0.19	0.46
Midparent	0.20	0.52

SOURCE: After Burks, 1928, p. 278.

however, by subsequent research in which larger numbers of MZ twins reared apart were tested. The results of these more recent studies (Burt, 1966; Shields, 1962), both conducted in Great Britain, are also shown in Table 11.9. In these two studies there is very little evidence indeed that the environmental differences of separated twins were influential. Another analysis with a similar conclusion is that of Vandenberg and Johnson (1968) who analyzed 37 published reports of IQ's of separated MZ twins in terms of the twins' age at separation. The proposition being examined was that twins who were separated earlier, thus having had a shorter period of shared environment, should be less similar than those separated later. The obtained results were that twins separated before they were one year old had an average IQ difference of 5.5 points and that those separated when older than one year actually had a larger IQ difference of 9.6 points.

Several major efforts have also been devoted to the study of the intelligence of adopted children. Perhaps the best known is that of Burks (1928) who compared the correlations between IQ of adopted children and mental age of adoptive parents to those between control children and their biological parents. Results are shown in Table 11.10. A very similar outcome was obtained by Leahy (1935), whose results are shown in Table 11.11. Substantial evidence for an influence of home environment on IQ score, however, was obtained by Freeman and associates (1928) who found that the adopted child's IQ performance was highly correlated ($r = 0.48$) with a rating of the adequacy of the adoptive home. Interpretation of this result is complicated by the fact that many adoptive agencies practice selective placement in that those children judged, by whatever criteria available, to be "better endowed" are given for adoption to "better" adoptive homes. Because the extent and effectiveness of selective placement at work in any particular sample are difficult to determine, the actual influence of home environment in these studies cannot be confidently assessed.

Measures of the resemblance of parent and offspring and of siblings constitute another source of information on the genetics of intelligence. A large number of studies, many of them reported in the late 1920's and 1930's,

Table 11.11

*Correlations between IQ of children and parental
IQ for adopted and control children*

IQ	Adopted children	Control children
Father	0.15	0.51
Mother	0.20	0.51
Midparent	0.18	0.60

SOURCE: After Leahy, 1935, p. 282.

have been published. Although many of these studies individually suffered from one or another difficulty, such as small sample size or bias in sampling, the combined weight of their evidence is quite impressive. Erlenmeyer-Kimling and Jarvik (1963) have reviewed and summarized these studies and the results are presented in Figure 11.4, along with correlations between twins and between parents and their adopted children. Particularly impressive is the relationship between the median values of the various studies and the coefficients of relationship of the individuals investigated. Thus, for those individuals with a coefficient of relationship of 0.50, the median reported phenotypic correlations cluster quite closely around this same value. A very simple genetic model (additive gene action, random mating, no environmental effects) will yield the expectation of a phenotypic correlation of 0.50. However, the empirical results can hardly be regarded as evidence substantiating the simple model. Strong assortative mating is known to exist with respect to intelligence, for example, and environmental agencies are presumed to be effective in modifying the IQ phenotype. The extent to which these departures from the assumptions may cancel each other out, and their relative influences on the displayed phenotypic resemblance of relatives, remain to be disentangled. The empirical result itself, however, appears to be a most robust one.

Although the principal thrust of research has been with respect to global IQ measurements, evidence has long existed that intelligence may consist of subfactors with different genetic bases. Willoughby (1928), for example, examined the correlations of parents and offspring, of siblings, and of husband and wife with respect to performance on a variety of cognitive tests such as arithmetical reasoning, sentence meaning, geometric forms, symbol digit, symbol series, completion, etc. Wide ranges in the obtained correlation coefficients were found for the various tests. Since that time, several investigators (Blewett, 1954; Husén, 1963; Thurstone, Thurstone, and Strandskov, 1955; Vandenberg, 1968) have contributed important further information using twin-study methodology. Representative of the results of this approach is the summary in Table 11.12 provided by Vandenberg, which combines

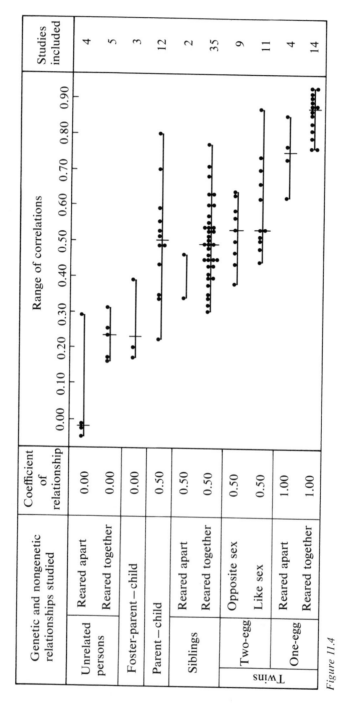

Figure 11.4

A summary of correlation coefficients compiled by L. Erlenmeyer-Kimling and L. F. Jarvik from various sources. The horizontal lines show the range of correlation coefficients in "intelligence" between individuals of various degrees of genetic and environmental relationship. The vertical lines show the medians. (After Erlenmeyer-Kimling and Jarvik, "Genetics and Intelligence: A Review," Science, 142, 1477–1479, 1963. Copyright © 1963 by the American Association for the Advancement of Science.)

Table 11.12

Combined results of ten twin studies showing the relative importance of heredity for seven abilities

Ability	F
Verbal	2.53
Word fluency	2.47
Perceptual speed	2.26
Spatial	2.25
Memory	2.18
Number ability	1.91
Reasoning	1.65

SOURCE: From Vandenberg, 1968, p. 157.

the results of ten different twin studies showing the relative importance of heredity for seven specific abilities. The reported F values are the ratios of the DZ within-pairs variance to the MZ within-pairs variance, so the larger the number the stronger the evidence for a hereditary contribution to variance with respect to the named trait. These results would appear to be of great importance for subsequent research on the genetics of intelligence. If the global IQ measurement is an assessment of a heterogeneous collection of different abilities, it would appear that the most rapid advancement in our knowledge could be made by concentrating research on those specific abilities themselves.

That there may indeed be a biological distinctness to the factors of intelligence is suggested by the specific cognitive defect that accompanies Turner's syndrome (see Chapter 7). First reports on affected individuals suggested a possible slight mental retardation, but further observations by Shaffer (1962) revealed that their verbal IQ on the Wechsler Adult Intelligence Scale was actually rather superior, with the performance IQ showing a substantial decrement. Further analysis of factor scores revealed a pattern of scores on perceptual organization that had previously been identified as a "brain damage" pattern. Money (1963) confirmed these observations and described the pattern of deficit as "space-form blindness." Subsequently, Alexander, Walker, and Money (1964) showed Turner's patients to be severely deficient in a test that requires right-left orientation as projected onto a map. These results clearly show that spatial ability may be influenced largely independently of verbal abilities, and are suggestive of the possibility of using other syndromes for a more detailed dissection of intelligence.

Table 11.13

*Correlations among parents and offspring for
quantitative reasoning and spatial visualization*

Correlation	N	Quantitative reasoning	Spatial visualization
Mother–son	50	0.62	0.41
Father–daughter	63	0.21	0.36
Mother–daughter	64	0.25	0.22
Father–son	51	0.08	0.03
Mother–father	99	0.07	0.05

SOURCE: From Stafford, 1965, p. 184.

Stafford (1965) has provided further evidence on the independence of genetic bases for separate abilities. On tests of spatial ability and quantitative reasoning, it is frequently observed that male performance, on the average, exceeds that of females. This sort of sexual difference suggests the possibility of sex linkage. Note that, in general, X-linkage implies that the following relationships should exist among various correlations: Mother–son should approximately equal father–daughter, which should exceed mother–daughter, which should exceed father–son, which should be approximately equal to mother–father (in the absence of assortative mating), which should equal zero. Table 11.13 gives the results of Stafford's investigation. From this table it may be seen that the observed pattern of correlations for spatial visualization conforms more closely to that expected of a sex-linked character than does that for quantitative reasoning.

The remarkable progress in identifying single loci and chromosome anomalies underlying mental retardation has been reviewed previously. The conditions of severe general retardation naturally attracted research interest early, because of the obvious incapacity of the affected individuals. In recent years, increasing attention is being given to children with IQ's in the normal or superior range who are not culturally or educationally disadvantaged, but who nevertheless suffer from some specific learning disability. Such children may constitute 5 percent or more of the school-age population (McCarthy and McCarthy, 1969). Clearly, a problem as large as this is worthy of intense research effort, including studies on the possibility of genetic etiology for some or all of the conditions included in the learning-disability category. One of the most interesting of these disorders is *dyslexia*, which is characterized by inability or severely reduced ability to read. That dyslexia tends to be familial has been noted since the condition was first delineated as a syndrome, and the published observations to this effect (reviewed by Critchley, 1970) are numerous. (Critchley calls attention to the fact that a publication from Germany in 1936 recommended sterilization for sufferers from "word-blindness.") The most substantial research project to date has been that of Hallgren (1950) who examined 276 Swedish dyslexic cases and their families. The incidence of dyslexia in relatives of the probands, including a few MZ and DZ twins, provided strong evidence of a hereditary factor, and Hallgren concluded that a dominant allele at a single autosomal locus was responsible.

This brief summary should demonstrate that, although much remains to be learned, there already exists a robust body of knowledge about the inheritance of intellectual abilities. From the earliest, there have been associated social issues.

We have already seen that the heritability of intelligence is very substantial and that assortative mating for various measures of intellectual functioning is high (see Chapter 9). Burt (1961) has also called attention to the fact that intelligence varies according to occupational class. The strength of this relationship is illustrated in Table 11.14, where six occupational classes are identified and where the table entries refer to relative incidence per 1,000 employed persons in the population. Given that social mobility exists, persons may move among socioeconomic strata partly on the basis of IQ, a highly heritable character. Most persons tend to marry within their occupational class and thus to marry someone more similar genetically than they would if mating were entirely at random; consequently, allelic frequencies at loci influencing intelligence may come to differ in the different socioeconomic strata. That is to say, even if initially IQ was randomly distributed in a population, the process just outlined would generate a social structure in which class membership was based substantially upon genotype (see Eckland, 1967). This process doesn't work with perfection, of course. It may be seen in Table 11.14 that there is considerable overlapping of intellectual capacity among the various occupational classes. In any case the process would not result in a fixed caste system, because there would continue to be mobility between classes. Due to genetic segregation, children are not uniformly like their parents and many possess IQ's more characteristic of some other class. Burt (1961) has provided specific information on this point, which is shown in Table 11.15. There it may be seen that fairly large percentages of children have measured intelligence levels appropriate to some occupational class other than that of their fathers.

Figure 11.5 shows the relationship between occupational class and mean IQ for parents and their offspring separately. It is obvious that, if the relationship that appears in the parental generation is to be maintained in the next generation, it will be necessary for there to be substantial migration of offspring of relatively high IQ to classes higher than that into which they were born. If the relative size of the different classes remains approximately the same, this would also imply a migration of individuals of relatively low IQ from the higher occupational classes downward to lower classes.

Table 11.14

Distribution of intelligence among adults according to occupational class

Class	50–60	60–70	70–80	80–90	90–100	100–110	110–120	120–130	130–140	140+	Total	Mean IQ
I Higher Professional							2	13	2	1	3	139.7
II Lower Professional				1	8	16	56	38	15	1	31	130.6
III Clerical			2	11	51	101	78	14	3		122	115.9
IV Skilled		5	15	31	135	120	17	2	1		258	108.2
V Semiskilled		18	52	117	53	11	9				325	97.8
VI Unskilled	1									—	261	84.9
Total	1	23	69	160	247	248	162	67	21	2	1000	100.0

SOURCE: After Burt, 1961, p. 11.

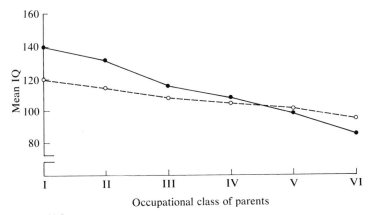

Figure 11.5

Relationship between occupational class and IQ for parents (solid line) and for offspring (broken line). (After Burt, 1961, p. 11.)

The particular specific component or components of intelligence that help qualify one for membership in a particular social or occupational stratum may differ widely from individual to individual. One person might be qualified by a high level of verbal ability, for example; another by high spatial ability. Because of assortative mating there will be an accumulation of alleles for these separate factors within the particular strata. The offspring will receive genes for high or moderate or low level of the various factors, on the average, according to the class membership of their parents. Tryon (1957) suggested that this process might account for the evidence of a "general intelligence" factor in intellectual functioning. At least in part, then, the evidence that such a general factor exists may be due to a particular social system and some degree of genetic determination of the factors that are relevant for placement within that system.

Table 11.15

Percentage of children whose intelligence is below, above or equivalent to that of their occupational class

Class	Below	Equivalent	Above	Number
I–III	75.5	16.8	7.7	156
IV–V	34.8	34.3	30.9	583
VI	—	42.9	57.1	261
Total population	32.1	33.5	34.4	1000

SOURCE: After Burt, 1961, p. 13.

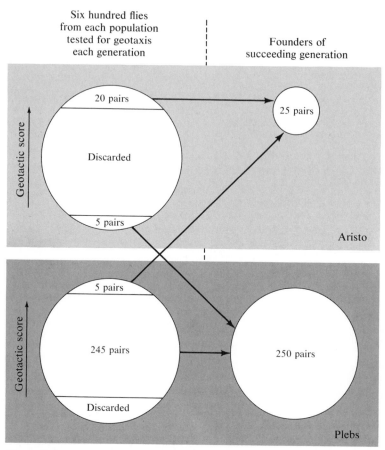

Figure 11.6

Experimental Drosophila *model for testing the genetic consequences of social mobility. (After Dobzhansky, 1968, pp. 140–141.)*

Although experimental studies of the genetic consequences of social mobility among classes in the human population are necessarily lacking, some evidence has been obtained using animal models. Dobzhansky and Spassky (1967) and Dobzhansky (1968) have employed *Drosophila* populations in such research, but only a small portion of their extensive work will be reviewed here. In one of their studies, an "Aristo" or "elite" population was started every generation by a select group of only 25 parental pairs, whereas a companion "Plebs" population was started with 250 pairs (see Figure 11.6). Each generation, 600 flies from each population were tested for geotactic behavior, a character previously discussed in Chapter 7. In the Aristo population, selection was quite rigorous, with only the 40 most positively geotactic individuals (20 pairs) being retained as parents. From the larger Plebs population, however, 490 individuals were saved. In addition, in each generation the populations exchanged 5 pairs: The 10 "best" (most positively geotactic) individuals from the Plebs were sent to the Aristo population, whereas the Aristo population sent its 10 "worst" individuals to the Plebs.

Since the most positively geotactic Aristo and Plebs individuals were utilized as the founders of the next generation for the Aristo population, it is reasonable to expect that the response in this group should have been toward positive geotaxis. However, what about the Plebs? Selection within the Plebs population was relatively weak and, moreover, the "dregs" from the Aristo population were incorporated into the founding population of the Plebs each generation. Somewhat surprisingly, both populations changed toward positive geotaxis, although the response was more marked in the Aristo population, i.e., downward social mobility transferred the selectional improvements of the "elite" population to the much larger Plebs population. That this transfer of minus variants from a plus population resulted in a plus transfer to the recipient population may have been due to the rather low within-group heritability for the character (less than 5 percent).

Although these populations were intended to serve as a model for studying the genetic consequences of social mobility in a class society, extrapolation of the results to human populations is clearly unwarranted. Nevertheless, these results do indicate that social mobility among classes may have important genetic consequences and, furthermore, that the theoretical formulations of Eckland (1967) previously discussed may be subjected to empirical verification.

One of the concerns of the eugenics movement was the differential reproduction of different social classes. To a considerable extent, of course, the concern was really over differential reproduction of individuals of different intelligence levels. Although this problem has been raised and expressed in various ways (see Falek, 1971, for review), one of the most salient presentations of the issue was made by Cattell (1936) when he reported that parents with lower IQ scores tended to have larger families than did parents with higher IQ scores. If this trend persisted and if there were any genetic basis at all for intelligence, it seemed obvious that the result would be a net decline in the average IQ of the species. The actual magnitude of the reproductive differential was such as to lead to a prediction of a rate of loss of about $1-1\frac{1}{2}$ IQ points per generation. However, when the IQ's of 10-year-old children in England were later measured in a follow-up study (Cattell, 1950) an increase of slightly more than one IQ point was found, rather than the decline that had been predicted. Even more massive evidence suggesting an increment rather than a decrement was provided by a large-scale survey of Scottish children. The IQ test results of nearly 71,000 Scottish children aged 11 in 1947 were compared with those of approximately 87,000 11-year-old

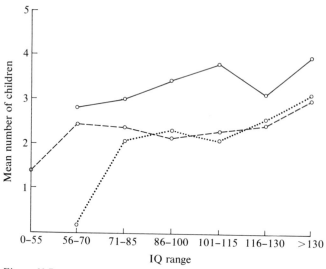

Figure 11.7

Differential fertility and intelligence. Data from Higgins, Reed, and Reed are plotted by the dashed line; those from Bajema by the dotted line; and those from Waller by the solid line. (After Falek, 1971, p. 555.)

Scottish children tested in 1932. The mean IQ gain was in excess of two IQ points (Maxwell, 1954).

These results were quite puzzling in view of the described reproductive advantage of those with lower IQ's. A number of explanations were attempted (see Burt, 1952) and such factors as an hypothesized increase in children's test sophistication between 1932 and 1947 were carefully examined. The clearest answer, however, appears to have been provided by the research of Higgins, Reed, and Reed (1962), Bajema (1962), and Waller (1971). Noting that the earlier results had not taken into consideration unmarried or otherwise nonreproducing individuals, these investigators were able to show that, although reproducing individuals of the lower IQ levels tend to have large families, there is a large percentage of individuals at this level who do not reproduce at all. When these "zero-family-size" individuals are included in the analysis, a strikingly different picture appears. Figure 11.7 shows the results of these three investigations with "zero" families included. There is clearly an over-all trend for higher reproductive rate of the higher IQ individuals. Cattel's paradox has evidently been resolved.

Another old problem related to intelligence and social class concerns race. Although widely deplored, racial discrimination in some form or another has continued around the world, justified usually on the grounds that those discriminated against are innately inferior in some fashion. In the

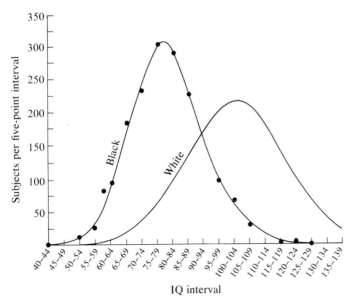

Figure 11.8

Distribution of IQ in black and white Americans. The distribution on the left was obtained from tests on 1,800 black school children from the South; that on the right is based on a "normative" sample of white Americans. (From Kennedy et al., Monographs of the Society for Research in Child Development, 1963, 28, Ser. 90.)

United States, assimilation of ethnic groups that, at one time or another, were regarded as undesirable (the Irish, and the eastern and southern Europeans, for example) has reduced to some extent the bitterness that accompanied the restrictive immigration laws. Assimilation of the Afro-American population, however, has not been as rapid, and the issue of equal opportunity for black and white citizens is currently a burning and urgent one. A key issue has been the question of differences in intelligence between Afro-American and Caucasian populations. The evidence can be stated simply enough. Although the distributions of IQ scores overlap extensively, the average for the blacks is about 10–20 IQ points below that of the whites. Illustrative of the data obtained are those shown in Figure 11.8, comparing measured IQ's of black elementary school children in the southeastern United States to a normative sample of white children.

In the abstract, the possibilities are clear enough. If two populations have been to some considerable extent reproductively isolated for an appreciable period of time, they will likely have come to have different allelic frequencies for a number of genes. There is no reason to believe that genes affecting intelligence are different from any others in this respect. Thus, two groups that have been largely separated reproductively might well have different frequencies of genes relating to intelligence. The difficulty, of course, is in determining if the observed average difference in IQ's is due to genetic differences or to the manifestly unequal environmental opportunities that have been available to the two groups. Thus, to assess the possibility that the means of the black and white populations in the United States differ because of differences in the gene pools of the populations, the phenotypic difference in means must be adjusted to compensate for the undeniable environmental disadvantage of the blacks. The question then becomes, how much should the compensation be? We simply do not know. The range of possible outcomes, if the appropriate compensation were made, extends from that of an increase in the apparent IQ difference, through equality, to a difference in the opposite direction, with the black population having a higher average IQ than the white.

Another approach to this problem is to examine the relationship of within-group and between-group heritabilities (DeFries, 1972). Does the finding of a high within-group heritability imply that an observed difference between groups is also highly heritable? In 1969, Jensen reviewed the evidence concerning the genetics of intelligence and concluded that intelligence, as measured by conventional IQ tests, was a highly heritable character within Caucasian populations. Based upon this evidence for a high within-group heritability (assumed to be about 0.8), Jensen hypothesized that genetic factors may be strongly implicated in the observed difference between the means of Caucasians and Afro-Americans.

Recall that it was shown in Chapter 9 that the broad-sense heritability for performance on IQ tests may be less than 0.52. Recent evidence from black twins in Philadelphia (Scarr-Salapatek, 1971) suggests that the heritability may be even less within the Afro-American population. Although the within-group heritability may be less than that indicated by Jensen, the question about the implication of within-group heritability nevertheless still remains, i.e., does the finding of a nonzero within-group heritability necessarily imply that an observed group difference is to some extent heritable? In order to answer this question, it is necessary to express the between-group heritability as a function of the within-group heritability.

When a population is composed of two or more groups, both the genetic and phenotypic variances may be partitioned into components within and between groups. The ratio of the additive genetic variance within groups to the phenotypic variance within groups yields the within-group heritability (h_w^2):

$$h_w^2 = h^2 \frac{(1-r)}{(1-t)},$$

where h^2 is the population heritability, t is the intraclass phenotypic correlation of members of the same group and r is an analogous genetic correlation.

The ratio of the between-group additive genetic variance to the between-group phenotypic variance yields the between-group heritability (h_f^2):

$$h_f^2 = h^2 \left[\frac{1 + (n-1)r}{1 + (n-1)t} \right].$$

When the number within groups is large,

$$h_f^2 \cong h^2 \left(\frac{r}{t} \right).$$

Thus, we may express the between-group heritability as a function of the within-group heritability as follows:

$$h_f^2 \cong h_w^2 \left[\frac{(1-t)r}{(1-r)t} \right].$$

If genotype-environment correlations are absent, this expression may be used to estimate the heritability of group differences. For example, if $h_f^2 = 0.2$, this would indicate that 20 percent of the observed difference between two groups would remain if the two groups were reared for one or more generations in exactly the same environment. If a positive correlation exists between heredity and environment, this expression would tend to underestimate the heritable nature of the group difference. If it is negative, the expression would overestimate the extent to which the difference was heritable. In fact, if the correlation between environment and heredity were large and negative, it may be shown that the group with the lower mean phenotypic value could actually have a higher mean genotypic value.

From the reported difference between the means of the Caucasian and Afro-American groups and assuming a standard deviation of 15 within each group, an estimate of $t = 0.2$ is obtained. No estimate of r is currently available. However, when h_f^2 is plotted as a function of h_w^2 (Figure 11.9) for various values of r and the obtained value of t, several generalizations emerge. Although h_f^2 is a monotonic increasing function of h_w^2 for given values of r and t, this relationship is greatly dependent upon the relative magnitude of r. When r is low, a relatively large change in h_w^2 is accompanied by a small change in h_f^2. When r is high, a small change in h_w^2 results in a large change in h_f^2. Thus, the finding of a relatively high within-group heritability does not necessarily imply that an observed difference between group means is also highly heritable.

One of the most insidious aspects of the whole problem has been the confusion of the idea of differences with the idea of superiority and inferiority. Implicit in much of the discussion of this problem of race differences in intelligence has been acceptance of a unidimensional model of intelligence, with a high score on this global IQ index being related to "superiority" in

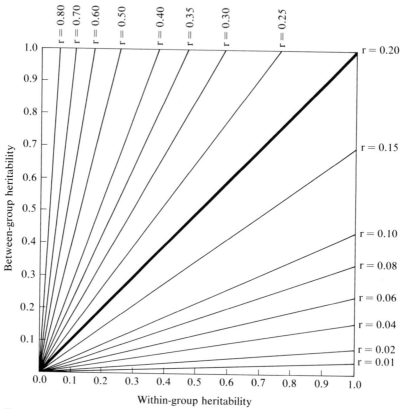

Figure 11.9

Between-group heritability expressed as a function of within-group heritability and the genetic correlation of members of the same group (r). See text for explanation.

some very general and pervasive sense. This view oversimplifies the issue inordinately. The domain of intelligence has clearly been shown to be multi-dimensional, consisting of a variety of special aptitudes that are to a large extent independent of each other. Thus, an individual's intellectual prowess must be characterized by a profile, and not by a single number. Regrettably, little of the available data on race differences has taken this finer-grained approach. Also requiring examination is the notion that intelligence is "linearly" good; that is to say, the higher the IQ, the better. It has already been noted in Chapter 10 that there is some reason to believe that the type of thing we measure as intelligence today (in the form of global IQ score) has not been very closely related to Darwinian fitness for any appreciable stretch of evolutionary time in man. To relate the index to superiority, there-fore, requires considerable qualification. In any case, the whole notion of superiority is an evasive one. The question quite properly can be put: superior for what?

Table 11.16

*Concordance with regard to adult criminality in
monozygotic and same-sexed dizygotic twin pairs*

Investigator and year		Monozygotic pairs		Dizygotic pairs		Concordance (%)	
		Conc.	Disc.	Conc.	Disc.	MZ	DZ
SEXES DIFFERENTIATED							
Rosanoff, Handy & Plesset (1941)	males	29	9	5	18	76	22
	females	6	1	1	3	86	25
SEXES UNDIFFERENTIATED							
Lange (1929)		10	3	2	15	77	12
Le Gras (1933)		4	–	–	5	100	0
Kranz (1936)		20	11	23	20	65	53
Stumpfl (1936)		11	7	7	12	61	37
Borgström (1939)		3	1	2	3	75	40
Yoshimasu (1965)		14	14	–	26	50	0

SOURCE: After Slater and Cowie, 1971, p. 114.

Finally, specific attention should be called to the overlap of the distributions in Figure 11.8. So extensive is this overlap that group membership does not allow accurate or efficient prediction of IQ. It seems clear that on matters where IQ is a relevant consideration, a person must be evaluated in his own right and not on the basis of race membership.

Criminality, Psychopathy, and Other Antisocial Behavior. Criminality, immorality, prostitution, and vice were other targets of concern of the early day eugenicists. Investigation of these behaviors from the genetic perspective is made particularly difficult because they are defined in legal rather than in psychometric or psychiatric terms. We have already noted difficulties enough with respect to the measurement of intelligence or the assessment of schizophrenia. For those traits defined in terms of legal codes the problems of ascertainment and the difficulties of comparing results from studies conducted in different countries or at different times are probably even more severe (see Rosenthal, 1970). Nevertheless, evidence of a possible genetic influence is accumulating. The possibility that individuals with an XYY karyotype may have an increased predisposition to crimes of violence has already been discussed (see Chapter 7). Slater and Cowie (1971) have summarized the results of several twin studies on criminality in Table 11.16, where the over-all indication is of a higher concordance for MZ than for DZ twins.

Table 11.17

Arrest records of the biological offspring (reared by adoptive parents) of female criminal offenders compared to those of controls

Arrest records	Probands	Controls	p[a]
Number of subjects checked for records	52	52	
Subjects with records	8	2	0.046
Total number of arrests	18	2	
Subjects arrested as adults	7	2	0.084
Subjects with convictions	7	1	0.030
Subjects with 2 or more arrests	4	0	0.059
Subjects incarcerated for an offense[b]	5	0	0.028
Total time incarcerated[b]	$3\frac{1}{2}$ yrs	0	

SOURCE: After Crowe, 1972, p. 602.

[a]Fisher's exact.

[b]This includes 2 subjects sent to the training school as juveniles, which accounts for $1\frac{1}{2}$ years.

Crowe (1972) has recently conducted an adoptees study method examination of the inheritance of criminal behavior, in which arrest records of biological offspring (reared by adoptive parents) of female criminal offenders were compared to those of controls. The biological mothers of the probands in this study were 41 offenders (37 convicted felons and 4 misdemeanants) who had been incarcerated in the state of Iowa and had given up babies for adoption. Their offspring (27 males and 25 females) ranged in age from 15 to 45 years, with a mean age of 25.6 years, at the time of this study. A control group consisted of 52 adoptees matched for age, sex, race, and age at the time of the adoptive decree. In order to test for possible differences in incidence of antisocial behavior between the proband and control groups, data were obtained from arrest records of the Iowa Bureau of Criminal Investigation, an agency that maintains records of every adult arrest in the state in which the subject was officially charged, as well as those of juvenile offenders sent to state training schools.

The principal results of this study are summarized in Table 11.17. From this table it may be seen that eight probands accounted for a total of 18 arrests, in contrast to only two arrests among controls. Seven probands were actually convicted of offenses, versus one control. Finally, five probands were incarcerated for a total of three and one-half years, whereas none of the control subjects was found to have been incarcerated.

These results clearly suggest that biological children of female criminal offenders may be more likely than controls to be arrested, receive a conviction, and be incarcerated for an offense. Since the probands and control subjects were both separated from their mothers at an early age (less than

Table 11.18

Empiric risk of psychopathy and criminality in siblings of chronic alcoholics

Parents	N	Percent psychopathic (brothers & sisters)	Percent criminal (brothers only)
Neither alcoholic	252 brothers 265 sisters	12.8	16.3
One alcoholic	97 brothers 100 sisters	20.8	18.6

SOURCE: Åmark, 1951, p. 111.

18 months) and reared in adoptive homes, it is tempting to ascribe these differences to biological factors. However, the possibility of environmental differences between the homes in which the children were reared cannot be discounted. For example, it is not known if the adoptive parents of the probands knew of the mothers' prison records and what effect, if any, this might have had on their attitudes toward the children. Nevertheless, these results, in conjunction with those from the twin studies, are highly suggestive and additional research in this very important area is clearly warranted.

The psychiatric terms psychopathy or sociopathy, referring generally to antisocial or immoral behavior, overlap the legal definition of criminality to a considerable extent, but are not synonymous with it. Of particular interest is the fact that psychopathy is frequently found among relatives of probands with some other psychiatric diagnosis. Heston (1970), for example, reported a higher relative incidence of antisocial personalities and felons among biological offspring (reared by adoptive parents) of schizophrenics than among controls (see Table 11.4). Similarly, Åmark (1951) reported on psychopathy and criminality in siblings of alcoholics, with the results shown in Table 11.18. These results raise the possibility that psychopathic behavior may be either a form of variable expression of major genes that are usually expressed as one of the other defined syndromes, or a quantitatively graded response due to possession of a number of alleles in a polygenic system insufficient for full manifestation of, say, schizophrenia, but sufficient to cause a deviation from behavioral norms. Much additional research will be required to illuminate this important issue.

The possibility of a genetic component in criminality has raised a number of interesting legal problems (Shah, 1970): In 1968, the defense attorney for a man on trial for murder in Paris presented an unusual defense. He claimed that his client possessed an extra Y chromosome and, thus, was not criminally responsible for this act. Although convicted, a reduced sentence was imposed, presumably due to the evidence presented concerning the chromosomal anomaly. At about the same time in Australia, a jury acquitted a man charged with murder on the grounds of legal insanity after a defense witness testified that the man had an extra Y chromosome. Also in 1968, Richard Speck, convicted murderer of eight nurses in Chicago, was reported to have an extra Y chromosome. Although this purported anomaly was reportedly going to be raised in an appeal, it was later announced that Speck had a normal chromosomal complement. Nevertheless, there now appears to be some precedent for diminished responsibility before the law of XYY individuals, at least in France and Australia. Should other countries adopt this precedent?

As discussed in Chapter 7, the relationship between the 47, XYY karyotype and predisposition to violent acts of crime is not well established. However, even if it is eventually found that XYY individuals are no more predisposed to criminality than are those with a normal chromosomal complement, the question of diminished responsibility for individuals genetically predisposed to criminal behavior remains. In this chapter, we have seen evidence that clearly supports the hypothesis of a possible genetic basis for antisocial behavior. In earlier chapters we saw evidence from laboratory animals that clearly demonstrates that aggressive behavior is heritable, although subject to experiential factors. Breeders of game cocks and fighting bulls have long known that there are heritable differences in such behavior. There is no reason to believe that heritable differences in aggressive behavior do not exist in man. Above some threshold, society may define aggressiveness as abnormal or criminal. What, then, should be society's stance with respect to those having a strong and partly genetically based predisposition to commit aggressive or criminal acts? Fortunately, as we have seen, a heritable basis for a pathological condition by no means precludes the possibility of environmental treatment or rehabilitation.

Mutagens.   The earlier concern over "poisoning the germ plasm" has undergone considerable change, due to increased understanding of the molecular processes of mutagenesis and to some new hazards that have accompanied technological advances. The principal disquietude has been with respect to the possibility that mutations produced by the utilization or testing of atomic weapons or the use of atomic energy would have serious consequences for the human gene pool. A great deal of research has been done in an effort to assess the seriousness of this hazard. The concern became generalized to radiation sources in general, and resulted in increased caution being taken in the diagnostic and therapeutic use of X-rays by dentists and by roentgenologists, and a banning of the fluoroscopes formerly used for fitting of shoes; perhaps of greatest importance, it contributed to the test-ban treaty that has reduced the amount of radiation contamination from testing and development of weapons. It now appears that the mutational hazard to the gene pool is probably low. Background radiation, from natural sources, provides 4–10 rems per generation (a *rem* is a unit of radiation intensity). Another 2–3 rems is estimated to be provided by occupational and medical radiation sources. The radiation released by atomic testing amounts to between 0.05 and 0.1 rems (Lerner, 1968). This is not to say, of course, that the potential hazard is small. An atomic war might lead to incalculable increases in mutational events.

The possibility that some chemical pollutants may be mutagenic, in addition to their other undesirable properties, is just beginning to be examined.

The New Social Biology

From the preceding sections, it is clear that the basic problems that aroused the early eugenicists have not gone away; nor have they been ignored. We have seen that these problems are subjects of intense and increasing research effort. The opprobrium attached to eugenic action programs, particularly the deformed version of the Nazis, has prompted a desire for a new label to describe the academic pursuit of these issues. From the examples of two prestigious journals, the former *Eugenics Quarterly* that became *Social Biology* and the former *Eugenics Review* that became the *Journal of Biosocial Science*, it appears that the term "social biology," or some variant thereof, is emerging as an appropriate label to describe this interdisciplinary effort. As indicated below, behavioral genetics constitutes a central core of this new social biology.

Genetic Counseling. Apart from the new academic look, there is, as well, a new perspective on action programs concerning genetic conditions that influence behavior. Although it is not done with a primarily eugenic intent, society still does "segregate" mentally retarded individuals (and also convicted criminals, of course) but, in general, the old coercive aspects of negative eugenics find few champions today. Instead, the emphasis is on the provision of information through genetic counseling, with the individuals involved free to adopt whatever course of action they choose. Couples seeking counsel either prior to marriage or prior to beginning their families usually do so because of some affected relative. Couples who have already had an affected child wish to know the chances of a subsequent child being normal. For an increasing number of conditions, accurate statements can be made concerning the probability of an affected child being born.

The ability to detect heterozygous unaffected carriers of some deleterious alleles means that it is not always necessary for an affected child to be produced in order for the prospects for future children to be assessed. Even for the more complex polygenic situations, the knowledge of empiric risks for relatives of affected individuals is constantly being improved.

A relatively new technique that increases the potential of genetic counseling enormously is *amniocentesis*. Study of a sample of the amniotic fluid from a pregnant woman can provide information concerning a number of chromosomal anomalies and single-gene conditions with metabolic consequences such as PKU. Such information about the condition of a fetus can be obtained at a sufficiently early stage to make therapeutic abortion possible. Amniocentesis thus opens up new possibilities in genetic counseling. Heretofore, probability statements only could be given to prospective parents who were "at risk" with respect to some genetic anomaly. Their decision to reproduce or not to reproduce had to be made, therefore, on the basis of the "odds" that the next child would be an abnormal one. With amniocentesis, for many conditions, it is possible for at-risk parents to attempt successive pregnancies until the tests indicate that the child will be normal. Table 11.19 provides a list of disorders for which prenatal diagnosis can be accomplished by amniocentesis at present.

The benefits of these genetic diagnostic techniques need not be restricted to immediate relatives of affected individuals. If a population at risk can be identified on other grounds, general screening can be undertaken. For example, pregnant women over 40 might be observed routinely, by amniocentesis, for chromosome anomalies of the fetus. As another example, when a relatively high frequency for a deleterious allele is known to be characteristic of a given population, persons in that population might be screened for heterozygosity. Such a program has recently been launched to identify heterozygotes for Tay-Sachs disease in Jewish populations in several communities.

It should also be mentioned that, in addition to the deliberate utilization of genetic information in decisions about reproduction by individuals, changes in the reproductive pattern of an entire population may influence the frequency of genetic anomalies. Vogel (1972), for example, has suggested that a trend away from reproduction by older women can account for the observed steady reduction in Down's syndrome births in Germany since 1950. The over-all effect of this trend in reproductive age is not clear, however. Cavalli-Sforza and Bodmer (1971) have noted that reproduction at younger ages will result in higher fertility of those individuals predisposed to develop a disease that has a later onset, e.g., schizophrenia. Alcoholism might be suggested as another condition for which reproduction at younger ages might have the effect of increasing the frequency of relevant alleles in the next generation.

Therapy. The therapies that are presently available for treatment of genetic diseases include surgical repair, prostheses, drug therapy, and diet therapy (Fuhrmann, 1972). All of these are euphenic in that they result in an improvement of the phenotype by manipulation of the environment, and all are symptomatic in that they may bring about changes in the phenotype but they leave the genotype unaltered.

Many exciting new prospects for "genetic engineering" have been generated by the spectacular advances in molecular biology. Among these potential techniques are the introduction of synthesized human genes, directed mutation, the introduction of viruses or virus-like particles containing genetic information into human systems, the transplantation of nuclei, and the introduction into human systems of cells containing additional genetic information from other mammalian species. None of these measures is possible at the present time, although they may become so in the future. It is important to note, however, that most of the envisioned techniques would probably change the genotype of somatic cells, but not of reproductive cells.

Positive Eugenics.   From the above discussion it may be seen that the principal effects of therapy and genetic engineering are to modify the pheno-type but not the genotype, or to modify the somatic genotype but not the germinal one. These measures will reduce human suffering and grief; how-ever, they will have little actual eugenic influence in the sense of decreasing the frequency of deleterious alleles. In fact, therapy will likely increase the reproductive fitness of individuals afflicted with genetic diseases, and, thus, may actually increase slightly the frequency of undesirable alleles in the population. This possibility hardly needs to cause immediate alarm, however, at least as it applies to rare recessive conditions, because most individuals homozygous for recessive alleles are progeny of the matings of two hetero-zygotes. The reproduction of homozygotes will cause only a slight increase in allelic frequency in the next generation. Crow (1967), for example, has calculated the impact of successful therapy on the incidence of a genetic disease. Assuming the present frequency of the recessive allele for PKU to be 0.01, he concluded if the fertility of recessive homozygotes suddenly became equivalent to that of normal persons, it would require 40 genera-tions, or about 12 centuries, just to double the incidence of the disease.

Some authors have argued the importance of eugenic measures not so much from a concern over possible deterioration of the gene pool as from a conviction that the opportunity should be seized to improve it. It is sug-gested that biotechnological progress in preservation of sperm and in artifi-cial insemination, developed largely for the purpose of improving farm animals, could now be applied as positive eugenic measures in man. Muller (1965) has particularly championed this idea in his discussions of "germinal choice." His suggestion is to preserve sperm of eminent men and, presum-ably after their death, make this germinal material available to couples who wish to have and rear a "genetically superior" child. A possible future varia-tion on this theme could involve the obtaining of ova from "superior" women, fertilizing them and implanting them in "ordinary" women's uteri where they would undergo development and birth. Although these procedures are not without unresolved difficulties, the problem is not so much a technical one as a philosophical one. The central issue becomes one of deciding which traits are the good ones to be selected. In lists of desirable attributes, intelli-gence usually appears, but we have seen that even the assumption that IQ

Table 11.19

Disorders for which prenatal diagnosis is possible or likely

Disorder	Tests utilized
Acatalasia	Catalase assay
Aminoacidopathies (see also specific individual diseases) Individual traits such as prolinemias and iminoacidopathies, phenylala-ninemias (several), tyrosinemias (several), methyl histidinurias, phospho-ethanolaminuria, Fanconi syndromes	Many procedures, including partition chromatography, column chroma-tography, electrophoresis, many chemical screening tests, and enzyme assays when indicated
Ceramide lactosidosis	Ceramide lactosidase
Chromosomal disease (and carrier detection in some conditions)	Cytogenetics (macro and micro methods)
Cystinosis	Cystine accumulation (chemical analysis) and quantitative amino acid analyses
Fabry's disease	Ceramide trihexosidase α-galactosidase
Fucosidosis	α-fucosidase
Galactosemias	Galactose-1-phosphate, uridyl transferase and galactokinase electrophoresis; chromatography of urine for reducing sugars
GM1 gangliosidosis (generalized) types I & II	β-galactosidase
Gaucher's disease	β-glucosidase and chemical analyses; thin layer chromatography to identify glucocerebroside of cell culture
Glycogen storage disease type II (Pompe's disease)	α-1,4 glucosidase assay (and other enzymes of glycogen metabolism) and glycogen content
Glycogen storage disease type IV	Brancher enzyme assays (and other enzymes of glycogen metabolism) and glycogen content and structure
Homocystinuria	Screening via cyanide-nitroprusside, quantitation of homocystine and methionine in blood and urine by column chromatography, and cystathionine synthetase assay
Hunter's syndrome	β-galactosidase; histochemical stains on cultured skin fibroblasts and cultured white blood cells, chemical analysis of mucopolysaccharides of skin fibroblast cultures, and cloning for X-linked mode of inheritance

Disorder	*Tests utilized*
Hurler's syndrome	β-galactosidase; histochemical stains on cultured skin fibroblasts and cultured white blood cells and chemical analyses of mucopolysaccharides of skin fibroblast cultures
Juvenile GM2 gangliosidosis	Hexosaminidase A
Krabbe's disease	Galactose cerebrosidase
Lesch-Nyhan syndrome (and variants)	Hypoxanthine-guanine, phosphoribosyl-transferase and other enzymes in purine pathway, chemical methods and screening
Lysosomal acid phosphatase deficiency	Lysosomal acid phosphatase
Maple syrup urine disease (branched chain ketoaciduria)	Branched chain keto acid decarboxylase, and quantitative chromatography
Metachromatic leukodystrophy (infantile, juvenile)	Arylsulfatase assay
Methylmalonic acidemia	Methylmalonic acid (quantitative assay of metabolic intermediates), and methylmalonic acid mutase assay
Muscular dystrophies (on the basis of sex determination alone)	Creatinine phosphokinase; acetylcholine esterase and choline acetyl transferase
Orotic aciduria	Orotidine mono P. decarboxylase, orotidine mono P. pyrophosphorylase, ornithine transcarbamylase, of orotic acid and other pyrimidines in urine
Refsum's disease	Phytanic acid A-hydroxylase and gas liquid chromatography (quantitative assay on metabolic intermediates)
Sandhoff's disease	Hexosaminadase A & B
Tay-Sachs disease	Hexosaminidase A & B, *N*-acetyl hexosaminidase isozyme A
Wolman's disease	Acid lipase
Xeroderma pigmentosa	DNA repair

SOURCE: From *Laboratory Management*, 1972, 10(10), p. 27.

is linearly good needs to be carefully examined. Other traits that might be mentioned include cooperativeness, general vigor, sympathy, moral courage, reasonableness, and creativity. However, critics have pointed out that we cannot at present measure these attributes with any respectable degree of reliability or validity. Furthermore, we know nothing about their genetic correlates. Do the genes for general vigor pleiotropically influence aggression, so that if we select for general vigor we would also be increasing

Table 11.20

Expectations under eutelegenesis (sperm selection) as a function of the proportion of females participating in the program, and the proportion of males used as sperm donors: mean IQ in first generation assuming 50 percent heritability

Sperm donors		Percentage of females participating					
IQ	Percentage selected	1	5	10	20	50	100
115	38	100	100.2	100.4	100.8	100.9	103.8
130	6	100.1	100.4	100.8	101.5	103.8	107.5
145	0.36	100.1	100.6	101.1	102.2	105.6	111.2
160	0.008	100.1	100.8	101.5	103.0	107.5	115.0
175	0.0001	100.2	101.0	101.9	103.8	109.4	118.8

SOURCE: After Cavalli-Sforza and Bodmer, *The Genetics of Human Populations*. W. H. Freeman and Company. Copyright © 1971, p. 769.

aggressiveness? Is cooperativeness similarly part of a pleiotropic network that includes servility? Would it be a good thing or a bad thing to have more aggression or more servility? Good for what? Is reasonableness, as manifested by an ability to see both sides of a question, related to an inability to make decisions? It seems most unlikely that we could maximize all of the positive phenotypes, even if we knew what they were!

At another level, the question becomes one of selection of the donors. Fashions change a great deal, and the acclaimed of one generation do not always retain their luster in the perspective of historical judgment. An example of this is provided by Muller's own writings. In his earlier work, he suggested that the sperm of such men as Marx and Lenin would obviously be appropriate for a sperm bank. In later works, after a change in his political philosophy, these men were dropped from his suggested donor list.

In any case, the actual beneficial effects, either upon the gene pool of the species or upon phenotypes, would probably be very small. Cavalli-Sforza and Bodmer (1971) have computed the expected mean IQ of the species after one generation of "germinal choice" under different conditions of selection for sperm donors and with different percentages of the women in the population participating, assuming that the heritability of IQ is 0.50. The results are shown in Table 11.20. No matter how extreme the selection for IQ of the male donors, results are miniscule unless a large proportion of women participate in the program. On the other hand, some 5–10 thousand women per year are inseminated by artificial means. Proponents of the Mullerian view propose that the opportunity should be taken to be systematic in choice of donors. Programmatic use of sperm of men who achieved eminence in some domain or other could hardly lead to *less* successful results than the current rather haphazard and furtive methods of selecting donors.

We have already seen (Chapter 9) that the gene pool of a species that is of appreciable size is a very conservative system. Any dysgenic influences, short of the mutational burden that might accompany all-out nuclear warfare, would be able to affect gene frequencies only slowly. However, as Stern has recently put it,

> If the hopes and fears of the eugenics movement seem greatly exaggerated in the light of a numerical treatment of the problems, it should not be forgotten that the idealism which concerns itself with the genetic fate of future generations has a sound core. To say that the loss of supposedly desirable genotypes in one or even many generations of differential reproduction is small does not mitigate the fact that it *is* a loss which may be regrettable and, possibly, even have serious consequences. To state that reproductive selection against severe physical and mental abnormalities will reduce the number of the affected from one generation to the next by only a small percentage does not alter the fact that the small percentage may represent tens of thousands of individuals. Conversely, even a slight increase of desirable genotypes, through progressive eugenic measures, would be a social gain (1973, p. 793).

Moreover, it appears that problems of population size may be forcing us into a situation where the genetic consequences of various actions must be examined. Whether the response to burgeoning populations be *laissez faire*, sanctions against large families, granting every couple the right to produce one pair of children, the issuance of saleable reproductive rights to women, or lottery choice of those to be permitted to reproduce, the result will be a departure from the present mating pattern of the species. In addressing ourselves to the possible consequences of these changes, we need much more information than is currently available about the structure of the intellect and the heritabilities of its components, about the genetic architecture of other behavioral characteristics that are a part of the functioning human being, about the effect of nonrandom mating on behavioral traits and so on. We need also most urgently for the old nature-nurture dichotomous model to be replaced in the thinking of the citizenry at large, as well as in the perspectives of the social and behavioral scientists, with a model of genotype-environment co-action. No matter how well intentioned it may be, a naive environmentalistic position is a vulnerable one and one that may lead to ignorant decisions. No less vulnerable, of course, nor less generative of ignorance is a naive hereditarian view.

Crow, in an evaluation of eugenics, has said of the issues:

> Will future generations regard our generation somewhat as we do the pioneers who destroyed our forests and wildlife — as geneticists without the wisdom and courage to look to the future? Or, on the other hand, will they

regard this generation as one which prudently refrained from rushing to act too soon in ignorance?

I have one conviction; it is high time that the social implications of our expanding genetic knowledge be discussed. Early eugenics was crude, over-simplified, and got confused in various dubious (and in some cases disastrous) political movements. I hope we are ready for a more mature consideration of eugenics and euphenics as complementary possibilities. It may well be that the second century of Mendelism will mark the beginning of a serious and informed consideration of the extent to which man can and should influence his biological future, with full deliberation on both the opportunities and the risks. (1967, p. 372)

Epilogue

If there is a central message from behavioral genetics, it is with respect to individual differences. With the trivial exception of members of identical multiple births, each one of us is a unique genetic experiment never tried before and never to be repeated again. Here is the conceptualization on which to build a philosophy of the dignity of the individual! Human varia-bility is not simply imprecision in a process that, if perfect, would generate unvarying representatives of a Platonic ideal. Individuality is the quintes-sence of life; it is the product and the agency of the whole grand sweep of evolution.

It is important not to confuse biological identity with the political concept of equality. The problem is to an extent a semantic one. The Greeks had not one, but many words for equality, distinguishing, for example (Hutchins and Adler, 1968, p. 305):

> *isonomia:* equality before the law
> *isotimia:* equality of honor
> *isopoliteia:* equality of political rights
> *isokratia:* equality of political power
> *isopsephia:* equality of votes or suffrage
> *isegoria:* equality in right to speak
> *isoteleia:* equality of tax or tribute
> *isomoiria:* equality of shares or partnership
> *isokleria:* equality of property
> *isodaimonia:* equality of fortune

Perhaps we confuse ourselves by using only the one word, equality.

A proper attention to individual needs, the provision of the environmental circumstances that will optimize the development of each person, is a utopian ideal and no more attainable than other utopias. However, we can approach this ideal more closely if we recognize the fundamental nature of individuality. Although the requisite knowledge will require much research effort and expense, it would seem to warrant a high priority. The human individuality that would be cultivated is the fundamental natural resource of our species.

References

Abdullah, S., Jarvik, L. F., Kato, T., Johnston, W. C., and Lanzkron, J. Extra Y chromosome and its psychiatric implications. *Archives of General Psychiatry*, 1969, 21, 497–501.

Ader, R., and Conklin, P. M. Handling of pregnant rats: Effects of emotionality on their offspring. *Science*, 1963, 142, 411–412.

Akesson, H. O., Forssman, H., and Wallin, L. Chromosomes of tall men in mental hospitals. *Lancet,* 1968, ii, 1040.

Alexander, D., Walker, H. T., Jr., and Money, J. Studies in direction sense. *Archives of General Psychiatry*, 1964, 10, 337–339.

Åmark, C. A study in alcoholism. Clinical, social-psychiatric and genetic investigations. *Acta Psychiatrica et Neurologica Scandinavica*, 1951, Suppl. 70.

Amin, A., Chai, C. K., and Reineke, E. P. Differences in thyroid activity of several strains of mice and F_1 hybrids. *American Journal of Physiology*, 1957, 191, 34–36.

Aristotle. Politics (Politica). *In* R. M. Hutchins (Ed.), *Great Books of the western world*. Vol. 9. *Aristotle* (Vol. II). Chicago: Encyclopedia Britannica, 1952. Pp. 531–532.

Bajema, C. Estimation of the direction and intensity of natural selection in relation to human intelligence by means of the intrinsic rate of natural increase. *Eugenics Quarterly*, 1963, 10, 175–187.

Baker, D., Telfer, M. A., Richardson, C. E., and Clark, G. R. Comparative studies of XYY and Klinefelter males characterized by antisocial characteristics. *Journal of the American Medical Association*, 1970, 214, 869–878.

Barnicot, N. A. Taste deficiency for phenylthiourea in African Negroes and Chinese. *Annals of Eugenics*, 1950, 15, 248–254.

Barr, M. L., and Bertram, E. G. A morphological distinction between neurones of the male and female, and the behaviour of the nucleolar satellite during accelerated nucleoprotein synthesis. *Nature*, 1949, 163, 676.

Bateson, W. *Mendel's principles of heredity.* Cambridge: University Press, 1909.

Beadle, G. W., and Tatum, E. L. Experimental control of developmental reaction. *American Naturalist*, 1941, 75, 107–116.

Belknap, J. K., Jr. Ethanol-induced sleep time in mice as a function of hepatic alcohol and aldehyde dehydrogenase activities. Unpublished doctoral dissertation, University of Colorado, 1971.

Benzer, S. Behavioral mutants of *Drosophila* isolated by countercurrent distribution. *Proceedings of the National Academy of Sciences*, 1967, 58, 1112–1119.

Bickel, H., Gerrard, J., and Hickmans, E. M. Influence of phenylalanine intake on phenylketonuria. *Lancet*, 1953, ii, 812–813.

Bigelow, R. *The dawn warriors: Man's evolution toward peace.* Boston: Little, Brown and Company, 1969.

Blewett, D. B. An experimental study of the inheritance of intelligence. *Journal of Mental Science*, 1954, 100, 922–933.

Böök, J. A. Oligophrenia. *In* A. Sorsby (Ed.), *Clinical genetics.* London: Butterworth, 1953. Pp. 322–331.

Böök, J. A. Genetical investigation in a north Swedish population: The offspring of first-cousin marriages. *Annals of Human Genetics*, 1957, 21, 191–221.

Borgström, C. A. Eine Serie von kriminellen Zwillingen. *Archiv fuer Rassen- und Gesellschaftsbiologie*, 1939, 33, 334–343.

Boring, E. G. A history of experimental psychology. (2nd ed.) New York: Appleton, 1950.

Brace, C. L., Nelson, H., and Korn, N. *Atlas of fossil man.* New York: Holt, Rinehart & Winston, 1971.

Broadhurst, P. L. Experiments in psychogenetics. Applications of biometrical genetics to the inheritance of behaviour. *In* H. J. Eysenck (Ed.), *Experiments in personality.* Vol. 1. *Psychogenetics and psychopharmacology.* London: Routledge & Kegan Paul, 1960. Pp. 1–102.

Broadhurst, P. L. Analysis of maternal effects in the inheritance of behaviour. *Animal Behaviour*, 1961, 9, 129–141.

Broadhurst, P. L. The biometrical analysis of behavioural inheritance. *Science Progress, Oxford*, 1967, 55, 123–139.

Brown, B. S. *Memo from the director.* National Institute of Mental Health, December 1970.

Bruell, J. H. Dominance and segregation in the inheritance of behavior in mice. *In* E. L. Bliss (Ed.), *Roots of behavior.* New York: Hoeber Medical Division, Harper & Row, 1962. Pp. 48–67.

Bruell, J. H. Heterotic inheritance of wheelrunning in mice. *Journal of Comparative and Physiological Psychology*, 1964a, 58, 159–163.

Bruell, J. H. Inheritance of behavioral and physiological characters of mice and the problem of heterosis. *American Zoologist*, 1964b, 4, 125–138.

Bruell, J. H. Behavioral heterosis. *In* J. Hirsch (Ed.), *Behavior-genetic analysis.* New York: McGraw-Hill, 1967. Pp. 270–304.

Bruell, J. H. Behavioral population genetics and wild *Mus musculus*. *In* G. Lindzey and D. D. Thiessen (Eds.), *Contributions to behavior-genetic analysis: The mouse as a prototype.* New York: Appleton-Century-Crofts, 1970. Pp. 261–291.

Buck vs. Bell (1927), 274 U.S. 200, 47 S.Ct. 584, 71 L.Ed. 1000.

Burks, B. S. The relative influence of nature and nurture upon mental development: A comparative study of foster parent-foster child resemblance and true parent-true child resemblance. *Nature and nurture: Their influence upon intelligence, Twenty-Seventh Yearbook of the National Society for the Study of Education,* 1928, Part 1, 219–316.

Burt, C. *Intelligence and fertility: The effect of the differential birthrate on inborn mental characteristics.* (2nd ed.) London: The Eugenics Society and Cassell and Company, 1952.

Burt, C. The inheritance of mental ability. *American Psychologist,* 1958, 13, 1–15.

Burt, C. Intelligence and social mobility. *British Journal of Statistical Psychology,* 1961, 14, 3–24.

Burt, C. The genetic determination of differences in intelligence: A study of monozygotic twins reared together and apart. *British Journal of Psychology,* 1966, 57, 137–153.

Carter, H. D. Family resemblances in verbal and numerical abilities. *Genetic Psychology Monographs,* 1932, 12, 1–104.

Casey, M. D., Blank, C. E., Street, D. R. K., Segall, L. J., McDougall, J. H., McGrath, P. J., and Skinner, J. L. YY chromosomes and antisocial behaviour. *Lancet,* 1966, ii, 859–860.

Caspersson, T., Farber, S., Foley, G. E., Kudynowski, J., Modest, E. J., Simonsson, E., Wagh, U., and Zech, L. Chemical differentiation along metaphase chromosomes. *Experimental Cell Research,* 1968, 49, 219–222.

Caspersson, T., Hultén, M., Lindsten, J., Therkelsen, Å. J., and Zech, L. Identification of different Robertsonian translocations in man by quinacrine mustard fluorescence analysis. *Hereditas,* 1971, 67, 213–220.

Caspersson, T., Lomakka, G., and Møller, A. Computerized chromosome identification by aid of the quinacrine mustard fluorescence technique. *Hereditas,* 1971, 67, 103–109.

Castle, W. E. The law of heredity of Galton and Mendel and some laws governing race improvement by selection. *Proceedings of the American Academy of Sciences,* 1903, 39, 233–242.

Cattell, R. B. Is national intelligence declining? *Eugenics Review,* 1936, 28, 181–208.

Cattell, R. B. The fate of national intelligence: Tests of a thirteen-year prediction. *Eugenics Review,* 1950, 42, 136–148.

Cavalli-Sforza, L. L., and Bodmer, W. F. *The genetics of human populations.* San Francisco: W. H. Freeman and Company, 1971.

Centerwall, S. A., and Centerwall, W. R. The discovery of phenylketonuria. *In* F. L. Lyman (Ed.), *Phenylketonuria.* Springfield, Ill.: Thomas, 1963. Pp. 3–10.

Chamberlain, H. S. *The foundations of the nineteenth century.* (Transl. by J. Lees) London and New York: J. Lane, 1910.

Changeux, J.-P. The control of biochemical reactions. *Scientific American,* 1965, 212, 36–45. Reprinted in *The molecular basis of life* (readings from *Scientific American*). San Francisco: W. H. Freeman and Company, 1968. Pp. 244–253.

Cohen, R., Bloch, N., Flum, Y., Kadar, M., and Goldschmidt, E. School attainment in an immigrant village. *In* E. Goldschmidt (Ed.), *The genetics of migrant and isolate populations.* Baltimore: Williams & Wilkins, 1963. Pp. 350–351.

Collins, R. L. A new genetic locus mapped from behavioral variation in mice: Audiogenic seizure prone (*asp*). *Behavior Genetics*, Plenum Publishing, Inc., 1970, 1, 99–109.

Collins, R. L., and Fuller, J. L. Audiogenic seizure prone (*asp*): A gene affecting behavior in linkage group VIII of the mouse. *Science*, 1968, 162, 1137–1139.

Cooper, R. M., and Zubek, J. P. Effects of enriched and restricted early environment on the learning ability of bright and dull rats. *Canadian Journal of Psychology*, 1958, 12, 159–164.

Court Brown, W. M. Males with an XYY sex chromosome complement. *Journal of Medical Genetics*, 1968, 5, 341–359.

Crew, F. A. E. Inheritance of educability. A first report on an attempt to examine Prof. McDougall's conclusions relating to his experiment for the testing of the hypothesis of Lamarck. *Proceedings of the Sixth International Congress of Genetics*, 1932, 1, 121–134.

Crick, F. *Of molecules and men.* Seattle: University of Washington Press, 1966.

Critchley, M. *The dyslexic child.* London: William Heinemann Medical Books, 1970.

Crow, J. F. Genetics and medicine. *In* R. A. Brink (Ed.), *Heritage from Mendel.* Madison: University of Wisconsin Press, 1967. Pp. 351–374.

Crow, J. F., and Felsenstein, J. The effect of assortative mating on the genetic composition of a population. *Eugenics Quarterly*, 1968, 15, 85–97.

Crow, J. F., and Kimura, M. *An introduction to population genetics theory.* New York: Harper & Row, 1970.

Crowe, R. R. The adopted offspring of women criminal offenders: A study of their arrest records. *Archives of General Psychiatry*, 1972, 27, 600–603.

Daly, R. F. Neurological disorders in XYY males. *Nature*, 1969, 221, 472–473.

Darwin, C. *On the origin of species by means of natural selection, or the preservation of favoured races in the struggle for life.* London: John Murray, 1859 (New York: Appleton, 1869.)

Darwin, C. *The variation of animals and plants under domestication.* New York: Orange Judd, 1868.

Darwin, C. *The descent of man and selection in relation to sex.* London: John Murray, 1871. (New York: Appleton, 1873.)

Darwin, C. *The expression of the emotions in man and animals.* London: John Murray, 1872.

Darwin, C. (Letter from C. Darwin to C. Lyell, 1858). *In* F. Darwin (Ed.), *The life and letters of Charles Darwin.* Vol. II. London: John Murray, 1888. P. 117.

Darwin, C. *Journal of researches into the natural history and geology of the countries visited during the voyage of H.M.S. Beagle round the world, under the command of Capt. Fitz Roy, R.N.* New York: Appleton, 1896.

Davenport, C. B. *Heredity in relation to eugenics.* London: Williams & Norgate, 1912.

Davis, H. *The works of Plato.* Vol. II. London: Bohn, 1849.

Dawson, N. J. Body composition of inbred mice (*Mus musculus*). *Comparative Biochemistry and Physiology*, 1970, 37, 589–593.

Day, E. J. The development of language in twins: II. The development of twins: Their resemblances and differences. *Child Development*, 1932, 3, 237–256.

DeFries, J. C. Prenatal maternal stress in mice: Differential effects on behavior. *Journal of Heredity*, 1964, 55, 289–295.

DeFries, J. C. Quantitative genetics and behavior: Overview and perspective. *In* J. Hirsch (Ed.), *Behavior-genetic analysis*. New York: McGraw-Hill, 1967. Pp. 322–339.

DeFries, J. C. Pleiotropic effects of albinism on open field behaviour in mice. *Nature*, 1969, 221, 65–66.

DeFries, J. C. Quantitative aspects of genetics and environment in the determination of behavior. *In* L. Ehrman, G. S. Omenn, and E. Caspari (Eds.), *Genetics, environment, and behavior: Implications for educational policy*. New York: Academic Press, 1972. Pp. 5–16.

DeFries, J. C., and Hegmann, J. P. Genetic analysis of open-field behavior. *In* G. Lindzey and D. D. Thiessen (Eds.), *Contributions to behavior-genetic analysis: The mouse as a prototype*. New York: Appleton-Century-Crofts, 1970. 23–56.

DeFries, J. C., Hegmann, J. P., and Weir, M. W. Open-field behavior in mice: Evidence for a major gene effect mediated by the visual system. *Science*, 1966, 154, 1577–1579.

DeFries, J. C., and McClearn, G. E. Social dominance and Darwinian fitness in the laboratory mouse. *American Naturalist*, 1970, 104, 408–411.

DeFries, J. C., and McClearn, G. E. Behavioral genetics and the fine structure of mouse populations: A study in microevolution. *In* Th. Dobzhansky, M. K. Hecht, and W. C. Steere (Eds.), *Evolutionary biology*. Vol. 5. New York: Appleton-Century-Crofts, 1972. Pp. 279–291.

DeFries, J. C., Thomas, E. A., Hegmann, J. P., and Weir, M. W. Open-field behavior in mice: Analysis of maternal effects by means of ovarian transplantation. *Psychonomic Science*, 1967, 8, 207–208.

DeFries, J. C., Weir, M. W., and Hegmann, J. P. Differential effects of prenatal maternal stress on offspring behavior in mice as a function of genotype and stress. *Journal of Comparative and Physiological Psychology*, 1967, 63, 332–334.

DeMars, R., Sarto, G., Felix, J. S., and Benke, P. Lesch-Nyhan mutation: Prenatal detection with amniotic fluid cells. *Science*, 1969, 164, 1303–1305.

Dobzhansky, Th. *Heredity and the nature of man*. New York: Harcourt, Brace & World, 1964.

Dobzhansky, Th. Genetics and the social sciences. *In* D. C. Glass (Ed.), *Genetics*. New York: The Rockefeller University Press and Russell Sage Foundation, 1968. Pp. 129–142.

Dobzhansky, Th. *Genetics of the evolutionary process*. New York: Columbia University Press, 1970.

Dobzhansky, Th., and Spassky, B. Effects of selection and migration on geotactic and phototactic behavior of *Drosophila*. *Proceedings of the Royal Society* (Series B), 1967, 168, 27–47.

Drets, M. E., and Shaw, M. W. Specific banding patterns of human chromosomes. *Proceedings of the National Academy of Sciences*, 1971, 68, 2073–2077.

Dugdale, R. L. *The Jukes*. New York: G. P. Putnam's Sons, 1877.

East, E. M., and Hayes, H. K. Inheritance in maize. *Bulletin of the Connecticut Agricultural Experiment Station*, 1911, 167, 1–142.

Eckland, B. K. Genetics and sociology: A reconsideration. *American Sociological Review*, 1967, 32, 173–194.

Edwards, J. H., Harnden, D. G., Cameron, A. H., Crosse, V. M., and Wolff, O. H. A new trisomic syndrome. *Lancet,* 1960, i, 787–789.

Ehrman, L. Mating success and genotype frequency in *Drosophila. Animal Behaviour,* 1966, 14, 332–339.

Ehrman, L. Simulation of the mating advantage of rare *Drosophila* males. *Science,* 1970a, 167, 905–906.

Ehrman, L. The mating advantage of rare males in *Drosophila. Proceedings of the National Academy of Sciences,* 1970b, 65, 345–348.

Ehrman, L. A factor influencing the rare male mating advantage in *Drosophila. Behavior Genetics,* 1972, 2, 69–78.

Eiseley, L. Charles Darwin, Edward Blyth, and the theory of natural selection. *Proceedings of the American Philosophical Society,* 1959, 103, 94–158.

Eldridge, R., Harlan, A., Cooper, I. S., and Riklan, M. Superior intelligence in recessively inherited torsion dystonia. *Lancet,* 1970, i, 65–67.

Ellis, W. *The politics of Aristotle.* London: Dent, 1912.

Elässer, G. Die Nachkommen geisteskranker Elternpaare. Stuttgart: Thieme, 1952.

Emerson, R. A., and East, E. M. The inheritance of quantitative characters in maize. *University of Nebraska Research Bulletin,* 1913, 2, 5–120.

Entrikin, R. K., and Erway, L. C. A genetic investigation of roller and tumbler pigeons. *Journal of Heredity,* 1972, 63, 351–354.

Eriksson, K., and Pikkarainen, P. H. Differences between the sexes in voluntary alcohol consumption and liver ADH-activity in inbred strains of mice. *Metabolism,* 1968, 17, 1037–1042.

Erlenmeyer-Kimling, L., and Jarvik, L. F. Genetics and intelligence: A review. *Science,* 1963, 142, 1477–1479.

Erlenmeyer-Kimling, L., and Paradowski, W. Selection and schizophrenia. *American Naturalist,* 1966, 100, 651–665.

Erlenmeyer-Kimling, L., Rainer, J. D., and Kallmann, F. J. Current reproductive trends in schizophrenia. *In* P. H. Hoch and J. Zubin (Eds.), *Psychopathology of schizophrenia.* New York: Grune & Stratton, 1966. Pp. 252–276.

Essen-Möller, E. Psychiatrische Untersuchungen an einer Serie von Zwillingen. *Acta Psychiatrica,* 1941, Suppl. 23.

Estabrook, A. H. *The Jukes in 1915.* Washington, D.C.: Carnegie Institution, 1916.

Eysenck, H. J., and Broadhurst, P. L. Experiments with animals: Introduction. *In* H. J. Eysenck (Ed.), *Experiments in motivation.* New York: Macmillan, 1964. Pp. 285–291.

Falconer, D. S. *Introduction to quantitative genetics.* New York: Ronald Press, 1960.

Falconer, D. S. Quantitative inheritance. *In* W. J. Burdette (Ed.), *Methodology in mammalian genetics.* San Francisco: Holden-Day, 1963. Pp. 193–216.

Falconer, D. S. The inheritance of liability to certain diseases estimated from the incidence among relatives. *Annals of Human Genetics,* 1965, 29, 51–71.

Falconer, D. S. Genetic consequences of selection pressure. *In* J. E. Meade and A. S. Parkes (Eds.), *Genetic and environmental factors in human ability.* Edinburgh: Oliver & Boyd, 1966.

Falek, A. Differential fertility and intelligence: Current status of the problem. *Social Biology,* 1971, 18(Suppl.), S50–S59.

Feuer, G., and Broadhurst, P. L. Thyroid function in rats selectively bred for emotional elimination: III. Behavioural and physiological changes after treatment with drugs acting on the thyroid. *Journal of Endocrinology*, 1962, 24, 385–396.

Fischer, M. Schizophrenia in twins. Paper presented at the Fourth World Congress of Psychiatry, Madrid, 1966.

Fischer, M., Harvald, B., and Hauge, M. A Danish twin study of schizophrenia. *British Journal of Psychiatry*, 1969, 115, 981–990.

Fisher, J., and Hinde, R. A. The opening of milk bottles by birds. *British Birds*, 1948, 42, 347–357.

Fisher, R. A. The correlation between relatives on the supposition of Mendelian inheritance. *Transactions of the Royal Society of Edinburgh*, 1918, 52, 399–433.

Fisher, R. A. *The genetical theory of natural selection.* Oxford: Clarendon Press, 1930.

Fisher, R. A. *The genetical theory of natural selection.* (2nd ed.) New York: Dover, 1958.

Følling, A., Mohr, O. L., and Ruud, L. *Oligophrenia phenylpyrouvica*, a recessive syndrome in man. *Norske Videnskaps/Akademi I Oslo, Matematisk-Naturvidenskapelig Klasse*, 1945, 13, 1–44.

Fonseca, A. F. da *Analise heredo-clinica das pertubacoes afectivas atraves de 60 pares de gemeos.* Oporto: Facultade de Medicina, 1959.

Freedman, D. G. Constitutional and environmental interactions in rearing of four breeds of dogs. *Science,* 1958, 127, 585–586.

Freeman, F. N., Holzinger, K. J., and Mitchell, B. C. The influence of environment on the intelligence, school achievement, and conduct of foster children. *Nature and nurture: Their influence upon intelligence, Twenty-Seventh Yearbook of the National Society for the Study of Education*, 1928, 27(Pt. 1), 103–217.

Fuhrmann, W. Therapy of genetic diseases in man and the possible place of genetic engineering. *In* G. Raspé (Ed.), *Advances in the biosciences.* Oxford: Pergamon Press, 1972. Pp. 387–396.

Fuller, J. L. Experiential deprivation and later behavior. *Science*, 1967, 158, 1645–1653.

Fuller, J. L., Easler, C., and Smith, M. E. Inheritance of audiogenic seizure susceptibility in the mouse. *Genetics*, 1950, 35, 622–632.

Fuller, J. L., and Sjursen, F. H., Jr. Audiogenic seizures in eleven mouse strains. *Journal of Heredity*, 1967, 58, 135–140.

Fuller, J. L., and Thompson, W. R. *Behavior genetics.* New York: Wiley, 1960.

Fuller, J. L., and Wimer, R. E. Neural, sensory, and motor functions. *In* E. L. Green (Ed.), *Biology of the laboratory mouse.* (2nd ed.) New York: McGraw-Hill, 1966. Pp. 609–628.

Galton, F. *Hereditary genius: An inquiry into its laws and consequences.* London: Macmillan, 1869.

Galton, F. *Inquiries into human faculty and its development.* London: Macmillan, 1883.

Galton, F. Eugenics: Its definition, scope and aims. Paper read before the Sociological Society at a Meeting in the School of Economics and Political Science, London University, May 16, 1904.

Garrod, A. E. The Croonian lectures on inborn errors of metabolism, I, II, III, IV. *Lancet,* 1908, ii, 1–7, 73–79, 142–148, 214–220.

Gates, R. R. The inheritance of mental defect. *British Journal of Medical Psychology*, 1933, 13, 254–267.

Gates, R. R. *Human genetics.* New York: Macmillan, 1946.

Geneticists press for regionalization. *Laboratory Management*, 1972, 10(10), 25–27.

Ginsburg, B., and Allee, W. C. Some effects of conditioning on social dominance and subordination in inbred strains of mice. *Physiological Zoology*, 1942, 25, 485–506.

Ginsburg, B. E., Cowen, J. S., Maxson, S. C., and Sze, P. Y.-L. Neurochemical effects of gene mutations associated with audiogenic seizures. Paper presented at the Second International Congress of Neuro-Genetics and Neuro-Ophthalmology, September 19, 1967.

Ginsburg, B. E., and Miller, D. S. Genetic factors in audiogenic seizures, *Colloques Nationaux du Centre National de la Recherche Scientifique, Paris*, 1963, 112, 217–225.

Gobineau, J. A. de *The moral and intellectual diversity of races.* (1853–1855) (Transl. by H. Hotz) Philadelphia: Lippincott, 1856.

Goodard, H. H. *The Kallikak family.* New York: Macmillan, 1912.

Goddard, H. H. *Feeblemindedness: Its causes and consequences.* New York: Macmillan, 1914.

Goddard, H. H. In defense of the Kallikak study. *Science*, 1942, 95, 574–576.

Goodman, R. M., Smith, W. S., and Migeon, C. J. Sex chromosome abnormalities. *Nature*, 1967, 216, 942–943.

Goodwin, D. W., Schulsinger, F., Hermanson, L., Guze, S. B., and Winokur, G. Alcohol problems in adoptees raised apart from alcoholic biological parents. *Archives of General Psychiatry*, 1973, 28, 238–243.

Gottesman, I. I. Heritability of personality: A demonstration. *Psychological Monographs*, 1963, 77(9, Whole No. 572).

Gottesman, I. I. Personality and natural selection. *In* S. G. Vandenberg (Ed.), *Methods and goals in human behavior genetics.* New York: Academic Press, 1965. Pp. 63–80.

Gottesman, I. I. Severity/concordance and diagnostic refinement in the Maudsley-Bethlem schizophrenic twin study. *In* D. Rosenthal and S. S. Kety (Eds.), *The transmission of schizophrenia.* Oxford: Pergamon Press, 1968. Pp. 37–48.

Gottesman, I. I., and Shields, J. Schizophrenia in twins: 16 years' consecutive admissions to a psychiatric clinic. *Diseases of the Nervous System*, 1966, 27, 11–19.

Gottesman, I. I., and Shields, J. A polygenic theory of schizophrenia. *Proceedings of the National Academy of Sciences*, 1967, 58, 199–205.

Gottesman, I. I., and Shields, J. Genetic theorizing and schizophrenia. *British Journal of Psychiatry*, 1972, 121, in press.

Green, E. L. Breeding systems. *In* E. L. Green (Ed.), *Biology of the laboratory mouse.* (2nd ed.) New York: McGraw-Hill, 1966. Pp. 11–22.

Griffiths, A. W., and Zaremba, J. Crime and sex chromosome anomalies. *British Medical Journal*, 1967, 4, 622.

Gruneberg, H. *The genetics of the mouse.* The Hague: Martinus Nijhoff, 1952.

Gun, W. T. J. The heredity of the Tudors. *Eugenics Review*, 1930a, 22, 111–116.

Gun, W. T. J. The heredity of the Stewarts. *Eugenics Review*, 1930b, 22, 195–201.

Haggard, E. A. *Intraclass correlation and the analysis of variance.* New York: Holt, Rinehart & Winston, 1958.

Hall, C. S. The inheritance of emotionality. *Sigma Xi Quarterly*, 1938, 26, 17–27.

Haller, M. H. *Eugenics.* New Brunswick, N.J.: Rutgers University Press, 1963.

Hallgren, B. Specific dyslexia ("congenital word-blindness"): A clinical and genetic study. *Acta Psychiatrica et Neurologica*, 1950, Suppl. 65.

Halperin, S. L. Human heredity and mental deficiency. *American Journal of Mental Deficiency*, 1946, 51, 153–163.

Hardin, G. *Biology: Its human implications.* San Francisco: W. H. Freeman and Company, 1949.

Harlow, H. F. The evolution of learning. *In* A. Roe and G. G. Simpson (Eds.), *Behavior and evolution.* New Haven: Yale University Press, 1958. Pp. 269–290.

Harris, H. Enzyme variation in man: Some general aspects. *In* J. F. Crow and J. V. Neel (Eds.), *Proceedings of the Third International Congress of Human Genetics.* Baltimore: Johns Hopkins Press, 1967. Pp. 207–214.

Harvald, B., and Hauge, M. Hereditary factors elucidated by twin studies. *In* J. V. Neel, M. W. Shaw, and W. J. Schull (Eds.), *Genetics and the epidemiology of chronic diseases.* Washington, D.C.: U.S. Dept. H.E.W., 1965.

Hays, W. L. *Statistics for psychologists.* New York: Holt, Rinehart & Winston, 1963.

Hegmann, J. P. Physiological function and behavioral genetics. I. Genetic variance for peripheral conduction velocity in mice. *Behavior Genetics*, 1972, 2, 55–67.

Henderson, N. D. Brain weight increases resulting from environmental enrichment: A directional dominance in mice. *Science*, 1970, 169, 776–778.

Henry, K. R. Audiogenic seizure susceptibility induced in C57BL/6J mice by prior auditory exposure. *Science*, 1967, 158, 938–940.

Henry, K. R., and Schlesinger, K. Effects of the albino and dilute loci on mouse behavior. *Journal of Comparative and Physiological Psychology*, 1967, 63, 320–323.

Heston, L. L. Psychiatric disorders in foster home reared children of schizophrenic mothers. *British Journal of Psychiatry*, 1966, 112, 819–825.

Heston, L. L. The genetics of schizophrenic and schizoid disease. *Science*, 1970, 167, 249–256.

Higgins, J. V., Reed, E. W., and Reed, S. C. Intelligence and family size: A paradox resolved. *Eugenics Quarterly*, 1962, 9, 84–90.

Hindley, C. B. Social class influences on the development of ability in the first five years. Paper presented at the Fourteenth International Congress of Applied Psychology, Copenhagen, 1961.

Hirsch, J. Studies in experimental behavior genetics: II. Individual differences in geotaxis as a function of chromosome variations in synthesized *Drosophila* populations. *Journal of Comparative and Physiological Psychology*, 1959, 52, 304–308.

Hirsch, J., and Erlenmeyer-Kimling, L. Sign of taxis as a property of the genotype. *Science*, 1961, 134, 835–836.

Hirsch, J., and Erlenmeyer-Kimling, L. Studies in experimental behavior genetics: IV. Chromosome analyses for geotaxis. *Journal of Comparative and Physiological Psychology*, 1962, 55, 732–739.

Hitler, A. *Mein Kampf.* (1925) (Transl. by R. Manheim) Boston: Houghton Mifflin, 1943.

Holmes, S. J. *A bibliography of eugenics.* Berkeley: University of California Press, 1924.

Honzik, M. P. Developmental studies of parent-child resemblance in intelligence. *Child Development,* 1957, 28, 215–228.

Hook, E. B. Behavioral implications of the human XYY genotype. *Science,* 1973, 179, 139–150.

Hopkinson, D. A., and Harris, H. Recent work on isozymes in man. *Annual Review of Genetics,* 1971, 5, 5–32.

Hotta, Y., and Benzer, S. Genetic dissection of the *Drosophila* nervous system by means of mosaics. *Proceedings of the National Academy of Sciences,* 1970, 67, 1156–1163.

Howells, W. *Mankind in the making: The story of human evolution.* (Rev. ed.) Garden City, N.Y.: Doubleday, 1967.

Hsia, D. Y.-Y. (Ed.) *Lectures in medical genetics.* Chicago: Year Book Medical Publishers, 1966.

Hsia, D. Y.-Y. *Human developmental genetics.* Chicago: Year Book Medical Publishers, 1968.

Hsia, D. Y.-Y. Phenylketonuria and its variants. *In* A. G. Steinberg and A. G. Bearn (Eds.), *Progress in medical genetics.* Vol. 7. New York: Grune & Stratton, 1970. Pp. 29–68.

Hsia, D. Y.-Y., Driscoll, K. W., Troll, W., and Knox, W. E. Detection by phenylalanine tolerance tests of heterozygous carriers of phenylketonuria. *Nature,* 1956, 178, 1239–1240.

Hsia, D. Y.-Y., Knox, W. E., Quinn, K. V., and Paine, R. S. A one-year controlled study of the effect of low-phenylalanine diet on phenylketonuria. *Pediatrics,* 1958, 21, 178–202.

Hunter, H. Chromatin-positive and XYY boys in approved schools. *Lancet,* 1968, i, 816.

Huntley, R. M. C. Heritability of intelligence. *In* J. E. Meade and A. S. Parkes (Eds.), *Genetic and environmental factors in human ability.* Edinburgh: Oliver & Boyd, 1966. Pp. 201–218.

Husén, T. *Tvillingstudier.* Stockholm: Almqvist & Wiksell, 1953.

Husén, T. Intra-pair similarities in the school achievements of twins. *Scandinavian Journal of Psychology,* 1963, 4, 108–114.

Hutchins, R. M., and Adler, M. J. The idea of equality. *In* R. M. Hutchins and M. J. Adler (Eds.), *The great ideas today.* Chicago: Encyclopædia Britannica, 1968. Pp. 302–350.

Huxley, J., Mayr, E., Osmond, H., and Hoffer, A. Schizophrenia as a genetic morphism. *Nature,* 1964, 204, 220–221.

Inouye, E. Similarity and dissimilarity of schizophrenia in twins. *Proceedings of the Third World Conference on Psychiatry,* 1961, 1, 524–530.

Jacob, F., and Monod, J. On the regulation of gene activity. *Cold Spring Harbor Symposia on Quantitative Biology,* 1961, 26, 193–209.

Jacobs, P. A., Brunton, M., Melville, M. M., Brittain, R. P., and McClemont, W. F. Aggressive behaviour, mental sub-normality and the XYY male. *Nature,* 1965, 208, 1351–1352.

Jacobs, P. A., Price, W. H., Court Brown, W. M., Brittain, R. P., and Whatmore, P. B. Chromosome studies on men in a maximum security hospital. *Annals of Human Genetics,* 1968, 31, 339–358.

Jensen, A. R. How much can we boost IQ and scholastic achievement? *Harvard Educational Review,* 1969, 39(1), 1–123.

Jinks, J. L., and Fulker, D. W. Comparison of the biometrical genetical, MAVA, and classical approaches to the analysis of human behavior. *Psychological Bulletin,* 1970, 73, 311–349.

Johnson, R. C., and Abelson, R. B. Intellectual, behavioral, and physical characteristics associated with trisomy, translocation, and mosaic types of Down's syndrome. *American Journal of Mental Deficiency,* 1969, 73, 852–855.

Jones, H. E. A first study of parent-child resemblance in intelligence. *Nature and nurture: Their influence upon intelligence, Twenty-Seventh Yearbook of the National Society for the Study of Education,* 1928, 27(Pt. 1), 61–72.

Kahn, E. Studien über Vererbung und Enstehung geistiger Störungen. IV. Schizoid und Schizophrenie im Erbgang. *Monographien aus dem Gesamtgebiete der Neurologie und Psychiatrie,* 1923, No. 36.

Kaij, L. Drinking habits in twins. *Acta Genetica et Medica,* 1957, 7, 437–441.

Kakihana, R., Brown, D. R., McClearn, G. E., and Tabershaw, I. R. Brain sensitivity to alcohol in inbred mouse strains. *Science,* 1966, 154, 1574–1575.

Kallmann, F. J. *The genetics of schizophrenia.* New York: J. J. Augustin, 1938.

Kallmann, F. J. The genetic theory of schizophrenia: An analysis of 691 schizophrenic twin index families. *American Journal of Psychiatry,* 1946, 103, 309–322.

Kallmann, F. J. The genetics of psychoses. An analysis of 1,232 twin index families. In *Congres International de Psychiatrie, Paris.* VI. *Psychiatrie Sociale Rapports.* Paris: Herman & Cie, 1950.

Kaplan, W. D., and Trout, W. E., III. The behavior of four neurological mutants of *Drosophila. Genetics,* 1969, 61, 399–409.

Keeler, C. Some oddities in the delayed appreciation of "Castle's law." *Journal of Heredity,* 1968, 59, 110–112.

Keeley, K. Prenatal influence on behavior of offspring of crowded mice. *Science,* 1962, 135, 44–45.

Kennedy, W. A., van de Riet, V., and White, J. C., Jr. A normative sample of intelligence and achievement of Negro elementary school children in the southeastern United States. *Monographs of the Society for Research in Child Development,* 1963, 28, No. 6.

Kessler, S., and Moos, R. H. The XYY karyotype and criminality: A review. *Journal of Psychiatric Research,* 1970, 7, 153–170.

Kettlewell, H. B. D. Insect survival and selection for pattern. *Science,* 1965, 148, 1290–1296.

Kety, S. S., Rosenthal, D., Wender, P. H., and Schulsinger, F. Mental illness in the biological and adoptive families of adopted schizophrenics. *American Journal of Psychiatry,* 1971, 128, 302–306.

Klein, T. W., and DeFries, J. C. Similar polymorphism of taste sensitivity to PTC in mice and men. *Nature,* 1970, 225, 555–557.

Kranz, H. *Lebensschicksale krimineller Zwillinge.* Berlin: Springer-Verlag OHG, 1936.

Kringlen, E. Schizophrenia in twins, an epidemiological-clinical study. *Psychiatry*, 1966, 29, 172–184.

Kringlen, E. *Heredity and environment in the functional psychoses. An epidemiological-clinical twin study.* Oslo: Universitetsförlaget, 1967.

Lack, D. Darwin's finches. *Scientific American*, 1953, 188(4), 66–72.

Lagerspetz, K. *Studies on the aggressive behaviour of mice.* Helsinki: Suomalainen Tiedeakatemia, 1964.

Lange, J. *Crime as destiny.* (1929) (Transl. by C. Haldane) London: George Allen & Unwin, 1931.

Leahy, A. M. Nature-nurture and intelligence. *Genetic Psychology Monographs*, The Journal Press, 1935, 17, 236–308.

Lederberg, J. Biological future of man. *In* G. Wolstenholme (Ed.), *Man and his future.* Boston: Little, Brown, 1963. Pp. 263–273.

Lee, R. B., and DeVore, I. (Eds.) *Man the hunter.* Chicago: Aldine, 1968.

Le Gras, A. M. Psychose und Kriminalität bei Zwillingen. *Zeitschrift fuer die Gesamte Neurologie und Psychiatrie*, 1933, 144, 198–222.

Lejeune, J. Gautier, M., and Turpin, R. Etude des chromosomes somatiques de neuf enfants mongoliens. *Comptes Rendus de l'Academie des Sciences, Paris*, 1959, 248, 1721–1722.

Lejeune, J., LaFourcade, J., Berger, R., Vialatte, J., Boeswillwald, M., Seringe, P., and Turpin, R. Trois cas de deletion partielle du bras court d'un chromosome 5. *Comptes Rendus de l'Academie des Sciences, Paris*, 1963, 257, 3098–3102.

Lerner, I. M. *Heredity, evolution, and society.* San Francisco: W. H. Freeman and Company, 1968.

Levin, B. R., Petras, M. L., and Rasmussen, D. I. The effect of migration on the maintenance of a lethal polymorphism in the house mouse. *American Naturalist*, 1969, 103, 647–661.

Levine, S., and Treiman, D. M. Differential plasma corticosterone response to stress in four inbred strains of mice. *Endocrinology*, 1964, 75, 142–144.

Lewontin, R., Kirk, D., and Crow, J. Selective mating, assortative mating, and inbreeding: Definitions and implications. *Eugenics Quarterly*, 1968, 15, 141–143.

Lindzey, G., Lykken, D. T., and Winston, H. D. Infantile trauma, genetic factors, *Annual Review of Psychology*, 1971, 22, 39–94.

Lindzey, G. Lykken, D. T., and Winston, H. D. Infantile trauma, genetic factors, and adult temperament. *Journal of Abnormal and Social Psychology*, 1960, 61, 7–14.

Lindzey, G., and Winston, H. Maze learning and effect of pretraining in inbred strains of mice. *Journal of Comparative and Physiological Psychology*, 1962, 55, 748–752.

Lindzey, G., Winston, H. D., and Manosevitz, M. Early experience, genotype, and temperament in *Mus musculus. Journal of Comparative and Physiological Psychology*, 1963, 56, 622–629.

Lush, J. L. Intra-sire correlations or regressions of offspring on dam as a method of estimating heritability of characteristics. *Thirty-third Annual Proceedings of the American Society of Animal Production*, 1940, 293–301.

Lush, J. L. Heritability of quantitative characters in farm animals. *Hereditas*, 1949, Suppl. Vol., 356–375.

Lush, J. L. Genetics and animal breeding. *In* L. C. Dunn (Ed.), *Genetics in the twentieth century.* New York: Macmillan, 1951. Pp. 493–525.

Luxenburger, H. Vorläufiger Bericht über psychiatrische Serienuntersuchungen an Zwillingen. *Zeitschrift fuer die Gesamte Neurologie und Psychiatrie,* 1928, 116, 297–326.

Lynch, C. B. Nest-building in mice: An evolutionary study of behavioral thermo-regulation. Unpublished doctoral dissertation, University of Iowa, 1971.

Lynch, C. B., and Hegmann, J. P. Genetic differences influencing behavioral temperature regulation in small mammals: I. Nesting by *Mus musculus. Behavior Genetics,* 1972, 2, 43–53.

Lynch, C. B., and Hegmann, J. P. Genetic differences influencing behavioral temperature regulation in small mammals: III. Inbreeding depression in *Mus musculus.* (unpublished)

Lyon, M. F. Twirler: A mutant affecting the inner ear of the house mouse. *Journal of Embryology and Experimental Morphology,* 1958, 6(Pt. 1), 105–116.

McCarthy, J. J., and McCarthy, J. F. *Learning disabilities.* Boston: Allyn & Bacon, 1969.

McClearn, G. E. Strain differences in activity of mice: Influence of illumination. *Journal of Comparative and Physiological Psychology,* 1960, 53, 142–143.

McClearn, G. E. The inheritance of behavior. *In* L. J. Postman (Ed.), *Psychology in the making.* New York: Knopf, 1963. Pp. 144–252.

McClearn, G. E. Genetics and motivation of the mouse. *In* W. J. Arnold (Ed.), *Nebraska symposium on motivation.* Lincoln: University of Nebraska Press, 1968. Pp. 47–83.

McClearn, G. E. Behavioral genetics. *Proceedings of the Twelfth International Congress of Genetics,* 1969, 3, 419–430.

McClearn, G. E., and DeFries, J. C. Genetics and aggression. *In* J. F. Knutson (Ed.), *Control of aggression: Implications from basic research.* Chicago: Aldine, in press.

McClearn, G. E. Wilson, J. R., and Meredith, W. The use of isogenic and hetero-genic mouse stocks in behavioral research. *In* G. Lindzey and D. D. Thiessen (Eds.), *Contributions to behavior-genetic analysis: The mouse as a prototype.* New York: Appleton-Century-Crofts, 1970. Pp. 3–22.

McDougall, W. *An introduction to social psychology.* London: Methuen, 1908.

McGaugh, J. L., and Cole, J. M. Age and strain differences in the effect of distribution of practice on maze learning. *Psychonomic Science,* 1965, 2, 253–254.

McKusick, V. A. *Mendelian inheritance in man.* (2nd ed.) Baltimore: Johns Hopkins Press, 1968.

Manning, A. *Drosophila* and the evolution of behaviour. *Viewpoints in Biology,* Butterworth and Co., Ltd., 1965, 4, 125–169.

Marinello, M. J., Berkson, R. A., Edwards, J. A., and Bannerman, R. M. A study of the XYY syndrome in tall men and juvenile delinquents. *Journal of the American Medical Association,* 1969, 208, 321–325.

Martin, M. A., and Hoyer, B. H. Adenine plus thymine and guanine plus cytosine enriched fractions of animal DNA's as indicators of polynucleotide homologies. *Journal of Molecular Biology,* 1967, 27, 113–129.

Masterton, B., Heffner, H., and Ravizza, R. The evolution of human hearing. *Journal of the Acoustical Society of America,* 1969, 45, 966–985.

Mather, K. The progress and prospect of biometrical genetics. *In* L. C. Dunn (Ed.), *Genetics in the twentieth century.* New York: Macmillan, 1951. Pp. 111–126.

Mather, K., and Jinks, J. L. *Biometrical genetics.* (2nd ed.) Ithaca, N.Y.: Cornell University Press, 1971.

Maxwell, J. Intelligence, fertility and the future: A report on the 1947 Scottish mental survey. *Eugenics Quarterly,* 1954, 1, 244–274.

Mayr, E. Behavior and systematics. *In* A. Roe and G. G. Simpson (Eds.), *Behavior and evolution.* New Haven: Yale University Press, 1958. Pp. 341–362.

Mayr, E. The emergence of evolutionary novelties. *In* S. Tax (Ed.), *The evolution of life.* Chicago: University of Chicago Press, 1960. Pp. 349–380.

Mayr, E. *Animal species and evolution.* Cambridge, Mass.: The Belknap Press of Harvard University, 1965.

Meckler, R. J., and Collins, R. L. Histology and weight of the mouse adrenal: A diallel genetic study. *Journal of Endocrinology,* 1965, 31, 95–103.

Meier, G. W. Differences in maze performances as a function of age and strain of house mice. *Journal of Comparative and Physiological Psychology,* 1964, 58, 418–422.

Meier, G. W., and Foshee, D. P. Genetics, age, and the variability of learning performances. *Journal of Genetic Psychology,* 1963, 102, 267–275.

Melnyck, J., Derencseny, A., Vanasek, F., Rucci, A. J., and Thompson, H. XYY survey in an institution for sex offenders and the mentally ill. *Nature,* 1969, 224, 369–370.

Mendel, G. J. Versuche über Pflanzen-Hybriden. *Verhandlungen des Naturforschunden Vereines in Bruenn,* 1866, 4, 3–47. (Transl. by Royal Horticultural Society of London. Conveniently available in recent texts, including: Dodson, E. O., *Genetics,* Philadelphia: Saunders, 1956; and Sinnott, E. W., Dunn, L. C., and Dobzhansky, Th., *Principles of genetics,* New York: McGraw-Hill, 1950.)

Merrell, D. J. *Evolution and genetics: The modern theory of evolution.* New York: Holt, Rinehart & Winston, 1962.

Merrell, D. J. Methodology in behavior genetics. *Journal of Heredity,* 1965, 56, 263–265.

Money, J. Cytogenetic and psychosexual incongruities with a note on space-form blindness. *American Journal of Psychiatry,* 1963, 119, 820–827.

Money, J. Two cytogenetic syndromes: Psychologic comparisons. I. Intelligence and specific-factor quotients. *Journal of Psychiatric Research,* 1964, 2, 223–231.

Money, J. Cognitive deficits in Turner's syndrome. *In* S. G. Vandenberg (Ed.), *Progress in human behavior genetics.* Baltimore: Johns Hopkins Press, 1968. Pp. 27–30.

Montagu, A. The reception of Darwin's theory. (Introduction to *Man's place in nature,* by T. H. Huxley.) Ann Arbor: University of Michigan Press, 1959. Pp. 2–3.

Montaigne, M. de. Of the resemblance of children to their fathers. (1588) *In* R. M. Hutchins (Ed.), *Great books of the western world.* Vol. 25. *Montaigne.* Chicago: Encyclopædia Britannica, 1952. Pp. 367–368.

Moor, L. Niveau intellectuel et polygohosomie: Confrontation du caryotype et du niveau mental de 374 malades dont le caryotype comporte un excess de chromosomes X ou Y. *Revue de Neuropsychiatrie Infantile et d'Hygiene Mentale de l'Enfance,* 1967, 15, 325–348.

Muller, H. J. Artificial transmutation of the gene. *Science,* 1927, 66, 84–87.

Muller, H. J. Means and aims in human genetic betterment. *In* T. M. Sonneborn (Ed.), *The control of human heredity and evolution.* New York: Macmillan, 1965. Pp. 100–122.

Munro, T. A. Phenylketonuria: Data on forty-seven British families. *Annals of Eugenics,* 1947, 14, 60–88.

Nadler, C. F., and Borges, W. H. Chromosomal structure and behavior. *In* D. Y.-Y. Hsia (Ed.), *Lectures in medical genetics.* Chicago: Year Book Medical Publishers, 1966. Pp. 25–58.

Napier, J. *The roots of mankind.* Washington, D.C.: Smithsonian Institution, 1970.

Neel, J. V. Lessons from a "primitive" people. *Science,* 1970, 170, 815–822.

Newman, H. H., Freeman, F. N., and Holzinger, K. J. *Twins: A study of heredity and environment.* Chicago: University of Chicago Press, 1937.

Nichols, R. C. The inheritance of general and specific ability. *National Merit Scholarship Research Reports,* 1965, 1, No. 1.

Nielsen, J. Y chromosomes in male psychiatric patients above 180 cm. tall. *British Journal of Psychiatry,* 1968, 114, 1589–1591.

Nielsen, J., Tsuboi, T., Sturup, G., and Romano, D. XYY chromosomal constitution in criminal psychopaths. *Lancet,* 1968, ii, 576.

Nilsson-Ehle, H. Einige Ergebnisse von Kruezungen bei Hafer und Weisen. *Botanische Notiser,* 1908–1909, 257–294.

Nirenberg, M. W. The genetic code: II. *Scientific American,* 1963, 208, 80–94.

Oakley, K. P., and Muir-Wood, H. M. The succession of life through geological time. (7th ed.) London: Trustees of the British Museum (Natural History), 1967.

O'Brien, J. S., Okada, S., Chen, A., and Fillerup, D. L. Tay-Sachs disease: Detection of heterozygotes and homozygotes by serum hexosaminidase assay. *New England Journal of Medicine,* 1970, 283, 15–20.

Ødegaard, Ø. The psychiatric disease entities in the light of genetic investigation. *Acta Psychiatrica Scandinavica,* 1963, 39(Suppl. 169), 94.

Okada, S., and O'Brien, J. S. Tay-Sachs disease: Generalized absence of a beta-D-N-acetylhexosaminidase component. *Science,* 1969, 165, 698–700.

Packard, A. S. *Lamarck, the founder of evolution.* New York: Longmans, Green & Co., 1901.

Parsons, P. A. *The genetic analysis of behaviour.* London: Methuen, 1967.

Partanen, J., Bruun, M., and Markkanen, T. *Inheritance of drinking behavior. A study on intelligence, personality, and use of alcohol of adult twins.* Helsinki: The Finnish Foundation for Alcohol Studies, 1966.

Patau, K., Smith, D. W., Therman, E., Inhorn, S. L., and Wagner, H. P. Multiple congenital anomaly caused by an extra autosome. *Lancet,* 1960, i, 790–793.

Pearson, K. *The life, letters and labours of Francis Galton.* London: Cambridge University, 1924.

Pearson, K. On the inheritance of mental disease. *Annals of Eugenics,* 1931, 4, 362–380.

Penrose, L. S. A study in the inheritance of intelligence. The analysis of 100 families containing subcultural mental defectives. *British Journal of Psychology, General Section, London,* 1933, 24, 1–19.

Penrose, L. S., and Smith, G. F. *Down's anomaly.* London: J. & A. Churchill, 1966.

Perris, C. Genetic transmission of depressive psychoses. *Acta Psychiatrica Scandinavica,* 1968, Suppl. 203, 45–52.

Petit, C. Le rôle de l'isolement sexuel dans l'évolution des populations de *Drosophila melanogaster*. *Bulletin Biologique de la France et de la Belgique*, 1951, 85, 392–418.

Petit, C., and Ehrman, L. Sexual selection in *Drosophila*. In Th. Dobzhansky, M. K. Hecht, and W. C. Steere (Eds.), *Evolutionary biology*. Vol. 3. New York: Appleton-Century-Crofts, 1969. Pp. 177–217.

Pittendrigh, C. Adaptations, natural selection, and behavior. *In* A. Roe and G. G. Simpson (Eds.), *Behavior and evolution*. New Haven: Yale University Press, 1958. Pp. 390–419.

Polani, P. E., Briggs, J. H., Ford, C. E., Clarke, C. M., and Berg, J. M. A mongoloid girl with 46 chromosomes. *Lancet*, 1960, i, 721–724.

Premack, A. J., and Premack, D. Teaching language to an ape. *Scientific American*, 1972, 227, 92–99.

Price, W. H., and Jacobs, P. A. The 47,XYY male with special reference to behavior. *Seminars in Psychiatry*, 1970, 2, 30–39.

Rasmuson, M. *Genetics on the population level*. Stockholm: Svenska Bokförlaget, 1961.

Reading, A. J. Effect of maternal environment on the behavior of inbred mice. *Journal of Comparative and Physiological Psychology*, 1966, 62, 437–440.

Reimer, J. D., and Petras, M. L. Breeding structure of the house mouse, *Mus musculus*, in a population cage. *Journal of Mammalogy*, 1967, 48, 88–89.

Remak, J. (Ed.) *The Nazi years: A documentary history*. Englewood Cliffs, N.J.: Prentice-Hall, 1969.

Ressler, R. H. Genotype-correlated parental influences in two strains of mice. *Journal of Comparative and Physiological Psychology*, 1963, 56, 882–886.

Reznikoff, M., and Honeyman, M. S. MMPI profiles of monozygotic and dizygotic twin pairs. *Journal of Consulting Psychology*, 1967, 31(1), 100.

Riss, W., Valenstein, E. S., Sinks, J., and Young, W. C. Development of sexual behavior in male guinea pigs from genetically different stocks under controlled conditions of androgen treatment and caging. *Endocrinology*, 1955, 57, 139–146.

Ritchie-Calder, P. R. *Leonardo and the age of the eye*. New York: Simon & Schuster, 1970.

Roberts, R. C. Some concepts and methods in quantitative genetics. *In* J. Hirsch (Ed.), *Behavior-genetic analysis*. New York: McGraw-Hill, 1967a. Pp. 214–257.

Roberts, R. C. Some evolutionary implications of behavior. *Canadian Journal of Genetic Cytology*, 1967b, 9, 419–435.

Robson, G. C., and Richards, O. W. *The variations of animals in nature*. London: Longmans, Green & Co., 1936.

Roderick, T. H. Selection for cholinesterase activity in the cerebral cortex of the rat. *Genetics*, 1960, 45, 1123–1140.

Rodgers, D. A., McClearn, G. E., Bennett, E. L., and Hebert, M. Alcohol preference as a function of its caloric utility in mice. *Journal of Comparative and Physiological Psychology*, 1963, 56, 666–672.

Roe, A. Children of alcoholic parents raised in foster homes. In *Alcohol, science and society*. New Haven: Journal of Studies on Alcohol, 1954. Pp. 115–127.

Romer, A. S. Phylogeny and behavior with special reference to vertebrate evolution. *In* A. Roe and G. G. Simpson (Eds.), *Behavior and evolution*. New Haven: Yale University Press, 1958. Pp. 48–79.

Roper, A. G. *Ancient eugenics*. Oxford: Blackwell, 1913.

Rosanoff, A. J., Handy, L. M., and Plesset, I. R. The etiology of child behavior difficulties, juvenile delinquency and adult criminality with special reference to their occurrence in twins. *Psychiatric Monographs (California)*, 1941, No. 1.

Rosanoff, A. J., Handy, L. M., Plesset, I. R., and Brush, S. The etiology of manic-depressive syndromes with special reference to their occurrence in twins. *American Journal of Psychiatry*, 1935, 91, 725–762.

Rosenthal, D. *Genetic theory and adnormal behavior*. New York: McGraw-Hill, 1970.

Rosenthal, D. Two adoption studies of heredity in the schizophrenic disorders. *In* M. Bleuler and J. Angst (Eds.), *The origin of schizophrenia*. Bern: Verlag Hans Huber, 1971. Pp. 21–34.

Rosenthal, D., Wender, P. H. Kety, S. S., Schulsinger, F., Welner, J., and Oster-gaard, L. Schizophrenics' offspring reared in adoptive homes. *In* D. Rosenthal and S. S. Kety (Eds.), *The transmission of schizophrenia*. London: Pergamon Press, 1968. Pp. 377–392.

Rosenthal, D., Wender, P. H., Kety, S. S., Welner, J., and Schulsinger, F. The adopted-away offspring of schizophrenics. *American Journal of Psychiatry*, 1971, 128, 307–311.

Rosenzweig, M. R. Effect of heredity and environment on brain chemistry, brain anatomy, and learning ability in the rat. *Kansas Studies in Education*, 1964, 14, 3–34.

Rothenbuhler, W. C. Genetic and evolutionary considerations of social behavior of honeybees and some related insects. *In* J. Hirsch (Ed.), *Behavior-genetic analysis*. New York: McGraw-Hill, 1967. Pp. 61–111.

Scarr-Salapatek, S. Race, social class, and I. Q. *Science*, 1971, 174, 1285–1295.

Scheinfeld, A. The Kallikaks after thirty years. *Journal of Heredity*, 1944, 35, 259–264.

Schlesinger, K., Boggan, W., and Freedman, D. X. Genetics of audiogenic seizures: I. Relation to brain serotonin and norepinephrine in mice. *Life Sciences*, 1965, 4, 2345–2351.

Schlesinger, K., Boggan, W. O., and Freedman, D. X. Genetics of audiogenic sei-zures: III. Time response relationships between drug administration and seizure susceptibility. *Life Sciences*, 1970, 9, 721–729.

Schoenfeldt, L. F. Hereditary-environmental components of the project TALENT two-day test battery. In *Proceedings of the Sixteenth International Congress of Applied Psychology*. Amsterdam: Swets & Zeitlinger, 1968.

Schuckit, M., Goodwin, D. W., and Winokur, G. The half-sibling approach in a genetic study of alcoholism. *In* M. Roff, L. N. Robins, and M. Pollack (Eds.), *Life history research in psychopathology*. Vol. 2. Minneapolis: University of Minnesota Press, 1972a. Pp. 120–127.

Schuckit, M., Goodwin, D. W., and Winokur, G. A study of alcoholism in half-siblings. *American Journal of Psychiatry*, 1972b, 128, 1132–1136.

Schull, W. J., and Neel, J. V. *The effects of inbreeding on Japanese children*. New York: Harper & Row, 1965.

Schulz, B. Kinder schizophrener Elternpaare. *Zeitschrift fuer die Gesamte Neuro-logie und Psychiatrie*, 1940, 168, 332–381.

Seegmiller, J. E., Rosenbloom, F. M., and Kelley, W. N. Enzyme defect associated with a sex-linked human neurological disorder and excessive purine synthesis. *Science*, 1967, 155, 1682–1683.

Selander, R. K. Behavior and genetic variation in natural populations. *American Zoologist,* 1970, 10, 53–66.

Sergovich, F. R. Chromosome studies in unselected neonates. *American Society of Human Genetics,* Program & Abstracts, Austin, Texas, 1969, 33.

Shaffer, J. W. A specific cognitive deficit observed in gonadal aplasia (Turner's syndrome). *Journal of Clinical Psychology,* 1962, 18, 403–406.

Shah, S. A. *Report on the XYY chromosomal abnormality.* (USPHS Pub. No. 2103) Washington, D.C.: U.S. Government Printing Office, 1970.

Sheppard, J. R., Albersheim, P., and McClearn, G. E. Aldehyde dehydrogenase and ethanol preference in mice. *Journal of Biological Chemistry,* 1970, 245, 2876–2882.

Shields, J. *Monozygotic twins.* London: Oxford University Press, 1962.

Shields, J. Summary of the genetic evidence. *In* D. Rosenthal and S. S. Kety (Eds.), *The transmission of schizophrenia.* Oxford: Pergamon Press, 1968. Pp. 95–126.

Shields, J., Gottesman, I. I., and Slater, E. Kallmann's 1946 schizophrenic twin study in the light of new information. *Acta Psychiatrica Scandinavica,* 1967, 43, 385–396.

Shire, J. G. M., and Spickett, S. G. Genetic variation in adrenal structure: Strain differences in quantitative characters. *Journal of Endocrinology,* 1968, 40, 215–229.

Skodak, M., and Skeels, H. M. A final follow-up of one hundred adopted children. *Journal of Genetic Psychology,* 1949, 75, 85–125.

Slater, E. Psychiatry. *In* A. Sorsby (Ed.), *Clinical genetics.* London: Butterworth, 1953a. Pp. 332–349.

Slater, E., with Shields, J. *Psychotic and neurotic illness in twins.* (MRC Special Rep. No. 278) London: H.M. Stationery Office, 1953b.

Slater, E., and Cowie, V. *The genetics of mental disorders.* London: Oxford University Press, 1971.

Smith, M. Similarities of marriage partners in intelligence. *American Sociological Review,* 1941, 6, 697–701.

Snedecor, G. W., and Cochran, W. G. *Statistical methods.* (6th ed.) Ames: Iowa State University Press, 1967.

Snyder, L. H. The inheritance of taste deficiency in man. *Ohio Journal of Science,* 1932, 32, 436–440.

Spuhler, J. N. Behavior and mating patterns in human populations. *In* J. N. Spuhler (Ed.), *Genetic diversity and human behavior.* Chicago: Aldine, 1967. 241–268.

Spuhler, J. N. Assortative mating with respect to physical characters. *Eugenics Quarterly,* 1968, 15, 128–140.

Srb, A. M., Owen, R. D., and Edgar, R. S. *General genetics.* (2nd ed.) San Francisco: W. H. Freeman and Company, 1965.

Staats, J. Standard nomenclature for inbred strains of mice: Third listing. *Cancer Research,* 1964, 24, 147–168.

Stafford, R. E. New techniques in analyzing parent-child test scores for evidence of hereditary components. *In* S. G. Vandenberg (Ed.), *Methods and goals in human behavior genetics.* New York: Academic Press, 1965. Pp. 171–186.

Stent, G. S. *Molecular biology of bacterial viruses.* San Francisco: W. H. Freeman and Company, 1963.

Stern, C. *Principles of human genetics.* (3rd ed.) San Francisco: W. H. Freeman and Company, 1973.

Stocks, P., and Karn, M. N. A biometric investigation of twins and their brothers and sisters. *Annals of Eugenics,* 1933, 5, 1–55.

Stumpfl, F. *Die Ursprünge des Verbrechens dargestellt am Lebenslauf von Zwillingen.* Leipzig: Georg Thieme Verlag, 1936.

Sturtevant, A. H. Experiments on sex recognition and the problem of sexual selection in *Drosophila. Journal of Animal Behavior,* 1915, 5, 351–366.

Sturtevant, A. H. *A history of genetics.* New York: Harper & Row, 1965.

Tanner, J. M. Regulation of growth in size in mammals. *Nature,* 1963, 199, 845–850.

Thiessen, D. D. The Wabbler-lethal mouse: A study in development. *Animal Behaviour,* 1965, 13, 87–100.

Thiessen, D. D., and Nealey, V. G. Adrenocortical activity, stress response and behavioral reactivity of five inbred mouse strains. *Endocrinology,* 1962, 71, 267–270.

Thompson, W. R., and Olian, S. Some effects on offspring behavior of maternal adrenalin injection during pregnancy in three inbred mouse strains. *Psychological Reports,* 1961, 8, 87–90.

Thompson, W. R., Watson, J., and Charlesworth, W. R. The effects of prenatal maternal stress on offspring behavior in rats. *Psychological Monographs,* 1962, 76 (38, Whole No. 557).

Thurstone, T. G., Thurstone, L. L., and Strandskov, H. H. *A psychological study of twins.* (Rep. No. 4) Chapel Hill, N.C.: Psychometric Laboratory, University of North Carolina, 1955.

Tienari, P. Psychiatric illnesses in identical twins. *Acta Psychiatrica Scandinavica,* 1963, 39(Suppl. 171), 1–195.

Tienari, P. Schizophrenia in monozygotic male twins. *In* D. Rosenthal and S. S. Kety (Eds.), *The transmission of schizophrenia.* Oxford: Pergamon Press, 1968. Pp. 27–36.

Tjio, J. H., and Levan, A. The chromosome number of man. *Hereditas,* 1956, 42, 1–6.

Tolman, E. C. The inheritance of maze-learning ability in rats. *Journal of Comparative Psychology,* 1924, 4, 1–18.

Tolman, E. C., Tryon, R. C., and Jeffress, L. A. A self-recording maze with an automatic delivery table. *University of California Publications in Psychology,* 1929, 4, 99–112.

Tredgold, A. E. *Mental deficiency (amentia).* London: Bailliere, Tindall & Cox, 1908.

Tredgold, A. E. *Mental deficiency (amentia).* (6th ed.) London: Bailliere, Tindall & Cox, 1937.

Tryon, R. Behavior genetics in social psychology. *American Psychologist,* 1957, 12, 453. (Abstract).

Tyler, P. A., A quantitative genetic analysis of runway learning in mice. Unpublished doctoral dissertation, University of Colorado, 1969.

Tyler, P. A. Coat color differences and runway learning in mice. *Behavior Genetics,* Plenum Publishing, Inc., 1970, 1, 149–155.

Vale, J. R., Vale, C. A., and Harley, J. P. Interaction of genotype and population number with regard to aggressive behavior, social grooming, and adrenal and gonadal weight in male mice. *Communications in Behavioral Biology,* 1971, 6, 209–221.

van Abeelen, J. H. F., and van der Kroon, P. H. W. *Nijmegen waltzer* — a new neurological mutant in the mouse. *Genetical Research*, 1967, 10, 117–118.

Vandenberg, S. G. The hereditary abilities study: Hereditary components in a psychological test battery. *American Journal of Human Genetics*, 1962, 14, 220–237.

Vandenberg, S. G. Hereditary factors in normal personality traits (as measured by inventories). *In* J. Wortis (Ed.), *Recent advances in biological psychiatry.* Vol. 9. New York: Plenum Press, 1967. Pp. 65–104.

Vandenberg, S. G. Primary mental abilities or general intelligence? Evidence from twin studies. *In* J. M. Thoday and A. S. Parkes (Eds.), *Genetic and environmental influences on behavior.* New York: Plenum Press, 1968. Pp. 146–160.

Vandenberg, S. G. What do we know today about the inheritance of intelligence and how do we know it? *In* R. Cancro (Ed.), *Intelligence: Genetic and environmental influences.* New York: Grune & Stratton, 1971. Pp. 182–218.

Vandenberg, S. G. Assortative mating, or who marries whom? *Behavior Genetics,* 1972, 2, 127–157.

Vandenberg, S. G., and Johnson, R. C. Further evidence on the relation between age of separation and similarity in IQ among pairs of separated identical twins. *In* S. G. Vandenberg (Ed.), *Progress in human behavior genetics.* Baltimore: Johns Hopkins Press, 1968. Pp. 233–260.

van Lawick-Goodall, J. *In the shadow of man.* Boston: Houghton Mifflin, 1971.

Vogel, F. Eugenic aspects of genetic engineering. *In* G. Raspé (Ed.), *Advances in the biosciences.* Oxford: Pergamon Press, 1972. Pp. 397–410.

von Guaita, G. Versuche mit Kreuzungen von verschiedenen Rassen der Hausmaus. *Bericht der Naturforschenden Gesellschaft, Freiburg,* 1900, 11, 131–138.

Waddington, C. H. *The strategy of the genes.* New York: Macmillan, 1957.

Waddington, C. H. *New patterns in genetics and development.* New York: Columbia University Press, 1962.

Wade, N. Creationists and evolutionists: Confrontation in California. *Science,* 1972, 178, 724–729.

Wallace, B. *Topics in population genetics.* New York: Norton, 1968.

Waller, J. H. Differential reproduction: Its relation to IQ test score, education, and occupation. *Social Biology,* 1971, 18, 122–136.

Washburn, S. L. Tools and human evolution. *Scientific American*, 1960, 203, 62–75.

Washburn, S. L. One hundred years of biological anthropology. *In* J. O. Brew (Ed.), *One hundred years of anthropology.* Cambridge: Harvard University Press, 1968. Pp. 97–118.

Washburn, S. L., Jay, P. C., and Lancaster, J. B. Field studies of old world monkeys and apes. *Science,* 1965, 150, 1541–1547.

Washburn, S. L., and Lancaster, C. S. The evolution of hunting. *In* R. B. Lee and I. DeVore (Eds.), *Man the hunter.* Chicago: Aldine, 1968. Pp. 293–303.

Watson, J. B. *Behaviorism.* New York: Norton, 1930.

Watson, J. D. *Molecular biology of the gene.* (2nd ed.) New York: W. A. Benjamin, 1970.

Watson, J. D., and Crick, F. H. C. Molecular structure of nucleic acids. A structure for deoxyribose nucleic acids. *Nature*, 1953a, 171, 737–738.

Watson, J. D., and Crick, F. H. C. Genetic implications of the structure of deoxyribonucleic acid. *Nature*, 1953b, 171, 964–967.

Weir, M. W., and DeFries, J. C. Prenatal maternal influence on behavior in mice: Evidence of a genetic basis. *Journal of Comparative and Physiological Psychology*, 1964, 58, 412–417.

Welch, J. P., Borgaonkar, D. S., and Herr, H. M. Psychopathy, mental deficiency, aggressiveness and the XYY syndrome. *Nature*, 1967, 214, 500–501.

White, R. W. *The abnormal personality*, New York; Ronald Press, 1948.

Whitney, G. D. Vocalization of mice: A single genetic unit effect. *Journal of Heredity*, 1969, 60, 337–340.

Whitney, G. Vocalization of mice influenced by a single gene in a heterogeneous population. *Behavior Genetics*, 1973, 3, 57–64.

Wictorin, M. *Bidrag till Räknefärdighetens Psykologi, en Tvillingsundersökning.* Goteborg: Elanders Boktryckeri, 1952.

Wiener, S., Sutherland, G., Bartholomew, A. A., and Hudson, B. XYY males in a Melbourne prison. *Lancet*, 1968, i, 150.

Willoughby, R. R. Family similarities in mental test abilities. *Nature and nurture: Their influence upon intelligence, Twenty-Seventh Yearbook of the National Society for the Study of Education*, 1928, 27(Pt. 1), 55–59.

Wilson, A. C., and Sarich, V. M. A molecular time scale for human evolution. *Proceedings of the National Academy of Sciences*, 1969, 63, 1088–1093.

Wilson, E. C., Respass, J. C., Hollifield, C., and Parson, W. Studies of alcohol metabolism in mice which preferentially consume ethanol. *Gastroenterology*, 1961, 40, 807–808.

Wilson, R. S. Twins: Early mental development. *Science*, 1972, 175, 914–917.

Wimer, R. E., Wimer, C., and Roderick, T. H. Genetic variability in forebrain structures between inbred strains of mice. *Brain Research*, 1969, 16, 257–264.

Winokur, G., Clayton, P. J., and Reich, T. *Manic depressive illness*. St. Louis: C. V. Mosby, 1969.

Witt, G., and Hall, C. S. The genetics of audiogenic seizures in the house mouse. *Journal of Comparative and Physiological Psychology*, 1949, 42, 58–63.

Woods, F. A. *Mental and moral heredity in royalty*. New York: Holt, 1906.

Woodworth, R. S. *Contemporary schools of psychology*. (Rev. ed.) New York: Ronald Press, 1948.

Wright, S. Systems of mating. *Genetics*, 1921, 6, 111–178.

Wright, S. Coefficients of inbreeding and relationship. *American Naturalist*, 1922, 56, 330–338.

Wright, S. *Evolution and the genetics of populations*. Vol. 1. *Genetic and biometric foundations*. Chicago: University of Chicago Press, 1968.

Wyatt, R. J., Murphy, D. L., Belmaker, R., Cohen, S., Donnelly, C. H., and Pollin, W. Reduced monoamine oxidase activity in platelets: A possible genetic marker for vulnerability to schizophrenia. *Science*, 1973, 179, 916–918.

Wynne-Edwards, V. C. Intergroup selection in the evolution of social systems. *Nature*, 1963, 200, 623–626.

Yanai, R., and Nagasawa, H. Age-, strain-, and sex-differences in the anterior pituitary growth hormone content of mice. *Endocrinologia Japonica*, 1968, 15, 395–402.

Yeakel, E. H., and Rhoades, R. P. A comparison of the body and endocrine gland (adrenal, thyroid and pituitary) weights of emotional and non-emotional rats. *Endocrinology*, 1941, 28, 337–340.

Yoshimasu, S. Criminal life curves of monozygotic twin-pairs. *Acta Criminol. et Med. Legal. Japon.,* 1965, 31, 5–6.

Zazzo, R. *Les jumeaux, le couple et la personne.* Vol. 2. Paris: Presses Universitaires de France, 1960.

Zirkle, C. The knowledge of heredity before 1900. *In* L. C. Dunn (Ed.), *Genetics in the 20th century.* New York: Macmillan, 1951. Pp. 35–57.

Author Index

Subject Index